>

LONG-TERM EXPERIMENTS IN AGRICULTURAL AND ECOLOGICAL SCIENCES

Long-term Experiments in Agricultural and Ecological Sciences

Proceedings of a conference to celebrate the 150th Anniversary of Rothamsted Experimental Station, held at Rothamsted, 14–17 July 1993

Edited by

R.A. Leigh
Head of Soils and Crop Sciences Division

A.E. Johnston
Lawes Trust Senior Fellow

*Institute of Arable Crops Research
Rothamsted Experimental Station
Harpenden, UK*

CAB INTERNATIONAL

CAB INTERNATIONAL
Wallingford
Oxon OX10 8DE
UK

Tel: Wallingford (01491) 832111
Telex: 847964 (COMAGG G)
E-mail: cabi@cabi.org
Fax: (01491) 833508

A catalogue entry for this book is available from the British
Library.

ISBN 0 85198 933 0

Typeset by MFK Typesetting Ltd, Hitchin, Herts
Printed and bound in the UK at the University Press, Cambridge

Contents

Preface

The conference for which this book is the proceedings was organized to celebrate the 150th anniversary of Rothamsted Experimental Station. The title of the conference was *Insight from foresight: The role of long-term experiments and databases in agricultural and ecological sciences* and reflected two aspects of Rothamsted's work. The first was the scientific insight that has been gained because of the foresight shown by the Station's founder Sir John Bennett Lawes and his collaborator, Sir Joseph Henry Gilbert, in retaining their long-term field experiments after their initial aims had been fulfilled. They did this because they recognized that future generations of scientists might see uses for them that they could not. This conviction has been amply rewarded as several of the following chapters indicate. The second aspect reflects the Station's interest and reputation in long-term research both in agriculture and ecology, the latter represented especially by the work of the Rothamsted Insect Survey.

What emerges in all chapters is that long-term experimentation and monitoring are vitally important in understanding the changes that are occurring in the environment and the way they interact with agriculture and natural ecosystems. This common theme emerges despite the diversity of topics covered and the very different reasons for starting many of these long-term studies.

One theme is that long-term data sets acquire a value that is often not predicted by the initiators of the work. Frequently, this is revealed because of the emergence of new techniques such as the use of ^{15}N in the Rothamsted experiments (Chapter 6 by Powlson) and the ability to analyse for organic pollutants (Chapter 9 by Jones *et al.*). Another example is the development of new methods of statistical or numerical analysis which allow additional information to be extracted from the observations (Chapters 10 (Barnett), 21 (Maberly *et al.*), and 18 (Woiwod and Harrington)). Often the availability of the long data sets are themselves the spur to the development of analytical and statistical tools as in the pioneering work of Fisher and Yates who, through their work on the Rothamsted data, devised current statistical approaches to data handling. Much attention today is given to modelling, often thought to be a reliable way of predicting future events, but models are only as good as the data on which they are based and reliable data for modelling biological systems need long-term experiments (Chapter 7 by Jenkinson *et al.*).

A second feature is that trends not obvious in short runs of results become obvious when much longer data sets are available and this often prevents erroneous conclusions being drawn. Examples are given in several Chapters (see those by Woiwod and Harrington (18); Greenwood *et al.* (19); Gamble (20)) of the way in which short-term declines in populations are found to be part of a longer cycle or can be linked to specific weather or environmental parameters, but only when long time series are analysed. In the case of slow changes, such as those that occur in soil organic matter, in the build up of pollutants in soil or in ecological systems subjected to a slowly changing environment, the changes are not measurable other than in long-term studies. In ecological studies long-term monitoring allows rare or episodic events to be detected and properly analysed. When seen as part of a long run of observations, the effects of these events can be appreciated and set in context, particularly when they lead to major ecological changes. The work at Rothamsted also highlights the importance of archiving samples from long-term experiments. The ability to go back to the 1840s and analyse time sequence samples of crops and soils has provided unique information on the accumulation of nutrients and pollutants in soil and their effects on crop quality. The importance of sample archiving cannot be stressed too strongly.

A final theme that permeates all chapters is the uncertainty that surrounds long-term research and the general low esteem in which it is viewed by many experimental scientists, in part perhaps because it does not lead to rapid publication of many research papers. As a result, those responsible for long-term experiments feel threatened because the experimentalists accuse them of conducting science without a proper scientific hypothesis to test. As the chapters in this book make clear, long-term experiments are often based on an initial hypothesis (this was so for the Rothamsted experiments) and then analysis of the data provides new hypotheses and insights that can be answered by further work that adds to the data set and starts a new cycle of hypothesis and measurement. Thus it becomes abundantly clear from the various chapters that short-term experimental approaches could not have provided the same information. If we are to secure the future of such long-term work it is essential that its importance is recognized both by the whole scientific community and by those who fund the work. The research outlined in the chapters that follow demonstrates what can be achieved when such recognition is given.

The Rothamsted 150th Anniversary Conference received generous financial support from The Lawes Agricultural Trust, The Tansley Fund of The New

Phytologist, The British Ecological Society, The Annals of Botany and Kemira Ince Ltd. Without their generosity we would not have had such a successful meeting. We are deeply grateful to them. Many individuals also contributed to the success of the meeting from which this book arose, not least the speakers and other participants. However, particular thanks must go to Mrs Deirdre Hughes who not only handled all the organization for the conference but was also responsible for the arduous task of collating and, in some cases, retyping the manuscripts of the chapters that follow. We also wish to thank all our colleagues at Rothamsted who helped in so many other ways.

Roger A. Leigh

A.E. Johnston

Note in proof
It is with regret that we must report that Dr John Gamble,
Director of the Sir Alister Hardy Foundation for Ocean Science
and author of Chapter 20, died on 10 August 1994.

Foreword

We are experiencing a period of intense reappraisal and change in the need for, and approaches to, agricultural research. Many of the important questions facing farming today such as the sustainability of different cultural systems, environmental impact, climate change and public perception of the industry and countryside, require long-term studies to provide meaningful information and answers. Such an approach has been pioneered and sustained by Rothamsted Experimental Station since its foundation and this book is a celebration of 150 years of continuous agricultural research and the inception of a unique set of field experiments – the Classical experiments – that have run continuously for all or most of that time. These experiments, established by the Station's founder, Sir John Bennett Lawes, FRS, and his scientific collaborator, Sir Joseph Henry Gilbert, FRS, originally aimed to study the nutrient requirements of arable crops. This objective was fulfilled relatively quickly but the experiments were continued and, with their associated archive of soil and crop samples, now provide a unique resource that is continuously providing new information and insights of relevance to the needs of late 20th-century agriculture and its interaction with the wider environment.

The benefits and insights gained from the Classical experiments are due largely to the foresight shown by Lawes and Gilbert and their successors in maintaining and continuing the experiments after their immediate objectives were achieved. The chapters in this book are based on the papers given at the Rothamsted 150th Anniversary Conference, held in July 1993, and all demonstrate how insight into long-term and often slow changes in environmental parameters or ecosystems require foresight, patience and a belief in the potential value of the information that eventually emerges.

The chapters discuss examples of the contributions that long-term field experiments and monitoring are making to agriculture, ecology and environmental sciences. They emphasize how such work has contributed to our understanding of changes in the wider environment and the responses of eco-systems to human and other activities. They should stand not only as a land-mark of progress, but as pointers to new studies necessary to sustain and improve productive agriculture worldwide in an attractive environment.

T. Lewis
Director of Research, Institute of Arable Crops Research,
Rothamsted Experimental Station

I

THE CONTRIBUTION OF LONG-TERM EXPERIMENTS TO AGRICULTURE AND FORESTRY

The Importance of Long-term Experimentation

T.R.E. Southwood
*Department of Zoology, University of Oxford,
Oxford OX1 2JD, UK.*

Time has always been a difficult concept in human reasoning; the utilitarian philosopher, John Stuart Mill summed up the views of medieval and modern philosophers when he said that time is not a factor which can be taken into account in the making of decisions, there are too many variables. The seasons' cycle was appreciated by the ancient Assyrians, it was the foundation of the agrarian society: autumn and harvest followed spring and sowing and one year followed the next. Others spoke of 'Time's Arrow' the uniqueness of the present moment, once passed never to be repeated. The Greek Sophists expressed this by saying one can never step into the same river twice. Between the assurance of the cyclical view and uncertainty of the linear view lay the desire to know something of the future. Will the season be good or bad, what is the mood of the gods or of God?

Two developments occurred between 200 and 100 years ago that have helped us to address that desire. First, there was an increasing move to replace memory with measurement. Weather was the great variable and the development of instrumentation encouraged the recording of temperatures, of rainfall and of other parameters. On the basis of these records – this monitoring of the weather – the science of meteorology developed, which led eventually to the great prize of weather forecasting. This indeed was foresight from insight!

Meteorologists measure the physical aspects of the atmosphere – and other scientists have measured chemical features. The carbon dioxide records from the Mauna Loa observatory now constitute a data set familiar to heads of governments throughout the world. It provides evidence on which national, indeed international, policy is being built. Although nobody would have questioned the gathering of the parallel set of data, that is the economic data which provided the information for the amount of fossil fuel that is burnt, it would not be surprising if the initial expenditure on routine carbon dioxide analysis had a difficult passage through the relevant policy and funding committees. Indeed, if you look at the results of the first two or three years of the carbon dioxide monitoring (a period after which so many research grants are now reviewed) one can imagine a conclusion being drawn that the variations within a year were much greater than those between years and that no further monitoring could be justified! It is indeed unusual for a consensus to be reached on the importance of initiating a monitoring programme (see Tinker, Chapter 22). In the 1920s G.M.B. Dobson, a Fellow of Merton College, Oxford, devised a way of measuring ozone in the upper atmosphere. I am told that at the time the study was regarded as a typical donnish eccentricity; Dobson covered a large part of the costs from his own personal resources. Now of course his work is the foundation of our knowledge of ozone levels in the upper atmosphere and these too are a global phenomenon of concern, not only to scientists, but to statesmen and the world's population.

The drive for these observations was largely the desire to use instruments to make precise measurements of the physical environment. The other impetus towards the development of long-term records and experiments was again a recognition of man's ability to change his environment. It was surely John Bennett Lawes' vision of the potential of an agricultural system scientifically managed by man that led him to establish his experiments at Rothamsted. With the hallmarks of an enquiring mind, he explored not only the impact of man, but the effects of the absence of human activity.

Long-term Studies in Ecology

Rothamsted's Broadbalk Wilderness, an area of a wheat field fenced off after harvest over a hundred years ago, is a remarkable demonstration of secondary succession. The part that was left untouched eventually reverted to woodland; this is the living proof of the theories of vegetational succession of ecologists such as Tansley, before these theories had been enunciated. These plots made a great impression on me when I first saw them in 1948. Later on in my career I reflected that we knew relatively little about the changes in the early stages of the secondary succession, and almost nothing about the changes in the fauna, the insects and the birds. Therefore, 29 years after first seeing Broadbalk I started a rather similar experiment at Imperial College Field Station at Silwood Park (Southwood *et al.*, 1979, 1988).

Ecological processes are often slow and it is only through long-term experiments such as these that they can be firstly detected and secondly understood. An example of this is the long-term decline of the partridge (*Perdix perdix*), a problem to which I was first introduced at Rothamsted, when an influential farmer wondered if the spraying of dinitro-*o*-cresol (DNOC) as a weedkiller on his cereal fields was having any impact on the survival of partridges for he felt that the numbers of birds were falling. Fortunately data had been collected on partridge populations over a number of years (Blank *et al.*, 1967) which hinted at the beginning of the decline. Other work showed that the application of herbicides reduced the population of weeds in the fields, with the loss of insects associated with them (Southwood and Cross, 1969). Dr G.R. Potts and his colleagues have made many further studies and have confirmed the decline of the partridge population and its association with the application of herbicides to cereals, as well as some other factors (Potts, 1986).

Insect population can also be surveyed by light traps, an approach C.B. Williams pioneered at Rothamsted. The Second World War inevitably produced a gap in the records and there was another gap shortly before and after his retirement, but the two periods of his catches taken with those from the continuous trapping since 1959 (when the Rothamsted Insect Survey was commenced; see Woiwod and Harrington, Chapter 18, this volume) and compared with the Geescroft traps' catches, show the impact of agricultural intensification. As in the study of partridges, this commenced in the mid 1950s and continues right through to today (Lewis, 1992).

The rare event is another type of phenomenon where long-term experiments are essential. I was just about to embark on the long-term study of the viburnum whitefly (*Aleurotrachelus jelinekii*) on three bushes at Silwood Park, when the very severe winter of 1962–63 caused considerable mortality both to the leaves on the bushes and to the whiteflies themselves. The result was that for many years the populations were 'recovering' and had the study been made without knowledge of this event for just a few years, a false picture would have been gained of the insect's population dynamics (Southwood *et al.*, 1989).

In some situations variability is such that a long run of data is necessary to determine the meaning. A good example of this is any attempt to measure the primary productivity of deserts, which are greatly influenced by rainfall (Franklin, 1987). Long-term observations are also required to separate a trend from a very variable background. Rising carbon dioxide levels and the fall in ozone concentrations in the atmosphere are in this category. The significance of anthropogenic sulphur emissions on the acidification of rivers and lakes in Scotland and southern Scandinavia is another example. The question was posed: were the changes just part of the natural cycle reflecting the state of maturity of the surrounding coniferous woodlands, or were they unique and reflection of man's industrial activity? In this instance nature conveniently recorded the information in the sediments of the lakes and palaeontological work has permitted a reconstruction (Battarbee and Renberg, 1990).

The Value of Long-term Experiments in Agriculture

One of the fundamental concepts of ecology is that the population of an animal in a particular area is determined by the carrying capacity of that habitat for it. Man is unique in that he has increased the carrying capacity of his habitat through agriculture. However, as population grows towards that carrying capacity, agriculture must be sustainable. We now know that the people who left behind those mysterious heads on Easter Island developed a population level based on an agriculture that was not sustainable. The famous Broadbalk winter wheat experiments, now in their 151st year, show what is sustainable with different fertilizers on Rothamsted soil. In the variation of yields over that period one can see the impact of weeds, the effect of changes in the availability of agricultural labour, the advent of herbicides and the impact of high-yield cultivars (Taylor, 1989; Jenkinson, 1991; Johnston, Chapter 2). Such long-term data sets also allow us to determine the impact on agriculture of extraneous factors such as pollution. The changes in the pH values of the soils of Park Grass and Geescroft Wilderness over more than 100 years show the effect of woodland vegetation in increasing the rate of acidic deposition (Johnston *et al.*, 1986).

Long-term Field Sites: Laboratories for Experiments and Data Sets for the Development of Analytical Tools

Long-term experiments such as those at Rothamsted should not be viewed as museum pieces, but rather as living laboratories providing opportunities for experimentation in which the effects of manipulation may be separated from other variables. A good example of this is provided by the work on Broadbalk on the effects of fungicides and of a break in cropping on wheat yields (Jenkinson, 1991). But it is possible not only to interpret variations in yields, but also to quantify other aspects of the whole system such as the nitrogen cycle (Jenkinson and Parry, 1989). Once a system is understood in this amount of detail, it is possible to manipulate almost any part and to be confident that one is able to assess the corresponding impacts.

An entirely different product of long-term experiments has been the development of many of the statistical techniques used today. Ronald Fisher's period at Rothamsted and the challenges the analysis of the experiments presented, provided a combination which effectively laid the foundation of much of modern statistics. When studying the variations in the size of populations of insects trapped over long periods at Rothamsted, Roy Taylor showed that the relationship between the variance and the mean are related by a power law with two constants one of which is characteristic of the species. This has been an important contribution to statistical ecology, and aids in the analysis of the changes in the relationships between species over time (Taylor, 1986). The

Rothamsted Insect Survey, provides a fundamental benchmark for pest species, against which we can assess changes due to agricultural practice or to widespread climatic variations (Woiwod and Harrington, Chapter 18).

In the last few decades, Rothamsted has developed efficient statistical methods to analyse large data sets, methods that owe their origins to the seed-bed of long-term experiments, but which are now used worldwide. With these statistical tools and this background, models can be developed which permit the inclusion of time into the decision-making processes: thus the dictum of John Stewart Mill no longer applies in this area.

Conclusions for the Future

At the beginning, I pointed out that whereas the monitoring of physical conditions (i.e. the weather) and various economic activities were accepted as valid activities, doubts were often voiced concerning requests that biological and chemical parameters should be followed. In part no doubt this scepticism is based on the almost unlimited variety of the parameters that might be measured and their inherent variability. Well-established long-term experiments and their related sets of measurements provide the framework for such monitoring. Politicians and others debate our changing environment, but it is only on the basis of such long-term information that we can ascertain what changes are occurring and, most importantly, gain insight into the mechanisms of change. I hope that tendencies to short-termism in funding scientific research will not lead to the loss of this knowledge, which is essential for our long-term prosperity and perhaps for our survival. We need both insight and foresight, otherwise we could all follow the fate of the inhabitants of Easter Island.

References

Battarbee, R.W. and Renberg, I. (1990) The surface water acidification project (SWAP) paleolimnology programme. *Philosophical Transactions of the Royal Society, London* 357, 227–232.

Blank, T.M., Southwood, T.R.E. and Cross, D.J. (1967) The ecology of the partridge: I. Outline of population processes with particular references to chick mortality and nest density. *Journal of Animal Ecology* 36, 549–556.

Franklin, J.F. (1987) Importance and justification of long-term studies in ecology. In: Likens, G.E. (ed.) *Long-term Studies in Ecology: Approaches and Alternatives*. Springer-Verlag, New York, pp. 1–19.

Jenkinson, D.S. (1991) The Rothamsted long-term experiments: are they still of use? *Agronomy Journal* 83, 2–10.

Jenkinson, D.S. and Parry, L.C. (1989) The nitrogen cycle in the Broadbalk wheat experiment: a model for the turnover of nitrogen through soil microbial biomass. *Soil Biology and Biochemistry* 21, 535–541.

Johnston, A.E., Goulding K.W.T. and Poulton, P.R. (1986) Soil acidification during more than 100 years under permanent grassland and woodland at Rothamsted. *Soil Use and Management* 2, 3-10.

Lewis, T. (1992) 150 years of research at Rothamsted: practice with science exemplified. *Journal of the Royal Agricultural Society of London, England* 153, 1-12.

Potts, G.R. (1986) *The Partridge: Pesticides, Predation and Conservation.* Collins, London.

Southwood, T.R.E. and Cross, D.J.M. (1969) The ecology of the partridge. III. Breeding success and the abundance of insects in natural habitats. *Journal of Animal Ecology* 38, 497-509.

Southwood, T.R.E., Brown, V.K. and Reader, P.M. (1979) The relationships of plant and insect diversities in sucession. *Biological Journal of the Linnean Society* 12 (4), 327-348.

Southwood, T.R.E., Brown, V.K., Reader, P.M. and Mason, E. (1988) Some characteristics of the primary trophic level of a secondary succession. *Proceedings of the Royal Society of London,* B 23, 11-44.

Southwood, T.R.E., Hassell, M.P., Reader, P.M. and Rogers, D.J. (1989) Population dynamics of the viburnum whitefly *(Aleurotrachelus jelinekii). Journal of Animal Ecology* 58, 921-942.

Taylor, L.R. (1986) Synoptic dynamics, migration and the Rothamsted Insect Survey. *Journal of Animal Ecology* 55, 1-38.

Taylor, L.R. (1989) Objective and experimentation in long-term Research. In: Likens, G.E. (ed.) *Long-Term Studies in Ecology: Approaches and Alternatives.* Springer-Verlag, New York, pp. 20-69.

2 The Rothamsted Classical Experiments

A.E. JOHNSTON
Institute of Arable Crops Research, Rothamsted Experimental Station, Harpenden, Hertfordshire AL5 2JQ, UK.

Introduction

J.B. Lawes (later Sir John) inherited the Rothamsted estate, which included an old manor house and a farm of about 100 ha, when only eight years of age but he did not assume responsibility for its management until he was 20, in 1834. As a landowner who was also a practising farmer, he was well aware of the need to keep agriculture financially viable. Gradually he became interested in crop nutrition and on 1 June 1843 he appointed J.H. Gilbert (later Sir Henry), a chemist by training, to assist him in field and laboratory experiments at Rothamsted on the nutrition of farm crops and animals.

It is impossible to conceive that in 1843 Lawes and Gilbert could have had the remotest idea that some of their experiments would still be continued in the 1990s. Today they are the oldest, continuous agricultural field experiments in the world. Their importance is recognized worldwide, not as fossilized monuments to the achievements of two great scientists, but because they still provide scientific data of immeasurable value. It is worth considering, therefore, some of the reasons for starting these experiments, their successes and failures, how they have been modified to maintain their usefulness and the lessons which can be drawn about the sustainability of agricultural systems.

J.B. Lawes

The family fortunes were at a low ebb when Lawes' father died in 1822. His mother let the manor house to a tenant whilst her son was educated, first at Eton, then at Oxford. In 1834 the tenant left the manor and Lawes' mother took up residence again. Lawes, then in his second year at Oxford, presumably saw it as his filial duty to return home and manage the estate. Probably this was no great hardship to him because his interests were undoubtedly in science, for which there was little formal training at that time. However, whilst at Oxford, he did attend lectures by Professor C.G. Daubeny whose research at Oxford from the early 1820s onwards ended a very barren period in the development of agricultural science. I suspect that Daubeny was a major, early formative influence on Lawes but there is no documentary evidence to support this.

Returning home, Lawes had a bedroom at the Manor converted into a laboratory and attempted to determine the 'active principles' of a number of medicinal plants. He was unsuccessful but a useful lesson was probably learnt, the need for rigorous analytical methods. He also attempted to develop commercially a newly proposed method for producing calomel, mercurous chloride, then used as a purgative in medicine. Again, without success but giving insight into 'trade', not then normally a part of a Victorian gentleman's training, but soon to be put to good use by him.

By the 1830s many farmers had found that an application of crushed bones increased yields of turnips (*Brassica rapa*). Lawes, and other farmers around Rothamsted, had little benefit from this practice. Like others, however, Lawes found that if bones were treated with sulphuric acid, to produce superphosphate of lime, this greatly increased yields of turnips grown on his soil. In 1842 Lawes took out a patent for the manufacture of superphosphate, not only from bones but from mineral phosphates and in 1843 he started production on a commercial scale at a factory at Deptford, London. This was the start of what was to become a successful manure business which financed the experiments at Rothamsted.

J.H. Gilbert

Gilbert's upbringing was very different from that of Lawes. He studied chemistry at University, first at Glasgow then at University College, London, where, after graduating, he worked first with Anthony Todd Thompson. He probably met Lawes in this period. Early in 1840 he went to Giessen in Germany as a pupil of Professor J. Liebig. He was awarded a doctorate within 12 months, apparently the criteria were different from those of today! Returning to England he was at University College, London in 1841 and then went to an industrial appointment in Manchester.

Lawes and Gilbert

By the early 1840s Lawes was becoming ever more interested in crop nutrition. Wisely deciding that he could not be both a fledgling industrialist with a business in London and an up and coming agricultural researcher at

Rothamsted 40 km away, Lawes decided that he needed an assistant at Rothamsted. Gilbert was appointed to help set up and supervise the experiments that Lawes was keen to start. Taking up his appointment on 1 June 1843, there started one of the longest (57 years) and most productive scientific partnerships on record. In agricultural science the names of Lawes and Gilbert are forever linked and together they are regarded as the founding fathers of the scientific method in agriculture. Together they published some 150 scientific papers together with many popular articles, perhaps 300, for farmers (Dyke, 1991).

The Years Before 1843

As early as 1837, Lawes had started to do some simple experiments on the nutrition of farm crops in pots and small plots in the field. At that time the consensus view was that plants derived their carbon, hydrogen and oxygen from carbon dioxide in the air and water taken up by roots from soil. Following the meticulous research of de Saussure the relative proportions of the mineral elements in plants were also known. These elements, e.g. phosphorus, potassium, calcium, magnesium, silicon, chlorine, were found in the ash of plants and were also known to be present in farmyard manure (FYM) then the main source of nutrients returned to soil to maintain its fertility. There was uncertainty about the source and the importance of nitrogen which was, of course, not found in plant ash but was present in both plants and FYM.

Lawes tested different forms of nitrogen, as ammonium and nitrate salts, applied to a cultivar of cabbages, used for animal feed, and to turnips. The results were published in the *Gardeners' Chronicle* in 1842 and 1843. He was clearly at a loss to explain why the fertilizers he tried did not increase yield in proportion to their nitrogen content but he did comment that 'ammonium phosphate was one of the most powerful manures known'. Herein probably is the explanation, the soils were so deficient in phosphorus that crops did not respond to nitrogen without it. In fact, he concluded the second article by noting that applying ammonia to a soil deficient in phosphorus was useless and where minerals were in adequate supply the ammonia supplied by the atmosphere would be insufficient for the wants of a crop.

Lawes had probably made a number of important decisions about how the field experiments he envisaged would be conducted before Gilbert's appointment. Perhaps the most important was to have separate experiments for each of the principal arable crops then usually grown in rotation. Although not all experiments were started in the same year, Table 2.1 shows that the aim of covering all relevant crops was eventually realized, but only once Lawes and Gilbert had developed their ability to plan, design and supervise large-scale field experiments and the farm staff had been trained in the appropriate experimental skills.

Table 2.1. The principal field experiments started by Lawes and Gilbert at Rothamsted and the arable crops each investigated.

Field	Year experiment started	Crop
Barnfield	1843	Turnips
Broadbalk	1843	Winter wheat
Geescroft	1847	Beans
Hoosfield	1849	Clover
Hoosfield	1852	Spring barley
Geescroft	1869	Oats
Agdell	1848	Crops in rotation
Park Grass	1856	Permanent pasture

In each experiment the same crop was to be grown year after year in succession on the same plots. This decision identifies Lawes as a scientist because it flew in the face of the accepted agricultural practice of crop rotation which had been gradually perfected over the previous century. Such rotations were obligatory on all tenanted farms and were perceived to maintain the productivity of a given area of land to the benefit of the tenant whilst maintaining soil fertility to the benefit of the landowner (Johnston, 1991b). Lawes justified his decision to grow crops in monoculture on the grounds that this would highlight not only the characteristic nutrient requirements of each crop but would also emphasize differences between crops. Possibly he had some suspicion that there could be benefits from growing crops in rotation which would be difficult to explain. Only in one experiment, that on Agdell (Table 2.1), were crops grown in rotation and Lawes and Gilbert frequently compared yields of each crop in this experiment with yields of the same crop in monoculture.

Another important decision identifies Lawes as a farmer. He decided to have large plots, initially on Broadbalk there were 20 plots on about 4.4 ha. He believed that fellow farmers would only accept the effects of different treatments if they could be seen to be consistent over a large area. Because the original plots were large, it has been possible to subdivide them to test changes in treatment. This was first done by Lawes and Gilbert but it has been done much more in recent years and has been of immeasurable value.

The Start of the Classical Experiments

Lawes must also have carefully considered the treatments to be tested because the experiment on turnips on Barnfield was started very soon after Gilbert arrived on 1 June. These treatments included an unmanured control and FYM. The rest were probably based not only on the experience he had already gained but they also reflected Liebig's views expressed in 1840. In his now famous report to the British Association for the Advancement of Science entitled 'Organic Chemistry in its Application to Agriculture and Physiology', Liebig stressed the need to supply the mineral elements (Liebig, 1840). The availability of simple, soluble inorganic salts of many of these mineral elements and of nitrogen made it possible to test them singly and in various combinations and compare the yields obtained with those given by FYM. The treatments initially applied to the winter wheat experiment on Broadbalk, started in autumn 1843, were very similar to those which had been used on Barnfield.

In both experiments there were few tests of nitrogen, eight out of 23 plots on Barnfield and five out of 21 plots on Broadbalk, and the maximum amount used was only 15 kg N ha^{-1}. The effects of nitrogen on Barnfield were far from clear. The method of applying fertilizer and drilling the seed brought the ammonium salts and the turnip seed into close proximity and the osmotic effect of the salt impaired germination. Where plants survived they were individually larger than on plots with minerals only but overall there were only small increases in yield because plant population was diminished. By contrast, winter wheat on Broadbalk responded well to the very modest application of readily available nitrogen. Compared to the unmanured crop, minerals alone increased yield by 10%, minerals plus nitrogen by 35%. I suspect that Lawes had not used cereals in his earlier experiments, probably because it was not usual to apply fertilizer to wheat in the four course rotation where it benefited from nitrogen residues from the preceding leguminous crop. Whether the benefits from giving readily available nitrogen was a surprise or not, Lawes and Gilbert were quick to appreciate the significance of the result. For the next crop of winter wheat drilled in autumn 1844 nitrogen was tested on 14 of the 21 plots and at four rates, 12, 24, 36 and 48 kg N ha^{-1}.

Lawes and Gilbert went on to prove the greater need for nitrogen than for minerals for all crops other than legumes. A brief summary of some early results is given in Table 2.2. However, because the fertility of the soils at that time was generally low, they were especially deficient in phosphorus, yields with NPK were always largest.

They also showed that inorganic fertilizers applied at appropriate rates could give the same yield as their generous application of FYM, 35 t ha^{-1} (Table 2.3). Thus they could say something about the relative importance of the nutrients in FYM. More important, at a time when the urban population was expanding rapidly, they could say with some confidence that crop yields could be maintained by the application of plant nutrients in readily available inor-

Table 2.2. Yields, t ha^{-1}, of winter wheat, spring barley, mangolds and beans in the early years of the field experiments at Rothamsted.

Experiment field name	Crop	Years	Treatment			
			None	PK	N	NPK
Broadbalk	Winter wheat grain	1852–61	1.12	1.29	1.63	2.52
Hoosfield	Spring barley grain	1872–81	0.85	1.10	1.66	2.62
Barnfield	Mangolds roots	1876–84	9.5	11.3	25.6	46.0
Geescroft	Field beans[a] grain	1847–62	1.11	1.50	—	1.64

[a] *Vicia faba*

Table 2.3. Yields (t ha^{-1}) of winter wheat and spring barley, grain at 85% dry matter, and roots of mangolds and sugar beet at Rothamsted.

Experiment field name	Crop	Period	Yield with	
			FYM[a]	NPK fertilizers[b]
Broadbalk	Winter wheat	1852–56	2.41	2.52
		1902–11	2.62	2.76
		1955–64	2.97	2.85
		1970–75	5.80	5.47
Hoosfield	Spring barley	1856–61	2.85	2.91
		1902–11	2.96	2.52
		1952–61	3.51	2.50
		1964–67(PA)[c]	4.60	3.36
		1964–67(MB)[c]	5.00	5.00
Barnfield	Mangolds[d]	1876–94	42.2	46.0
		1941–59	22.3	36.2
	Sugar beet	1946–59	15.6	20.1

[a] FYM 35 t ha^{-1} containing on average 225 kg N.
[b] N fertilizer per ha; barley, 48 kg; wheat, 144 kg; root crops, 96 kg.
[c] PA Plumage Archer given 48 kg N ha^{-1}, MB Maris Badger given 96 kg N ha^{-1}.
[d] Mangolds *Beta vulgaris*.

ganic salts if the large amounts of FYM they were testing were not available. Subsequently demand for fertilizer nitrogen increased to such an extent that by the end of the century Sir William Crookes, in his Presidential Address to the British Association in 1898, warned that the world's known reserves of combined nitrogen were being rapidly exhausted.

The Development of the Classicals

The Classicals or Classical experiments is the name now given to those experiments which were started by Lawes and Gilbert and which still continue. Mention was made earlier to the fact that Lawes was interested in animal nutrition and the pioneering experiments undertaken at Rothamsted laid a sure foundation for this branch of agricultural science. In fact in the early years more effort was probably expended on these experiments than on the crop experiments. However, research on animal nutrition gradually diminished and ended in the 1890s after Lawes and Gilbert had published some 50 scientific papers and articles on animal nutrition (see Dyke, 1991; Pirie, 1991).

The development of some of the Classicals, both in the early years and more recently, can be used to illustrate the value of long-term or continuing experiments. In recent times major changes were introduced on Agdell in 1952, on Exhaustion Land in 1957, on Barnfield in 1960 and on Hoosfield and Broadbalk in 1968. None of the changes was undertaken lightly and all followed much discussion and formal approval by the Lawes Agricultural Trust committee (see Leigh *et al.*, Chapter 14). The changes have made the Classicals more relevant to current research needs and to farming practice. All have enhanced the value of the experiments.

Barnfield

Yields of turnips, both tops and roots, in the second year, 1844, confirmed those in 1843; ammonium sulphate again decreased plant population and yields with fertilizers were not appreciably better than on the unmanured plot. For the third year the design of the experiment was modified. The number of plots running the length of the field was decreased to eight to test mineral treatments only – strip treatments. Ammonium sulphate and two other nitrogen fertilizers, sodium nitrate and rape cake (an organic manure used to supply nitrogen) were introduced that year, but applied at right angles to the strip treatments, on what became known as sections (Warren and Johnston, 1962). Germination was no longer seriously impaired and yields of roots of turnips and other root crops (swedes, mangolds and sugar beet) grown later were all increased by nitrogen in the presence of phosphorus.

A further modification in 1856 introduced FYM as a strip treatment, so that combinations of nitrogen fertilizer and FYM were then tested on some plots. Until 1968 this was the only Classical experiment with this treatment. The results, however, were interesting. Yields were always largest where FYM and nitrogen fertilizer were applied together (Table 2.4). To test whether 96 kg N ha⁻¹ as fertilizer was too little, the plots were subdivided in 1968 to test four amounts of nitrogen on four arable crops (Table 2.5). Potatoes, sugar beet,

Table 2.4. Yields (t ha⁻¹) of root crops grown on Barnfield, Rothamsted 1941–59.

	Mangolds[a]		Sugar beet	
Treatment[b]	No N	96 kg N	No N	96 kg N
PK	6.8	36.2	4.5	20.1
FYM	22.3	50.2	15.6	27.9

[a] Mangolds *Beta vulgaris*.
[b] Rates per ha: P, 33 kg; K 225; FYM 35 t.

Table 2.5. Yields of potatoes, sugar beet, spring barley and spring wheat on Barnfield, 1968–73.

		Fertilizer N applied[a]			
Crop	Manuring[b]	0	1	2	3
Potatoes	FYM	24.2	38.4	44.0	44.0
tubers	PK	11.6	21.5	29.9	36.2
Sugar beet	FYM	27.4	43.5	48.6	49.6
roots	PK	15.8	27.0	39.0	45.6
Barley	FYM	4.18	5.40	5.16	5.08
grain	PK	1.85	3.74	4.83	4.92
Wheat	FYM	2.44	3.73	3.92	3.79
grain	PK	1.46	2.97	3.53	4.12

[a] N rates, 0, 1, 2, 3 = 0, 72, 144, 216 kg ha⁻¹ to root crops and 0, 48, 96, 144 kg ha⁻¹ to cereals.
[b] Applied since 1843.

spring wheat and spring barley were grown in rotation, the two root crops and the two cereals in the same year by dividing the strips longitudinally; the large plots that Lawes originally set out allowed such subdivision. Although the largest rates of nitrogen were well in excess of those then used for these crops, yields of potatoes and sugar beet were always larger on the soils which had received FYM, 35 t ha⁻¹, each year since 1856 (Table 2.5). By 1968 these soils had about 2.5 times as much total organic matter as soils given fertilizers only.

This was probably the first suggestion that extra organic matter in Rothamsted soil could have a beneficial effect on yield that could be exploited by using extra nitrogen.

Except for three years, 1853-55, when barley was grown, various root crops were grown each year until 1959 (Warren and Johnston, 1962); then between 1960 and 1974 various arable crops were grown. In 1975 all plots except those on the no nitrogen section were sown to grass. Mineral manures continued to be applied but not FYM, and initially (1976-81) the same amount of nitrogen was applied to all plots. On average, yields of dry matter were larger, by about 2 t ha^{-1}, on soils which had received FYM. This result was not easy to explain because both FYM- and PK-treated soils had about the same amount of readily soluble P and K which was well in excess of that at which grass would be expected to respond to either nutrient. In 1983, a test of four amounts of N was introduced and optimum yields now differed by only about 1 t ha^{-1}, but more nitrogen was needed to get maximum yields on soils low in organic matter than on soils with more organic matter. This suggests that some nitrogen was being diverted from production of aboveground biomass to soil organic matter where there was much carbon available from grass root residues and soil organic matter was not at the equilibrium value for the cropping system.

Broadbalk

Lawes and Gilbert changed N treatments frequently in the first few years often on half plots made by dividing the long narrow plots longitudinally, so that the half plot occupied one land (defined below). This was to consolidate the finding that it was essential to supply readily assimilable or plant available nitrogen to get acceptable yields and that the nitrogen added in fertilizer had little or no residual value. They later regretted these frequent changes and after 1852 most of the plot treatments remained unaltered. There was then no change to the basic design of the experiment for more than 80 years because there were no major problems with growing wheat year after year other than the need to control weeds. Also, the design was simple and the arrangement of the plots provided a living histogram of crop response to nitrogen, applied either as FYM or as fertilizers, on either side of the unmanured plot and that given minerals only. This was, and still is, so striking a visual demonstration of the importance of supplying nitrogen in readily available forms that it is utterly convincing.

One modification in 1849 did, however, have important consequences. In Britain the traditional method of managing heavier textured soils was a ridge and furrow system. The soil over a width of about 3.5-4 m was ploughed towards the centre from each side to create a shallow ridge at the centre and leave a furrow at the sides. The ground between the furrows was called a land.

The width of a land would vary somewhat because the field width was divided into whole lands. The lands were parallel to the slope. Surface water ran into the furrows, down the slope and was taken away in ditches. This explains why initially the plots on Barnfield and Broadbalk were long and narrow. In 1843 Lawes laid out the plots on the existing lands; two lands for the Broadbalk plots, one for those on Barnfield.

In 1849 tile drains, of the horseshoe and sole pattern, were put below the centre furrow of each plot to further aid the removal of surplus water and maintain reasonable soil conditions for cultivation and drilling until late in the year. In 1866 Lawes and Gilbert realized that because there was one tile drain per plot, nutrient losses in drainage could be related to treatment if the drainage was collected and analysed. They went on to show very little loss of phosphorus. Where potassium, sodium and magnesium were applied very little potassium and magnesium were lost; instead much calcium appeared in the drainage water. This provided evidence from a field experiment to support the concept developed some years earlier by Thomas Way, that cation exchange occurred in soil (Way, 1850, 1852). Although ammonium sulphate was applied to many plots, nitrate rather than ammonia was found in the drainage. No reason was offered, microbial nitrification had still to be discovered. However, in retrospect the analyses of the drainage waters between 1878 and 1883 are of great interest. They showed there was a loss of mineral nitrogen even on plots given no nitrogen fertilizer and there was little increase where 48 kg N ha^{-1} was applied (Table 2.6). It is interesting to compare the mean concen-

Table 2.6. Nitrate – N (mg l^{-1}) in drainage water from plots growing winter wheat each year, 1878–83, Broadbalk, Rothamsted.

N applied[a] kg ha^{-1}	Period				
	March to May	June to harvest	Harvest to early Nov	November to March	Whole year
0	1.7	0.2	5.6	4.5	3.9
48	8.1	0.7	7.3	4.8	5.0
96	16.3	1.4	8.3	5.2	6.4
144	21.5	4.0	14.7	7.3	9.3
96[b]	5.7	2.9	7.4	26.4	19.4

[a] All nitrogen applied as ammonium sulphate in spring except [b] where all nitrogen applied in autumn.

tration of nitrate-nitrogen in the annual drainflow with the current EC maximum admissible concentration for nitrate in potable water (11.3 mg NO_3–N l[-1]) introduced over 100 years after these measurements were made. Where 144 kg N ha[-1] was applied in spring the concentration in the drain-flow approached the limit and exceeded it where 96 kg N ha[-1] was applied in autumn. When 144 kg N ha[-1] was applied as ammonium sulphate in spring, this amount was obviously in excess of that which could be used efficiently by the cultivar, Red Rostock, grown at that time. The ammonia had been nitrified and the excess nitrate retained in soil until the autumn rains caused through leaching with enhanced nitrate concentrations (Warrington, 1891).

The early results on Broadbalk showed that wheat could be grown continuously provided nutrients were applied and weeds controlled, at that time by hoeing. By the 1920s labour for this task was too expensive and weed competition became serious. In 1925 the experiment was divided into five sections across the length of the plots. Sections I, II, IV and V were each fallowed twice, in consecutive years, in the next four years, whereas the centre section III was fallowed in all four years. All sections were cropped in 1930 and in 1931 a regular scheme of fallowing was started in which each section was fallowed one year in five (Johnston and Garner, 1969). This change in management was essential to ensure the continuity of the experiment but it did have useful scientific spin-offs. It led to research on: (i) the mineralization of organic nitrogen in the fallow year and its benefit to the next crop of winter wheat (Garner and Dyke, 1969); (ii) the effects of wheat bulb fly (*Delia coarctata Fall.*), which lays its eggs in fallow soil and each emerging larva enters and eventually kills a tiller of wheat so that this pest is especially damaging where poor nutrition limits tillering (Johnson *et al.*, 1969); (iii) weed studies (Thurston, 1969).

The advent of weedkillers allowed a return to continuous winter wheat and in autumn 1955 section I due to be fallowed in 1956 was halved and one half reverted to continuous wheat with weedkillers. Various weedkillers were tested on this half section for eight years and shown to have no recognizable adverse effects. From 1964, weedkillers have been used as required on all the experiments with the exception of one half section (see below) (Johnston and Garner, 1969).

There were two major changes in 1968 on both Broadbalk and Hoosfield. One was to introduce a more flexible approach to the cultivar to be grown. The second was to introduce a three-year rotation consisting of a two-year break from cereals followed by wheat (Broadbalk) or spring barley (Hoosfield) on a part of each experiment. Comparing yields of wheat and barley after a break with those grown continuously allowed the effects of take-all (caused by the fungus *Gaeumannomyces graminis*) and the factors that might inhibit take-all to be studied against a background of a known history of cropping and manuring over many years (Glynne, 1969; Etheridge, 1969).

On Broadbalk this change was accommodated by dividing each of the five sections created in 1925 into two, making 10 sections (0–9). Sections 0 and 1 at the west end and 8 and 9 at the east end of the experiment remained in continuous wheat while sections 2 to 7 were used to test two rotations (i) beans, potatoes, wheat and (ii) fallow, wheat, wheat, the latter to maintain some continuity with the previous test of fallowing. Unfortunately, stem nematode was introduced with one batch of bean seed, it quickly spread and beans had to be replaced by fallow after three cycles of the rotation. Little new was learnt from the shorter rotation with fallow and since 1984 a five-course rotation; fallow, potatoes, wheat, wheat, wheat has been tested. The sixth section reverted to continuous wheat without pesticides but with weedkillers.

Yields of Cappelle Desprez in 1970–78 were discussed in detail by Dyke *et al.* (1983). They used a novel approach involving curve fitting and shifting to distinguish between the effects of nitrogen and other factors on the yields of wheat grown in the different rotations. During 1979–84 cv. Flanders was grown and Fig. 2.1 summarizes yields of both cultivars between 1970 and 1984. Yields of Cappelle grown on PK-treated soils responded little or not at all to more than 96 kg N ha⁻¹. Soils given FYM annually since 1843 had about 2.5 times more soil organic matter than PK-treated soils and yields were a little larger than the best given by fertilizers. Unlike the response by root crops on Barnfield, giving an extra 96 kg ha⁻¹ nitrogen fertilizer decreased grain yield although straw yield was increased. A greater incidence of foliar pathogens was observed on FYM plus N plots and when the cultivar was changed to Flanders in 1979 every effort was made to control mildew by the judicious use of fungicides. Grain yields now responded to the largest amount of nitrogen fertilizer on both PK-and FYM-treated plots and the first wheat after a two-year break yielded appreciably more than continuous wheat (Fig. 2.1). For farmers

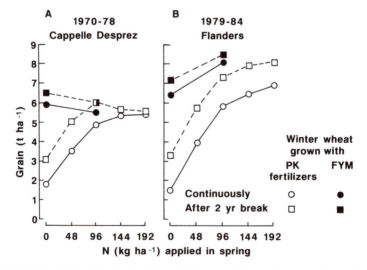

Fig. 2.1. Yields of two cultivars of winter wheat grown either continuously (circles) or after a two-year break (squares) and on soils with either 1.8% soil organic matter (open symbols) or 4.6% organic matter (solid symbols), Broadbalk, Rothamsted.

the practical lesson was clear; grow first wheats as frequently as possible and protect their yield potential by judicious use of fungicides.

In 1985 the cultivar was changed to Brimstone and, because the nitrogen response curves with Flanders had apparently not reached a plateau (Fig. 2.1), the nitrogen treatment on two PK-treated plots was modified to test 240 and 288 kg N ha^{-1}. Some recent results can be summarized as follows.

The lack of response by Cappelle to more than 96 kg N ha^{-1} in 1970–78 (Fig. 2.1) was generally attributed to lack of control of leaf fungal pathogens but growth regulators were not used in this period and occasional lodging with the largest amounts of nitrogen could have contributed. The inclusion of a test of 240 and 288 kg N ha^{-1} from 1985 for Brimstone led to the decision to apply a growth regulator annually but both this and the summer fungicides were tested by their omission from Section 6. The largest increase in grain yield from using both fungicides and growth regulators was about 0.5 t ha^{-1}. This was less than might have been expected from the poor response to nitrogen by Cappelle (Fig. 2.1) but perhaps reflects the fact the Brimstone was less susceptible to mildew than Cappelle whereas the effects of lodging on Cappelle might have been underestimated (R.D. Prew, personal communication).

There was a large benefit from using weedkillers; nearly 2 t ha^{-1} grain with most fertilizer N and over 3 t ha^{-1} on FYM plus N plots. Such large effects are only likely in long-term experiments where not using weedkillers over a long period allows large quantities of weed seed to be returned to the soil each year.

Maximum yields of first wheats after a two-year break were between 1.4 and 1.9 t ha^{-1} grain more than for continuous wheat. Once take-all decline had been broken, the build-up of take-all and the slower build-up of the factor responsible for take-all decline, resulted in the yield of the third wheat after the two-year break being less than that of continuous wheat.

For both Flanders and Brimstone, grown either continuously or in rotation, maximum yields with fertilizers (and more nitrogen than tested previously) exceeded those with FYM alone but largest yields were given by FYM plus N (Table 2.7). Presumably the available nitrogen supplied by the 35 t ha^{-1} FYM together with that mineralized from the extra soil organic matter in this plot, was not sufficient to meet the nitrogen requirement of a modern wheat cultivar where the yield potential was protected by the judicious use of weedkillers, fungicides and, when appropriate, pesticides. It is also probable that the extra soil organic matter contributed to yield potential in ways other than the supply of nitrogen; possible mechanisms were discussed by Johnston (1987).

Amounts of nitrogen removed in harvested grain and straw have been determined for various periods and percentage recovery of applied nitrogen calculated using the difference method (Table 2.8). Recovery of only about 30% in the 1850s improved little by the 1960s but in recent years recoveries by wheat grown continuously have exceeded 60%. This reflects the change in harvest index and the larger concentrations of nitrogen in grain. Extra grain

Table 2.7. Maximum yields of winter wheat, grain (t ha⁻¹) given by fertilizers, farmyard manure and farmyard manure plus fertilizer N, Broadbalk, Rothamsted.

| | Cultivar and years grown | | | |
| | Flanders 1979–84 | | Brimstone 1985–90 | |
Treatment	Continuously	In rotation	Continuously	In rotation
NPK[a]	6.93	8.09	6.69	8.61
FYM[b]	6.40	7.20	6.17	7.89
FYM+N[c]	8.13	8.52	7.92	9.36

[a] Maximum yield of Flanders given by 192 kg N ha⁻¹ and of Brimstone by 288 kg N ha⁻¹.
[b] 35 t ha⁻¹ FYM.
[c] 96 kg N ha⁻¹.

yield where wheat was grown in rotation (Table 2.8) boosted recoveries to over 80%. These nitrogen recoveries on Broadbalk tend to be larger than in many other experiments when determined by the difference method. This is because soil organic matter, after 150 years of continuous wheat and unchanged fertilizer applications, is in equilibrium. Thus on the plot given PK fertilizers only, soil nitrogen effectively does not contribute to the small (about 30 kg ha⁻¹) nitrogen content of the harvested crop. When this small amount of nitrogen is subtracted from the nitrogen content of the crop given fertilizer nitrogen the recovery of the added nitrogen is large. However, it is important to note that even with current large recoveries of the larger amounts of applied nitrogen not all of the smaller amounts were recovered. This indicates that there are loss processes removing nitrogen from the system irrespective of the amount applied (see Powlson, Chapter 6 and Jenkinson *et al.*, Chapter 7).

Hoosfield

The experiment on continuous spring barley started on Hoosfield in 1852. The original design, four strip treatments with minerals (O, KNaMg, P, PKNaMg) with nitrogen treatments at right angles built on the 1845 modification to the design of the Barnfield experiment. FYM, however, was not included as a strip treatment. Initially, 48 and 96 kg N ha⁻¹ as ammonium salts were tested together with rape cake as an organic nitrogen manure. The 96 kg N treatment invariably caused severe lodging and the amount was decreased to 48 kg in 1858. Sodium nitrate replaced ammonium salts in 1868. Before that, in 1864, this section had been halved to test sodium silicate (Warren and Johnston, 1967). Lawes and Gilbert introduced this test to see whether silicate – a main

Table 2.8. Percentage recovery of fertilizer nitrogen by winter wheat on Broadbalk, Rothamsted.

Wheat grown	N applied, kg ha⁻¹					
	48	96	144	192	240	288
Continuously						
1852–57	32	33	32	29	—	—
1966–67	32	39	36		—	—
1970–78	56	63	59	52	—	—
1979–84	69	83	76	69	—	—
1985–88	67	73	64	58	58	52
In rotation						
1970–78	71	68	56	46		
1979–84	83	88	80	71		
1985–88	69	81	82	72	64	59

constituent of straw – would improve its quality for ease of hand plaiting. At that time, straw plaiting was a major cottage industry for the wives and children of farm labourers, the straw plaits being sold to the hat factories in Luton, 8 km away. Silicate had little benefit on the quality of straw for plaiting but it did increase barley grain yields where phosphorus fertilizer was not applied. Exchange of silicate for phosphate ions is the most plausible explanation (Warren and Johnston, 1967).

Weeds were never the problem on Hoosfield that they were on Broadbalk because late autumn ploughing and spring cultivation controlled them effectively and the experiment has only been fallowed four times since 1852.

In 1968 all plots were subdivided to test 0, 48, 96 and 144 kg N ha⁻¹; yields of Maris Badger in 1966–67 (Table 2.3) showed that improved varieties would respond to more nitrogen. On the section which had tested rape cake, barley was grown in a three-course rotation, beans, potatoes, barley, and yields were compared with those of continuous barley. Yields were increased little by growing barley in rotation because take-all is less of a problem with spring barley than with winter wheat and the rotation was discontinued after 1979.

The nitrogen response curves on the PK fertilizer treated soils were very informative (Fig. 2.2). Nitrogen was used efficiently only when the soil was adequately supplied with phosphorus and potassium. This is an essential message for farmers today especially when financial constraints make them decrease inputs, which have little immediate visual impact, yet these inputs are essential to maintain soil fertility.

Yields of three cultivars grown between 1970 and 1990 on soils differing in organic matter by a factor of about 2.5, because of repeated applications of either FYM or PK fertilizer, are shown in Fig. 2.3. Yields of cv. Julia (1970–79)

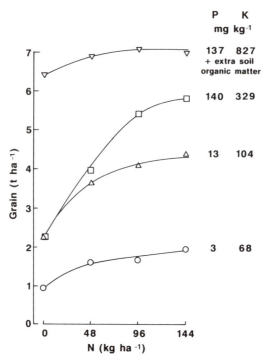

Fig. 2.2. Response by spring barley cv Triumph, to nitrogen on soils with different concentrations of readily soluble phosphorus and potassium and with extra soil organic matter (4.6 *vs* 1.8%), Hoosfield, Rothamsted.

followed the long-standing pattern of identical yields with FYM and NPK fertilizers, here 96 kg N ha⁻¹. In 1980–85 cv. Georgie required 144 kg N ha⁻¹ to give the same yield as FYM but there was an indication of a potential to respond to further fertilizer nitrogen on FYM-treated soil. In 1986–90 cv. Triumph yielded more on FYM-treated soil than with fertilizers and there was an appreciable response to nitrogen fertilizer on FYM-treated soil.

This pattern of larger yields on soils with more organic matter where additional nitrogen is supplied is now emerging quite strongly, especially for spring grown crops which, to achieve large yields, must develop quickly and therefore require good soil physical conditions (Johnston, 1986, 1991a). This observation may well be of importance to European agriculture in the future because many of the 'new' crops such as linseed, sunflowers, maize, are sown in spring.

The effects of FYM on the accumulation of soil organic matter in this experiment are discussed by Jenkinson *et al.* (Chapter 7). Lawes and Gilbert halved the FYM plot in 1872 and stopped FYM applications on one half. Plant-available nitrogen, mineralized from the extra organic matter accumulated during 20 years, declined quite rapidly at first and then more slowly (Jenkinson

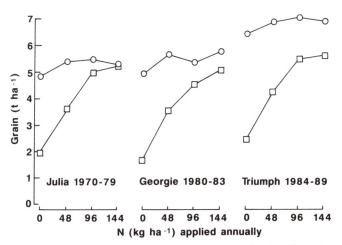

Fig. 2.3. Yields of three cultivars of spring barley grown on soils with different amounts of soil organic matter, Hoosfield, Rothamsted. Soils given farmyard manure (○) since 1852 now contain 4.6% organic matter; those given PK fertilizers (□) now contain 1.8% organic matter.

et al., Chapter 7, Fig. 7.4). Compared to the PK-treated soil, yields were doubled by the mineralized nitrogen in the first 10 years and for the following 20 years were increased a little (Warren and Johnston, 1967). Residues of phosphorus and potassium had also accumulated during this 20-year period and 100 years later they are still available to plants. When all the plots were subdivided in 1968 to test four amounts of nitrogen, spring barley was able to utilize the residues and yields increased dramatically between 1970 and 1974 (Table 2.9) (Johnston and Poulton, 1992). After another 20 years although barley yields were less than where FYM was being applied each year, the residues of phosphorus and potassium were still giving larger yields than soils without any residues (Table 2.9). The effectiveness of the residues after such a long period is of great interest; clearly they had not become irreversibly 'fixed' in soil. While no nitrogen was applied the soil was not stressed to supply P and K but once nitrogen was given in 1968 the residues were still plant available. This result suggests that on these and similar soils a manuring policy that seeks to build up PK residues to enhance soil fertility is justified (see below). Johnston and Poulton (1977) also discussed in detail the history of the Exhaustion Land experiment and the evaluation of P and K residues accumulated from both fertilizers and FYM during 1856–1901.

Park Grass

The Park Grass experiment, started in 1856, was the last of Lawes and Gilberts' major field experiments. The results from the experiments on arable crops, started periodically from 1843 (Table 2.1), had highlighted differences between crops representative of the great botanical families, especially in the amounts and ratios of nitrogen, phosphorus and potassium they required. Lawes and Gilbert were intrigued as to whether different species from these same families but found in the mixed sward of permanent grassland would respond similarly.

Table 2.9. Effect of P and K residues accumulated in soil between 1852 and 1871 on yields (t ha⁻¹) of spring barley in 1970–73 and 1988–91, Hoosfield, Rothamsted. (Adapted from Johnston and Poulton, 1992).

Nitrogen (kg ha⁻¹) since 1968	Unmanured since 1852		FYM applied 1852–71 only		FYM annually since 1852	
	1970–73	1988–91	1970–73	1988–91	1970–73	1988–91
None	1.46	0.98	1.07	2.20	5.38	5.50
48	2.01	1.43	3.20	3.24	5.86	5.94
96	2.12	1.73	5.27	3.48	5.62	6.06
144	2.26	1.66	4.94	3.63	5.38	6.00

A possible site was selected on part of the parkland to the south of the Manor House. The site had been in permanent grassland for some hundreds of years and had never received the large applications (up to 250 t ha⁻¹) of chalk which had been applied to the fields on which the experiments on arable crops had been sited. After two years of testing the uniformity of the site, plots ranging in size from 0.1 to 0.2 ha were established to test N, P and K singly and in various combinations and compare the effects with those of FYM. All fertilizers were applied each year, P K Na Mg in late winter, N once only in spring. Initially FYM (35 t ha⁻¹) was applied each year but it decomposed slowly and the residues accumulated on the surface and adversely affected yields. FYM was not applied between 1864 and 1904, for details of treatments testing FYM since 1905 see Warren and Johnston (1964).

In June each year the herbage on each plot was cut and made into hay on its own plot before being weighed and a sample taken for dry matter determination, chemical analysis and storing in the archive. This continued until 1959. From 1960 a part of each plot was cut by flail harvester at about the same time as previously, the herbage was weighed green and yields, unaffected by haymaking losses, were recorded as dry matter. The traditional hay making was maintained on the rest of the plot because in the past the shed seed might have helped maintain the botanical composition of the sward. The growth between haymaking and late autumn, the aftermath, was grazed by sheep until 1874, but after that it was cut, weighed green and yields recorded as dry matter;

Warren and Johnston (1964) give details of the changes in manuring and man-agement including the application of lime (see below).

The fertilizer treatments quickly changed the proportions of grasses, leg-umes and weeds in the herbage. Only two years from the start of the exper-iment, PK fertilizers had increased legumes from 5 to 20% (dry weight basis) and weeds were rare. Ammonium sulphate alone or with PK fertilizers elimin-ated the legumes and most of the weeds leaving a herbage with 90% or more grasses. In 1859, Lawes and Gilbert wrote '... the very varying degree in which they (the manures) respectively developed the different kinds of plants ... that the experimental ground looked almost as much as if it were devoted to trials with different seeds as with different manures'. The first botanical assessment appears to have been made by Professor Henfry in 1859. Much later, in 1896, Lawes commented that the experiment was 'the battlefield of the plants'. The early results led Lawes and Gilbert to realize that,unlike the wheat, barley and turnips in their other experiments, weight alone was not a satisfactory mea-sure of the worth of the crop and they gave more emphasis to the botanical composition. For reference to published work on botanical composition see Warren and Johnston (1964), and Thurston *et al.* (1976), and also Tilman *et al.* (Chapter 16).

To the practising farmer also yields were of little interest. Most grass fields are grazed more frequently than they are mown for hay and grazing has a major effect on species composition and most grassland receives more than one application of nitrogen each year. There are, however, some interesting aspects related to yield. There was a stepwise increase in yield on all plots from 1960 when the method of estimating yields was changed. However, examin-ation of the yields of herbage dry matter both before and after 1960 shows that yields have not declined with any treatment, all treatments have been sustain-able but at different levels of production. By the mid 1960s best yields on Park Grass were only about half those obtained on short-term grass leys. To test whether permanent grass could be as productive, microplots were established on plot 6 in 1965, to test up to 432 kg N ha^{-1} and more frequent harvesting. Annual yields, averaged over three years, exceeded those on intensively man-aged three-year grass leys because yields from the leys were small in the first establishment year. Recently there have been suggestions that increasing car-bon dioxide concentrations in the air would enhance plant growth. The rela-tive stability of the botanical composition of the unmanured plots and those getting nitrogen (as sodium nitrate) with P and K has allowed a test of this hypothesis (Jenkinson *et al.*, 1994). Using the statistical approach adopted, no increase in yield attributable to increased carbon dioxide could be detected. The authors concluded that the same was likely to be true for all ecosystems limited by either water and/or nitrogen.

After a very small test of lime (CaCO$_3$) in 1881, Lawes and Gilbert tested one application of lime, first to one half and then to the other half of most plots, between 1883 and 1896. There was little effect on either yield or botanical composition. In 1903, Hall halved most plots and started a test of 4 t ha^{-1} CaCO$_3$ applied cumulatively every fourth year (Warren and Johnston, 1964). In 1923, Crowther (1925) determined the pH of the 0–23 and 23–46 cm soil horizons on all plots. Where 96 or 144 kg N ha^{-1} had been applied as ammonium sulphate the pH of the surface soil was below 4.0 having declined from about 5.8 in 1856. The pH on the limed halves of the plots ranged from 4.1 to 4.8. By 1959 the pH of the unlimed soils had decreased only a little, that of the limed soils had increased further (Warren and Johnston, 1964). In 1965 the half plots were halved yet again to give subplots a, b, c, d and during the years that followed chalk was applied to try to achieve pH 7, 6 and 5 on subplots a, b and c, whilst subplot d remained without lime to assess the extent to which treatment and other acidifying inputs would decrease soil pH. On plots with a mat or thatch of only partially decomposed organic material it proved very difficult to raise the pH of the underlying mineral soil until the mat had been decomposed by increased microbial activity following liming (Johnston, 1972).

The combination of manurial treatment and the wide range (3.7–7.5) of soil pH gave major differences in botanical composition as shown in Table 2.10 which summarizes more detailed results given by Warren and Johnston (1964) (see also Thurston *et al.*, 1976 and Tilman *et al.*, Chapter 16). Today, as noted by Tilman, much interest centres on the large database on the botanical diversity on the different plots.

There have been other interesting studies. Johnston *et al.* (1986) compared changes in pH on the unmanured soils on Park Grass and Geescroft Wilderness. Left untended since 1886 the Geescroft Wilderness is now deciduous (mainly oak) woodland. The 0–23 cm depth of soil on Geescroft has become more acid (by 3 pH units) more quickly (in a 100 years) than the Park Grass soil (by 1 pH unit) in 150 years. This difference probably relates to the efficiency with which trees trap acidifying compounds in the atmosphere passing through the aboveground canopy and transferring them to soil.

Where superphosphate (supplying 33 kg P ha^{-1}) has been applied annually to the surface of the Park Grass plots for more than 130 years, P now enriches the subsoils below 23 cm. Similar enrichment of subsoils below 23 cm has not been detected where the same annual applications of superphosphate have been applied to the arable experiments for 140 years. Downwards movement of P has occurred on FYM-treated soils in the arable experiments. This suggests that low molecular weight organic P molecules may be involved in the downward movement of P in the aqueous phase (Johnston and Poulton, 1992). This has implications for the causes of enrichment of surface water with P which can play a role in eutrophication.

When similarly treated soils from the different Classical experiments were compared for their cadmium content there had been significant increases in soil cadmium from aerial deposition, especially since the 1940s. It was, however, only on the acid, organic soils of Park Grass that cadmium had accumulated from that applied in superphosphate. Increasing levels of cadmium, from

Table 2.10. Effect of manuring and soil on grass species in 1947–49, Park Grass, Rothamsted, given as % species present in more than 10% by weight of grass fraction.

Treatment	pH 3.7–4.1		pH 4.2–6.0		pH 6.0–7.5	
N1	*A. tenuis*	79	*D. glomerata*	36	*D. glomerata*	27
	F. rubra	16	*A. pratensis*	20	*F. rubra*	26
			F. rubra	13	*H. pubsescens*	22
			A. odoratum	12		
			H. lanatus	12		
N2P	*A. tenuis*	44	*F. rubra*	59		
	F. rubra	23	*A. pratensis*	28		
	H. lanatus	20				
	A. odoratum	10				
N2PK	*H. lanatus*	91	*A. pratensis*	38	*A. elatius*	48
			A. elatius	28	*D. glomerata*	14
			C. cristatus	15	*C. cristatus*	14
			D. glomerata	13	*A. pratensis*	13
			P. pratensis	10		
N3PK	*H. lanatus*	96	*A. pratensis*	72		
			A. elatius	18		
			P. pratensis	11		
			D. glomerata	10		

Fertilizers (kg ha^{-1}): N, 1, 2, 3; 48, 96, 144; P, 33; K, 200.
Agrostis tenuis, fine bent; *Alopecurus pratensis,* meadow foxtail; *Anthoxanthum odoratum,* sweet vernal; *Arrhenatherum elatius,* tall oat; *Cynosurus cristatus,* crested dogstail; *Dactylis glomerata,* cocksfoot; *Festuca rubra,* red fescue; *Helictotrichon pubescens,* downy oat; *Holcus lanatus,* Yorkshire fog; *Poa pratensis,* smooth stalked meadow grass.

aerial deposition, in neutral, low organic matter soils on Broadbalk and Hoos-field had not caused increases in cadmium in wheat and barley grain. On Park Grass, the herbage grown on acid soils (pH 5) has much enhanced concentrations of cadmium compared to those in herbage from limed soils with pH about 7.0 (Johnston and Jones, 1992).

Agdell

The experiment on Agdell started in 1848 and tested arable crops grown in a four-course rotation until 1951. To evaluate the residues of P and K accumulated during the previous 100 years, the six plots of the original experiment were divided in 1958 to grow grass or arable crops on each half plot. In 1964 each half plot was subdivided into eight and on four of these sub-sub-plots, four amounts of phosphorus were added while the other four received four amounts of potassium (Johnston and Penny, 1972 and Leigh *et al.*, Chapter 14). These applications of P and K were superimposed on the six levels of readily soluble P and K in soil from the manuring between 1848 and 1951. In addition, by 1970, soils which had been in grass since 1958 contained 2.4% organic matter whereas those which continued to grow arable crops had 1.5% organic matter. Thus there was a matrix of soils with 24 levels of readily soluble P and 24 levels of readily soluble K both at two levels of soil organic matter.

By further subdivision into microplots spring barley, potatoes and sugar beet, were grown in rotation to relate yields to readily soluble P in soil and response to fresh P fertilizer at each level of soil P (1970-72) and to readily soluble K and response to fresh K fertilizer (1973-76). The relationship between yields of potatoes and readily soluble P (Fig. 2.4) illustrates the results obtained. The marked difference in the shape of the response curve on the two soils again illustrates the importance of soil organic matter. In this case, the effect was almost certainly an improvement in soil structure because when all 48 soils were cropped with ryegrass in pots in the glasshouse, yields were all on one response curve which was similar in shape to that for potatoes on the soil with more organic matter. Soil organic matter, however, had no effect on the response of any of the three crops to readily soluble K in soil (Johnston and Goulding, 1990). This was probably because the soils testing K had received so much superphosphate that they contained more than 60 mg kg^{-1} readily soluble P. Figure 2.4 shows that at this concentration the P yields were the same on both soils so that differences in soil structure did not affect the acquisition of P and presumably also of K.

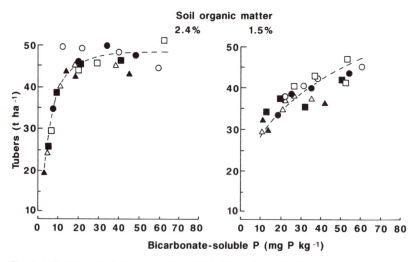

Fig. 2.4. Relationship between readily soluble P in soil and yields of potatoes grown on soils with either (a) 2.4% or (b) 1.5% soil organic matter, Agdell, Rothamsted.

The Productivity and Sustainability of Agricultural Systems

It is essential that the biological, chemical and physical factors which control the productivity of agricultural systems are identified. Once identified, it will be necessary to determine, for each factor, critical values below which yields of crops will decline appreciably, i.e. productivity will not be maintained or enhanced. Research discussed above suggests that we can define such critical values for readily soluble P and K and organic matter in Rothamsted soils. The values vary for different crops but soils should be maintained to meet the needs of the most demanding crop likely to be grown. The sustainability of the system can then be tested in long-term experiments.

Winter wheat has been grown on all or part of Broadbalk year after year for the last 150 years. Yields declined somewhat in the 1920s due to weed competition but, once weeds were controlled, initially by fallowing then by weed-killers, yields first recovered and then, with the introduction of high yielding cultivars and best management practices, they have increased appreciably (Fig. 2.5). This is a remarkable example of a sustainable system. So too are the Park Grass and Hoosfield experiments. Barley grain yields on Hoosfield, like those of wheat on Broadbalk, have increased in recent years. But judged by crop yield, cropping and manuring on Agdell were unsustainable (Table 2.11).

Table 2.11. Yields (t ha^{-1}) of turnip roots and winter wheat grain in the Agdell rotation experiment, Rothamsted, 1848–1951.

| | Crop and treatment | | | |
| | Turnip[a] roots | | Wheat grain | |
Year	None	NPK	None	NPK
1848–51	1.31	2.92	1.91	1.93
1852–83	0.24	3.47	1.46	1.96
1884–99	0.13	5.09	1.60	2.47
1900–19	0.11	3.98	0.97	1.37
1920–35	0.08	1.61	0.98	0.91
1936–51	0.04	0.54	1.27	2.07
Soil pH				
1953	8.2	5.6		

[a] Turnips *Brassica rapa.*

After the 1930s yields of turnips given NPK fertilizers were little better than those on soils without manuring for 100 years; winter wheat yields did not

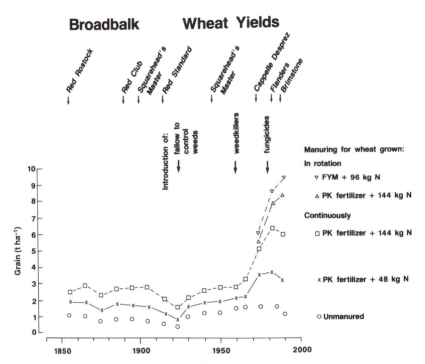

Fig. 2.5. Yields of winter wheat grown on Broadbalk, Rothamsted, from 1852 to 1990 with fertilizers and with farmyard manure showing the effects of changing cultivars and the introduction of weed control, fungicides and crop rotation to minimize effects of soil-borne pathogens.

decline so severely. On NPK-treated soil, the nitrogen was applied as ammonium sulphate and the soils became acid (pH$_{water}$ 5.6 in 1953). Once the acidity was recognized it could be, and later was, corrected for the experiments from 1958 onwards described above. However, acidity encouraged the development of 'finger-and-toe' caused by the fungus *Plasmodiophora* sp. which attacks the roots of brassicas and decreases yield. Once established, this fungus is difficult to eradicate and turnips could not be grown satisfactorily.

Another example of non-sustainability is the decline in yields of spring barley in the 1880s on the Classical Barley site at Woburn. The soil at Woburn is a sandy loam with an initial pH of about 6.0. Yields decreased initially due to the development of acidity where ammonium sulphate was used to supply nitrogen, but liming only increased yields slightly (Johnston and Powlson, 1994). In this case continuous cereal culture failed, perhaps because the light-textured soil favoured the survival and build-up of cereal cyst nematodes. An important part of the sustainability of agricultural systems is flexibility of cropping and a decline in any aspect of soil fertility which limits choice is to be deplored.

Long-term or continuing experiments are the best practical way of assessing the sustainability of an agricultural system (Johnston and Powlson, 1994; Powlson and Johnston, 1994). Comparing different husbandry systems often allows the possibility of explaining failure if this occurs. It also aids the quest for modifications that may be necessary to maintain, and where appropriate, increase production. In this respect it is interesting to ask why Lawes and Gilbert persisted with some of their experiments which have become the Classicals of today.

Why Did Lawes and Gilbert Continue?

I would suggest that, in part, the initial impetus came from their various controversies with Liebig (Johnston, 1991c). Of the three mentioned here perhaps the most important was the first. This was the controversy about the source from which plants get their nitrogen. Lawes and Gilbert (1855) eloquently summarized the cause of the controversy by putting into juxtaposition the two contradictory statements by Liebig:

'Cultivated plants receive the same quantity of nitrogen from the atmosphere as trees, shrubs, and other wild plants; BUT THIS IS NOT SUFFICIENT FOR THE PURPOSES OF AGRICULTURE'.
1st Edition, p. 85, 1840.

'Cultivated plants receive the same quantity of nitrogen from the atmosphere as trees, shrubs, and other wild plants; AND THIS IS QUITE SUFFICIENT FOR THE PURPOSES OF AGRICULTURE'.
3rd and 4th Editions, p.54, 1843.

Results from the field experiments at Rothamsted supported the 1840 statement, not that of 1843, yet Liebig stubbornly refused to retract his 1843 statement and the ensuing argument lasted over 20 years (Johnston, 1991c).

Then there was Liebig's frequently repeated statement that the cropping and manuring at Rothamsted would rapidly exhaust the soil of minerals, and thereby diminish its productivity. This did not happen. The third source of disagreement was Liebig's assertion that it was only necessary to replace the quantity of minerals actually removed in the harvested crops. Lawes and Gilbert showed that this was not true – to get good yields they had to apply more minerals than the crop contained. This was probably because at that time the soil contained little plant-available phosphorus and potassium.

I suspect that Lawes and Gilbert needed to go on showing that they were right on all counts including these three. And not only right, but right beyond all reasonable doubt. And so they kept the experiments going. Liebig was so wily and his apparent dislike of Lawes and Gilbert so intense that if the experiments had been stopped Liebig might have construed this as Lawes and Gilbert being in error whilst he was correct. Perhaps the benefits to agricultural science from the Rothamsted experiments owe more to Liebig than he has been given credit for in the past! There may also be a lesson for today, especially for those funding research. A healthy sense of competition, springing from a lack of consensus, can often lead to greater and more rapid progress than consensus and resulting complacency does.

It is also true that Lawes was a great educationalist. He wanted farmers to benefit by the results of Rothamsted research (see also Brown, Chapter 3, for his comment about the start of the Sanborn Plots). In his response to the Earl of Chichester on the presentation of the Testimonial Laboratory on 19 July 1855, Lawes said '. . . the object of these investigations (at Rothamsted) is not exactly to put money into my pocket but to give you the knowledge by which you may be able to put money into yours' (*The Agricultural Gazette*, 1856). What better way than living experiments with large plots to impart knowledge.

Taylor (1989), discussing long-term experiments in general, gives a somewhat different, and possibly inaccurate, picture of some aspects of the early history of these Rothamsted experiments. He quotes a reference (which I cannot trace), that at one stage Lawes ordered the termination of the Broadbalk experiment. G.V. Dyke (personal communication) in his very detailed perusal of all Lawes' papers and existing correspondence, can find no document to support Taylor's statement. Taylor's reference was to a book by a former Director of Rothamsted, E.J. Russell (1966), in which Russell does make, however, a very valid and important comment. It stemmed from the fact that Lawes financially supported the Rothamsted experiments from his own income. In part this derived from his manure business, until it was sold, and for a shorter period of time from a royalty of 10 shillings on every ton of superphosphate made by other manufacturers in England and Wales. This royalty (equivalent to about £15 a ton today) must have been a major source of income and Lawes fought two costly legal battles to uphold his patent and therefore the royalty income. Russell's comment was to raise the hypothetical question, 'What would have happened to the Rothamsted experiments if the case had gone against Lawes and the royalty income ceased?' Continued financial support for long-term experiments then, as now, is crucial.

Once the initial 20-year period had been completed and the nitrogen controversy had largely run its course, Lawes and Gilbert saw in the experiments the opportunity to solve other problems. In 1869 Hermann Liebig, son of Justus, wrote to Lawes and Gilbert as follows: 'The object for which they (the field experiments) were made is now attained, a continuation of them in the same way, is according to my opinion quite aimless, because they are not well adapted for the settlement of other questions.' He went on to suggest modifications to test ideas of his own. Lawes and Gilbert's reply was polite but adamant:

> We feel quite sure that if you possessed full and accurate knowledge
> of the experiments and the results to which you refer you would not

for a moment suppose ... that the objects for which they have been so many years carried on are now all attained and that to continue them on the same general plan any longer is useless. ... more would be lost than gained by disturbing the existing arrangement
(Letters in the Rothamsted Archive, now catalogued by Margaret Harcourt-Williams).

Later, Lawes came out strongly on the importance of continuous investigations in agriculture (Lawes, 1882). Then again in 1895, summarizing the results of the first 50 years, Lawes and Gilbert said that the experiments had become more valuable with time (Lawes and Gilbert, 1895). Today, in the 1990s, they have proved to be even more valuable than Lawes and Gilbert could ever have imagined not only in relation to agriculture but to environmental problems which are of concern in the final decades of the 20th century.

Sir John Lawes made a lot of money from his patent and manure business. He could readily have become a business tycoon and local squire but his intense interest in agriculture decreed otherwise. He maintained financially the Rothamsted experiments from his own resources and, using part of the proceeds from the sale of his manure business in 1872, he endowed Trustees in 1889 to ensure that the experiments he had started with Gilbert might continue. In 1935, money raised by public subscription allowed the Trustees to purchase the estate and security of tenure of the sites of the Classical experiments was assured. Today financial support comes from other sources but agricultural and ecological science, and the interface of both with the wider environment, reap great benefit from Lawes' generosity and Lawes and Gilbert's scientific objectivity and foresight. Rothamsted now has a well-documented database extending over a 150-year period, a unique archive of crop and soil samples over the same period which can be interrogated to answer questions Lawes and Gilbert could not have conceived, and continuing experiments which, I suggest, have at least as much to offer in the future as they have yielded in the past.

References

The Agriculture Gazette (1856) Mr Lawes of Rothamsted. *The Agricultural Gazette* January 1856.

Crowther, E.M. (1925) Studies on soil reaction, 3, 4, 5. *Journal of Agricultural Science, Cambridge* 15, 201–221, 222–231, 232–236.

Dyke, G.V. (1991) *John Bennett Lawes: The Record of His Genius*. Research Studies Press, Taunton; John Wiley, Chichester.

Dyke, G.V., George, B.J., Johnston, A.E., Poulton, P.R. and Todd, A.D. (1983) The Broadbalk Wheat experiment 1968–78. Yields and plant nutrients in crops grown continuously and in rotation. *Rothamsted Experimental Station Report for 1982*, Part 2, 5–44.

Etheridge, J. (1969) Take-all on Broadbalk Wheat, 1958-1967. *Rothamsted Experimental Station Report for 1968*, Part 2, 137-139.

Garner, H.V. and Dyke, G.V. (1969) The Broadbalk Yields. *Rothamsted Experimental Station Report for 1968*, Part 2, 26-49.

Glynne, M.D. (1969) Fungus diseases of wheat on Broadbalk, 1843-1967. *Rothamsted Experimental Station Report for 1968*, Part 2, 116-136.

Jenkinson, D.S., Potts, J.M., Perry, J.N., Barnett, V., Coleman, K. and Johnston, A.E. (1994) Trends in herbage yields over the last century on the Rothamsted Long-term Continuous Hay Experiment. *Journal of Agricultural Science, Cambridge* 122, 365-374.

Johnson, C.G., Lofty, J.R. and Cross, D.J. (1969) Insect pests on Broadbalk. *Rothamsted Experimental Station Report for 1968*, Part 2, 141-156.

Johnston, A.E. (1972) Changes in soil properties caused by the new liming scheme on Park Grass. *Rothamsted Experimental Station Report for 1971*, Part 2, 177-180.

Johnston, A.E. (1986) Soil organic matter, effects on soils and crops. *Soil Use and Management* 2, 97-105.

Johnston, A.E. (1987) Effects of soil organic matter on yields of crops in long-term experiments at Rothamsted and Woburn. *INTECOL Bulletin* 15, 9-16.

Johnston, A.E. (1991a) Soil fertility and soil organic matter. In: Wilson, W.S. (ed.) *Advances in Soil Organic Matter Research: The Impact on Agriculture and the Environment*. Royal Society of Chemistry, Cambridge, pp. 400.

Johnston, A.E. (1991b) Potential changes in soil fertility from arable farming including organic systems. *Proceedings No. 306 The Fertiliser Society*, Peterborough, 1-38.

Johnston, A.E. (1991c) Liebig and the Rothamsted experiments. In: Judel, G.K. and Winnewisser, M. (eds) *Symposium '150 Jahre Agrikulturchemie'*, Justus Liebig-Gesellschaft zu Giessen, Giessen, pp. 37-64.

Johnston, A.E. and Garner, H.V. (1969) The Broadbalk Wheat Experiment: Historical Introduction. *Rothamsted Experimental Station Report for 1968*, Part 2, 12-25.

Johnston, A.E. and Goulding, K.W.T. (1990) The use of plant and soil analyses to predict the potassium supplying capacity of soil. In: *Development of K-fertilizer Recommendations*. International Potash Institute, Berne, pp. 177-204.

Johnston, A.E. and Jones, K.C. (1992) The cadmium issue - long term changes in the cadmium content of soils and crops grown on them. In: Schultz, J.J. (ed.) *Phosphate Fertilizers and the Environment*. International Fertilizer Development Centre, Muscle Shoals, USA, pp. 255-270.

Johnston, A.E. and Penny, A. (1972) The Agdell Experiment, 1848-1970. *Rothamsted Experimental Station Report for 1971*, Part 2, 38-67.

Johnston, A.E. and Poulton, P.R. (1977) Yields on the Exhaustion Land and changes in the NPK contents of the soils due to cropping and manuring, 1852-1975. *Rothamsted Experimental Station Report for 1976*, Part 2, 53-85.

Johnston, A.E. and Poulton, P.R. (1992) The role of phosphorus in crop production and soil fertility: 150 years of field experiments at Rothamsted, United Kingdom. In: Schultz, J.J. (ed.) *Phosphate Fertilizers and the Environment*, International Fertilizer Development Centre, Muscle Shoals, USA, pp. 45-63.

Johnston, A.E. and Powlson, D.S. (1994) The setting up, conduct and applicability of long-term, continuing field experiments in agricultural research. In: Greenland, D.J. and Szabolcs, I. (eds) *Soil Resilience and Sustainable Land Use*. CAB International, Wallingford, pp. 395-421.

Johnston, A.E., Goulding, K.W.T. and Poulton, P.R. (1986) Soil acidification during more than 100 years under permanent grassland and woodland at Rothamsted. *Soil Use and Management* 2, 3-10.

Lawes, J.B. (1882) The future of agricultural field experiments. *Agricultural Students Gazette*, New Series 1, 33-35.

Lawes, J.B. and Gilbert, J.H. (1855) Reply to Baron Liebig's Principles of Agricultural Chemistry. *Journal of the Royal Agricultural Society of England* 16.

Lawes, J.B. and Gilbert, J.H. (1895) The Rothamsted experiments over fifty years. *Transactions of the Highland and Agricultural Society of Scotland Fifth Series* 7, 11-354.

Liebig, J. (1840) *Organic Chemistry in its Application to Agriculture and Physiology*. Taylor and Walton, London.

Pirie, N.W. (1991) The role of Rothamsted in making nutrition a science. *The Biochemist* 13, 8-13.

Powlson, D.S. and Johnston, A.E. (1994) Long-term field experiments - their importance in understanding sustainable land use. In: Greenland, D.J. and Szabolcs, I. (eds) *Soil Resilience and Sustainable Land Use*. CAB International, Wallingford, pp.367-394.

Russell, E.J. (1966) *A History of Agricultural Science in Great Britain*. George Allen and Unwin, London, pp. 493.

Taylor, L.R. (1989) Objective and experiment in long-term research. In: Likens, G.E. (ed.) *Long-term Studies in Ecology: Approaches and Alternatives*. Springer-Verlag, New York, pp. 20-70.

Thurston, J.M. (1969) Weed studies on Broadbalk. *Rothamsted Experimental Station Report for 1968*, Part 2, 186-208.

Thurston, J.M., Williams, E.D. and Johnston, A.E. (1976) Modern developments in an experiment on permanent grassland started in 1856; effects of fertilizers and lime on botanical composition and crop and soil analysis. *Annales Agronomique* 27, 1043-1082.

Warren, R.G. and Johnston, A.E. (1962) Barnfield. *Rothamsted Experimental Station Report for 1961*, 227-247.

Warren, R.G. and Johnston, A.E. (1964) The Park Grass Experiment. *Rothamsted Experimental Station Report for 1963*, 240-262.

Warren, R.G. and Johnston, A.E. (1967) Hoosfield Continuous Barley. *Rothamsted Experimental Station Report for 1966*, 320-338.

Warrington, R. (1891) Drainage and well water. Lecture 6 on Investigations at Rothamsted Experimental Station. *USDA Experiment Station Bulletin No.8*, 95-112.

Way, J.T. (1850) On the power of soils to absorb manure. *Journal of the Royal Agricultural Society of England* 11, 313-379.

Way, J.T. (1852) On the power of soils to absorb manure. *Journal of the Royal Agricultural Society of England* 13, 123-143.

3

The Sanborn Field Experiment

J.R. BROWN
University of Missouri, 144 Mumford Hall, Columbia, MO 65211, USA and the Missouri Agricultural Experiment Station, USA.

Those of us who have survived into our seventh decade of life on this planet have observed many changes in agriculture. As I was trying to decide how best to relate the 'Sanborn Field Experiment', I naturally needed to study the history of the field although many of the real benefits of the field have come to light during my association with the experiment. All these benefits, however, resulted from J.W. Sanborn's desire to learn and then demonstrate the results of that learning.

Another facet of the experiment was revealed to me by a response to a lament I made recently to Dr C.M. (Woody) Woodruff, a retiree now in his ninth decade. I had commented that many studies done by younger faculty repeat older reports. Woody pointed out to me that we should not deprive the younger folks of the opportunity to learn from their mistakes. Therein lies part of the 'Sanborn Field Experiment'.

History of Sanborn Field

At the time of the establishment of Sanborn Field, farming was relatively new in the central United States in comparison to the lengthy history of farming in Europe. The literature from the 1880s suggests, however, that yields even then were declining in the central US as fields originally in prairie grasses were farmed to domesticated plants (Anon., 1888).

J.W. Sanborn, Dean of Agriculture in Missouri in the 1880s, started the Rotation Field to address this problem of declining yields. It seemed he wished to demonstrate the value of crop rotation and barnyard manure (farmyard manure or animal manure). In modern terminology, he wished to demonstrate those farming practices that would return farming to a sustainable level.

©CAB INTERNATIONAL, 1994. From R.A. Leigh and A.E. Johnston (eds),
Long-term Experiments in Agricultural and Ecological Sciences.

In order to appreciate fully Sanborn's actions, one needs to consider the history of how the opportunity to initiate Sanborn Field came about. We know how Rothamsted came to be but the background of the long-term sites in the US is quite different.

One must presume that sometime after President Thomas Jefferson approved the Louisiana Purchase in 1803, migrants from the 'Colonies' headed west and progressively broke the prairie as they continued their westward movement. The exact date of a significant influx of immigrants into the Missouri territory is hard to define. In 1818, Henry Rowe Schoolcraft is reported to have commented on grasses in Missouri tall enough to hide a man on horseback (Nigh and Houf, 1993). Handwritten notes by Dean Sanborn (1888) concerning the site of Sanborn Field suggest it was an unimproved site. It is likely that the area had been grazed but it is unknown if it had a history of tilled crops. (This one experience from the Sanborn Field archives demonstrates the need for thorough records.)

The Louisiana Purchase and the decline in soil productivity was not enough on its own to cause initiation of Sanborn Field. The formation of the University of Missouri and the Agricultural Experiment Station in Columbia finished setting the stage. The Geyer Act of 1839 by the Missouri State Legislature established the University of Missouri. The Morrill Act passed by the Federal Congress in 1862 resulted in the formation, in 1870, of the College of Agriculture and Mechanic Arts as a division of the University of Missouri. Funding remained a problem but, in 1887, the Federal Congress passed the Hatch Act establishing agricultural experiment stations and made an annual appropriation of federal funds to the experiment stations (Longwell, 1970). The agricultural experiment station in Missouri was assigned to the College of Agriculture. Thus, Dean Sanborn finally had access to a mechanism and funds to initiate the action needed to determine the cause of the decline in soil productivity.

The Rotation Field, therefore, was established as a result of political reaction to perceived needs. In 1924, in honour of Dean Sanborn's foresight, the Rotation Field was renamed Sanborn Field.

The Studies of the First Half-century

In 1888, the Missouri farmer had a primary objective – to feed the family. This objective was achieved by growing crops on the farm both for human consumption and for animals providing meat, milk, eggs, etc. There was also a necessity for some off-farm purchases thus the need for saleable produce such as corn (*Zea mays* L.) and wheat (*Triticum aestivum* L.). The need to grow saleable produce meant draft animals were needed to expand the production beyond that resulting solely from manual labour. Draft animals required feed and produced manure. Therefore, several different crops were grown on the farm and manure had to be removed from the barns used to house the animals.

The initial Sanborn Field files of Dean Sanborn's contain no explanations for the selection and assignment of crop sequences and treatments on the Rotation Field (Sanborn, 1888, unpublished notes). The treatments would appear to demonstrate how crops could be fitted into a rotation to satisfy the above needs and to demonstrate the value of manure (Table 3.1). Particularly puzzling, however, are the multiple replications of manure treatments and few replications of the other treatments. (The files contain little attributable to Dean Sanborn after 1889 when he was forced to leave Missouri because of political conflicts.)

Table 3.1. Initial practices used on the Rotation (Sanborn) Field, 1888. (After Upchurch *et al.*, 1985.)

	Soil treatments (no. of plots)			
Cropping practice	None	Manure[a]	Fertilizer	Fertilizer + manure
Continuous cropping to				
Wheat (W)	2	6	1	0
Corn (C)	1	1	0	0
Oat[b] (O)	1	1	0	0
Clover[b] (Cl)	1	1	0	0
Timothy[b] (T)	1	1	0	0
Rotations				
W, Cl	1	2	0	0
C, W, Cl	1	3	0	0
C, O, W, Cl	1	4	0	0
C, O, W, Cl, T, T	1	6	1	1[c]

[a] Manure at 13.4 Mg ha^{-1}.
[b] Oat, *Avena sativa* L.; Cl, clover initially sweet clover, *Melilotus alba* Desr. and/or *M. officinalis* (L) Lam., red clover (*Trifolium repens*) was also used; Timothy, *Phleum pratense* L.
[c] Manure at 6.7 Mg ha^{-1} and fertility for a 'half-crop'.

The field was managed along the lines established by Dean Sanborn until the 1913 crop season; some changes were made starting with the 1914 season. Following a summary of 30 years of activity in 1921 (Miller and Hudelson, 1921), other changes in the management of many of the plots were made starting with the 1928 crop season followed by further changes in 1940 and 1950 (Upchurch *et al.*, 1985).

The changes made in the management combinations on the field in the first 50 years resulted in a reduction in the number of manured plots and a significant increase in fertilized plots (Tables 3.2 and 3.3). Continuous cropping with oats and cowpea proved to be impractical even under experimental conditions and growing clover continuously ceased because of disease. Cowpea was not well adapted to Missouri conditions.

Tables 3.1–3.3 indicate that the directors of the experiment found that manure was not a cure-all for nutrient problems as the shift was toward commercial fertilizer or to use of both manures and fertilizers. It is also evident that wheat was the crop of most interest during this period; in the second 50 years, corn replaced wheat as the key crop. This was, most likely, due to the development of corn hybrids during the 1930s, which caused corn to become more efficient at converting sunlight into commercial product.

Table 3.2. Cropping practices used on Sanborn Field between 1914 and 1927. (After Upchurch *et al.*, 1985.)

	Soil treatments (no. of plots)			
Cropping practice	None	Manure	Fertilizer	Fertilizer + manure
Continuous cropping to				
Wheat (W)	1	3	5	0
Corn (C)	1	1	0	0
Oat (O)	1	1	0	0
Timothy (T)	1	1	0	0
Cowpea[a] (Cp)	1	1	0	0
Rotations				
W, Cl	1	2	0	0
C, W, Cl	2	2	0	0
C, O, W, Cl	2	1	2	0
C, O, W, Cl, T, T	1	2	1	5

[a] Cowpea, *Vigna unquiculata* (L.) Walp.

Fertilizers used during the first part of the Sanborn Field Experiment were ordinary superphosphate, ammonium sulphate, and sodium nitrate. Some mixed fertilizers were used but these contained less than ten units of nutrients per ton.

Table 3.3. Cropping practices used on Sanborn Field between 1928 and 1939. (After Upchurch *et al.*, 1985.)

	Soil treatments (no. of plots)			
Cropping practice	None	Manure	Fertilizer	Fertilizer + manure
Continuous cropping to				
Wheat (W)	1	3	5	0
Corn (C)	1	1	0	0
Timothy (T)	1	1	0	0
Oat (O)	1	1	0	0
Rotations				
W, Cl	1	1	2	0
C, O, W, Cl	2	1	4	0
C, O, W, Cl, T, T	1	2	1	5
C, S, W(Cl)[a]	0	0	3	0

[a] C, S, W(Cl) is a three-year rotation including soyabean (*Glycine max* (L.) Merr.) and a clover green manure crop.

Early Results

The first published summary of results from Sanborn Field covered the initial 30 years (Miller and Hudelson, 1921). The results illustrated the value of crop rotation, the mixed blessings of farmyard manure and the value of the return of crop nutrients to the soil at least in the quantities removed by the crops. The authors also concluded that, as a rule, crops grown in rotation yielded more than did crops grown continuously, although there were exceptions (Table 3.4). At that time, manured plots yielded on a par with those receiving inorganic fertilizers (Table 3.5). One must remember, however, that the crop varieties were those that had evolved under relatively infertile conditions by modern standards and no hybrid vigour was involved in any crop variety.

Smith (1942) followed the same format as Miller and Hudelson (1921) in his summary of 50 years of cropping. It seems that by 1937 (50 crop years), the system had stabilized (Tables 3.6 and 3.7 compared with Tables 3.4 and 3.5, respectively). In fact, Smith's summary was much like that of Miller and Hudelson of 20 years earlier. Based upon these two reports, one might conclude that all the lessons had been learned from Sanborn Field and the experiment could have been terminated.

Fortunately for the continuation of Sanborn Field, the publication of Smith's summary was delayed until 1942. The field scientists responsible for Sanborn Field were swept up in the war effort and Dr W.A. Albrecht, the chairperson of the Soils Department, chose to continue the field per a revised plan

Table 3.4. Crop yields from untreated plots (1888–1917). (From Miller and Hudelson, 1921, Table 2.)

Cropping system	Crop yield (t ha⁻¹)					Average Annual Value[a] ($/ha)
	Corn	Oat	Wheat	Clover	Timothy	
6-year rotation	2.6	1.0	1.4	2.4	2.7	35.77
4-year rotation	2.4	1.0	1.6	2.9	—	44.02
3-year rotation	2.0	—	1.0	2.1	—	35.02
2-year rotation	—	—	1.2	3.3	—	39.69
C-continuous	1.3	—	—	—	—	31.91
O-continuous	—	0.6	—	—	—	18.67
W-continuous	—	—	0.6	—	—	24.01
Cl-continuous	—	—	—	2.7	—	26.85
T-continuous	—	—	—	—	2.9	28.48

[a] 1889 to 1918 average prices including valuation of stover and straw.

initiated in 1940. Several events occurred between 1930 and 1949 that made it fortuitous that Sanborn Field was not terminated. Dr William Albrecht had carried out some soil nitrogen work using Sanborn Field samples that opened new vistas in the evaluation of crop management effects on soils. Hybrid corn, rather than open pollinated varieties, became the accepted seed for field planting after the Second World War. The nitrogen fertilizer industry grew from the munitions industry after VE/VJ days ended the Second World War. Agriculture shifted from animal power to gasoline power enabling expanded grain production on land formerly used to sustain the animal power for agriculture. The

Table 3.5. Comparison of yields from plots untreated, manured and fertilized (1888–1917). (After Miller and Hudelson, 1921, Table 8.)

Cropping system	Yield (t ha⁻¹)		
	Untreated	Manured	Commercial fertilizer
Continuous wheat	0.6	1.2	1.3
6-year rotation			
Corn	2.6	2.8	2.6
Oat	1.0	0.9	1.4
Wheat	1.4	1.8	2.0
Clover	2.4	4.5	4.1
Timothy (2-yr avg.)	2.7	4.9	4.3

Table 3.6. Crop yields (t ha⁻¹) from untreated plots (1888–1937). (After Smith, 1942, Table 7.)

Cropping System	Corn	Oat	Wheat	Clover	Timothy
6-year rotation	2.4	1.1	1.1	2.3	2.8
4-year rotation	2.3	1.0	1.5	2.5	—
3-year rotation	2.0	—	0.9	1.9	—
Corn-continuous	1.2	—	—	—	—
Oat-continuous	—	0.7	—	—	—
Wheat-continuous	—	—	0.7	—	—
Timothy-continuous	—	—	—	—	2.8

Marshall Plan and other postwar programmes put greater emphasis on food production in the United States.

Although the 50-year report, therefore, suggested that Dean Sanborn's objectives had been met, it is fortunate that William Albrecht was able to keep the field operational through the Second World War and had the resource available for the study of postwar changes in agriculture.

Application of Science to Sanborn Field

Scientific agriculture blossomed after the Second World War. The scientists who left the campus to assist in the war effort returned with broadened outlooks and experiences. Farming practices changed due to the improvements in machinery and manufacturing resulting from the war effort. Ammonium nitrate and anhydrous ammonia were the stars of an expanding industry that provided relatively cheap fertilizer. The GI Bill, which provided a college education for the war veterans, created a generation of educated people who were

Table 3.7. Comparison of yields (t ha⁻¹) from plots untreated, manured and fertilized (1888–1937). (After Smith, 1942, Table 12.)

Crop sequences	Nutrient management		
	Untreated	Manure	Fertilizer
Continuous wheat	0.7	1.3	1.4
6-year rotation			
Corn	2.3	3.1	2.7
Oat	1.1	1.3	1.5
Wheat	1.0	1.8	2.1
Clover	2.3	5.0	3.9
Timothy (2-year avg.)	2.9	5.7	6.3

focused because of the maturity arising from their wartime experiences. At the Missouri Agricultural Experiment Station, the members of the Soils Department were in their heyday; Sanborn Field was an important aspect of postwar activities.

William Albrecht did considerable work with the soil samples collected from Sanborn Field from 1914 onward. The database which resulted was used to produce a soil nitrogen decay curve (Woodruff, 1950). This nitrogen work was continued and expanded by George Wagner who studied both soil nitrogen and carbon (see later).

Dr Albrecht, in 1945, was asked by Dr Benjamin Duggar to collect soil samples from different locations in Missouri. Dr Duggar was searching for organisms that produced antibiotics. A sample was taken from the untreated continuous timothy plot on Sanborn Field from which *Streptomyces aureofaciens* was isolated. This organism was found to produce the antibiotic called aureomycin which was released to the medical profession in 1948 by Lederle Laboratories (Brown, 1993).

In 1950, Dr George Smith initiated a new series of studies on Sanborn Field but also retained several plots in the original (1888) plan. This was the start of 40 years of relative stability during which the Missouri scientists charted changes in soil properties without the impact of 'another new plan'.

The postwar era also saw improvements in techniques, crop varieties, weather measurement, and concepts of the soil–plant–environment system. The availability of ^{32}P and ^{15}N, allowed an unprecedented expansion in soil research. Concepts of cation exchange, soil acidity and soil morphology became more sophisticated. Thus, in this postwar period, Dr Sanborn's namesake provided a resource used by fertile minds to explain soil changes resulting from management.

Post-Second World War Results

The nitrogen work initiated by Dr Albrecht during the 1930s and 1940s and the decay equation adapted by Dr Woodruff both suggested that the soil organic fraction under a given cropping system tended toward an equilibrium value. The results of Albrecht and Woodruff were based upon cropping systems from 1888 through 1949 when crop residues were removed to simulate the use of such residues for animal feed. Dr Smith's changes starting in 1950 included return of residues to the plot of origin (Table 3.8).

Conceptually, new equilibrium levels of soil organic matter should have been achieved under the post-1950 cropping systems. However, immediate large changes in soil properties would not be expected so little except yield results were reported from Sanborn Field until about 1960. Dr George Wagner replaced Dr Albrecht as the station soil microbiologist in the late 1950s and he started making use of the varied organic systems to be found on the individual

Table 3.8. Outline of the Sanborn Field management plan initiated in 1950. (After Upchurch *et al.*, 1985.)

	Nutrient management (no. of plots)		
Crop sequences	Untreated	Manure	Other[a]
Continuous cropping			
Wheat	1	2	1
Corn	1	1	2
Timothy	1	1	1
Clover	0	0	1
Alfalfa	0	0	1
Alfalfa-grass	0	0	1
Rotations			
C, W, Cl	1	0	6
C, O(Ls), W(Ls)[b]	0	0	3
C, S. W(Cl)	0	0	3
C, O, W, Cl	1	1	6
C, O, W, Cl, T, T	1	1	1

[a] Includes chemical fertilizers and combinations of fertilizer and manure.
[b] Ls refers to annual lespedeza, *Lespedeza stipulacea* (Maxim.). The parentheses designate a green manure crop.

Sanborn Field plots. C.M. Woodruff shifted from the theoretical to applied soil management and Sanborn Field became a focus for considerable activity. Such activity was noted and, in 1965, the field was designated a National Historic Landmark by the US National Park Service. (This was a non-financial designation but it served to prevent car park development.)

Experiences during this period showed that thermophilic bacteria counts in most Sanborn Field soils were less than mesophyllic counts (Fields *et al.*, 1974). A sample of soil from a well-fertilized plot annually seeded to corn was used successfully to detect soluble *Rhizobium japonicum* antigens (Kremer and Wagner, 1978). Monthly microbial counts showed that, when either crop residues or manures were incorporated into soil, the numbers of bacteria and actinomycetes were 10–40-fold those found in soils on the plots untreated since 1888 (Martynuik and Wagner, 1978).

Laboratory studies using low P soil from Sanborn Field demonstrated the value of adequate soil P to assure nodulation of soybean (Wagner *et al.*, 1978). The different cropping systems on Sanborn Field resulted in the proliferation of different organisms in the soil. For example, greater numbers of *Aspergillus* sp. were found where the plots were mouldboard ploughed than when no-till culture was used (Angle *et al.*, 1982). This finding was deemed a result of incorporation of corn residue into the soil and is one illustration of the impact of different crop management practices.

The dynamics of carbon have been studied using wheat, corn and soyabean (Buyanovsky and Wagner, 1983). Periods of greatest CO_2 in soil air under wheat occurred when crop residues were decomposing. In contrast, highest CO_2 concentrations in soil air under corn and soyabean were found when active root respiration occurred during plant growth. Net primary production available for synthesis of soil organic matter was greater under native prairie than continuous wheat. This observation offered at least a partial explanation for the decline in soil productivity reported in 1888 (Buyanovsky *et al.*, 1987).

Changes in soil organic matter (SOM) following several decades under different cultural systems have provided a resource for the study of carbon dynamics. Samples of soil from Sanborn Field have been used in recent years to study the influence of macroclimate on soil microbial biomass (Insam *et al.*, 1989), to study the SOM turnover using natural ^{13}C abundance (Balesdent *et al.*, 1988), and for the evaluation of the pool size of stable SOM (Hsieh, 1992).

Other studies have included partaking in a national study on the accumulation of Cd in plants from the phosphatic fertilizers containing less than 10 ppm Cd; Cd accumulation was negligible (Mortvedt, 1987).

Although the lack of initial soil samples has limited many of the usual attempts to demonstrate erosion and absolute profile changes, the selection of a study location in an area representative of a major soil catena has proved useful. The presence of a well-developed argillic horizon in the dominant soil on Sanborn Field has been used as a baseline by Gantzer *et al.* (1990) to evaluate degree of erosion (Table 3.9).

Considerable criticism has been made of Sanborn Field, as with other long-term study sites, because of the lack of replication and randomization as well as the failure of the founders to consider spatial variability. Such critics do not recognize that the century-old sites were started prior to R.A. Fisher's first edition of *The Design of Experiments* in 1935 (Fisher, 1951) and prior to the development of the science of soil. What the founders did not know could not be used! The study by Gantzer *et al.* (1990) did demonstrate that the past 100 years of management had resulted in differences between plots (Table 3.9). The differences in topsoil depth, clay content, and concentration of organic matter, although relatively small, have profound effects upon the tillage properties of the respective plots. The lack of organic matter on the unfertilized corn plot causes the clay properties (stickiness, tendency to puddle, etc.) to be profound whereas the greater organic matter content on other plots masks these negative effects of soil clays. These effects using hydraulic conductivity as an indicator were addressed by Anderson *et al.* (1990) (Table 3.9).

The management of Sanborn Field over time has also influenced the nutrient status of both topsoil and the entire soil profile. Manure is usually considered an organic waste product and its value as a soil amendment has most often been based on the organic properties and nitrogen content. However,

Table 3.9. Influence of long-term management on Sanborn Field surface soils, 1988. (From Gantzer *et al.*, 1990 and Anderson *et al.*, 1990.)

Management	Surface soil depth[a] (cm)	Clay (%)	Organic matter (%)	Hydraulic conductivity (m hr^{-1})
Continuous corn				
Unfertilized	18	27	1.0	0.058
Manured	21	30	2.4	1.089
6-year rotation				
Unfertilized	31	15	1.4	0.004
Manured	29	18	2.9	0.058
Fertilized	34	18	2.0	0.005
Timothy				
Unfertilized	45	17	2.0	0.262
Manured	44	17	3.4	1.037

[a] Depth to the B$_t$ horizon.

phosphorus, always a major component of animal excreta, is now receiving the attention it deserves because of its effects on eutrophication if leached into water courses.

Sanborn Field soil samples have been taken periodically since 1914 and have been stored with the exception of those for 1949. Jirapunvanich (1987) used samples from selected plots to evaluate three different phosphorus extractants. In this study, the fluctuation of extractable soil P over time on the manured plots became evident. In general, extractable P declined before the Second World War even on the manured plots (Table 3.10). After the Second World War, the extractable P increased on the manured plots. We suspect that manure used before the Second World War contained considerable amounts of animal bedding, which were low in P. After the Second World War, we know the manure came from locations where animals were confined and where little or no bedding was used. The conclusion is that manure is an unbalanced, variable source of nutrients for crops and modern manure management may cause build-up of P in soils.

Curiosity about the balance of nutrients seemed to have started during the first 25 years of Sanborn Field. One of the changes made, starting with the 1928 crop season, was to terminate the two-year wheat–clover rotation that was manure based and start a three-year rotation of corn–soyabean–wheat with a green manure clover crop during the wheat year (Tables 3.2 and 3.3). The unique feature of this change was to supply phosphorus and potassium to one plot, withhold phosphorus on one plot and withhold potassium on the third

Table 3.10. Extractable phosphorus (μg P g^{-1}) over time in selected Sanborn Field surface soil samples. (From Jirapunvanich, 1987.)

Management	Date of sample				
	1915	1928	1938	1962	1980
Continuous corn					
Unfertilized	12	11	11	10	4
Manured	50	38	38	46	91
6-year rotation					
Unfertilized	12	NA	12	8	4
Manured	38	50	38	35	36
Continuous timothy					
Unfertilized	16	18	14	10	4
Manure	44	58	41	51	58

plot. This rotation received no manure after 1928 and, until 1950, the only N available was residual from the soyabean and clover. Starting in 1950, all three plots received N for a 'full' corn crop and 37 kg N ha^{-1} at winter wheat seeding. The impact of this management on yields during the 1970–1990 crop years showed the great need for P for the winter wheat crop, the relative sensitivity to lack of K for the corn crop, and the insensitivity of soyabean to the treatments (Table 3.11). The effect of 61 years of depletion of P by cropping is clear in Table 3.12. The K profile (not shown) shows depletion of subsoil K in the upper argillic horizon.

Summary

The Sanborn Field Experiment has been an experience for many staff and students. It has been a living experience especially for the many college students who are one or more generations removed from the farm. The lessons, some of

Table 3.11. Influence of nutrient omissions on grain yields from Sanborn Field, 1970–90.

Fertility programme	Crop yield (t ha^{-1})		
	Corn	Soyabean	Wheat
Full fertility (FF)	7.6	2.0	4.2
FF minus K	7.1	2.1	4.4
FF minus P	7.8	2.1	3.0

Table 3.12. The distribution of extractable P[a] in soil profiles after 61 years of cropping (1928-88).

Depth increment (cm)	Extractable P (μg P g^{-1})	
	Full treatment	Minus P
0–10	35	4
10–20	18	3
20–30	6	2
30–40	8	2
40–50	13	2
50–60	17	2
60–70	18	3
70–80	16	3
80–90	12	5
90–100	12	8

[a] P extractable in 0.025 M-HCl and 0.03 M-NH$_4$F.

which were summarized in popular fashion by Brown (1993), have added to our understanding of the soil–plant–environment continuum.

Recent activity on C and N balances suggests new uses for the Sanborn Field resource. In spite of its limitations of being non-replicated, too small, and surrounded by buildings and small parks, Sanborn Field will continue to be used to document the effects of management on the soil resource.

References

Anderson, S.H., Gantzer, C.J. and Brown, J.R. (1990) Soil physical properties after 100 years of continuous cultivation. *Journal of Soil Water Conservation* 45, 117-121.

Angle, J.S., Dunn, K.A. and Wagner, G.H. (1982) Effect of cultural practices on the soil population of *Aspergillus flavus* and *Aspergillus parasiticus*. *Soil Science Society of America Journal* 46, 301-304.

Anon. (1888) Announcement. *Missouri Agricultural Experiment Station Bulletin 1*.

Balesdent, J., Wagner, G.H. and Mariotti, A. (1988) Soil organic matter turnover in long-term field experiments as revealed by carbon-13 natural abundance. *Soil Science Society of America Journal* 52, 118-124.

Brown, J.R. (1993) 100 years - Sanborn Field. *Missouri Agricultural Experiment Station Publication* MX 390.

Buyanovsky, G.A. and Wagner, G.H. (1983) Annual cycles of carbon dioxide level in soil air. *Soil Science Society of America Journal* 47, 1139-1145.

Buyanovsky, G.A., Kucera, C.L. and Wagner, G.H. (1987) Comparative analyses of carbon dynamics in native and cultivated ecosystems. *Ecology* 68, 2023-2031.

Fields, M.L., Chen Lee, P.P. and Wang, D. (1974) Relationships of soil constituents to spore counts and heat resistance of *Bacillus stearothermophilus*. *Canadian Journal of Microbiology* 20, 1625-1631.

Fisher, R.A. (1951) *The Design of Experiments*, 6th edn. Hafner, New York.

Gantzer, C.J., Anderson, S.H., Thompson, A.L. and Brown, J.R. (1990) Estimating soil erosion after 100 years of cropping on Sanborn Field. *Journal of Soil Water Conservation* 46, 641-644.

Hsieh, Y. (1992) Pool size and mean age of stable soil organic carbon in cropland. *Soil Science Society of America Journal* 56, 460-464.

Insam, H., Parkinson, D. and Domsch, K.H. (1989) Influence of macroclimate on soil microbial biomass. *Soil Biology and Biochemistry* 21, 211-221.

Jirapunvanich, M. (1987) Extractable phosphorus in historical Sanborn Field soil samples. Unpublished MS Thesis, Library, University of Missouri, Columbia.

Kremer, R.J. and Wagner, G.H. (1978) Detection of soluble *Rhizobium japonicum* antigens in soil by immunodiffusion. *Soil Biology and Biochemistry* 10, 247-255.

Longwell, J.H. (1970) The Centennial Report 1870-1970 of the College of Agriculture. *Missouri Agricultural Experiment Station Special Bulletin* 883.

Martynuik, S. and Wagner, G.H. (1978) Quantitative and qualitative examination of soil microflora associated with different management systems. *Soil Science* 125, 343-350.

Miller, M.F. and Hudelson, R.R. (1921) Thirty years of field experiments with crop rotation, manure and fertilizers. *Missouri Agricultural Experiment Station Bulletin* 182.

Mortvedt, J.J. (1987) Cadmium levels in soils and plants from some long-term soil fertility experiments in the United States of America. *Journal of Environmental Quality* 16, 137-142.

Nigh, T. and Houf, L. (1993) In the footsteps of Schoolcraft. *Missouri Conservationist* 54, 5, 9-15.

Smith, G.E. (1942) Sanborn Field - Fifty years of field experiments with crop rotations, manure and fertilizers. *Missouri Agricultural Experiment Station Bulletin* 458.

Upchurch, W.J., Kinder, R.J., Brown, J.R. and Wagner, G.H. (1985) Sanborn - Field historical perspectives. *Missouri Agricultural Experiment Station Research Bulletin* 1054.

Wagner, G. H., Kassein, G.M. and Martynuik, S. (1978) Nodulation of annual medicago by strains of *R. meliloti* in a commercial inoculant as influenced by soil phosphorus and pH. *Plant and Soil* 50, 81-89.

Woodruff, C.M. (1950) Estimating the nitrogen delivery of soil from the organic matter determination as reflected by Sanborn Field. *Soil Science Society of America Proceedings* 14, 208-212.

Long-term Field Trials in Australia

P. GRACE AND J.M. OADES
Cooperative Research Centre for Soil and Land Management, Waite Campus, Urrbrae, South Australia 5064.

Introduction

The areas of Australia suitable for intensive agricultural production are primarily determined by climatic factors affecting water availability. Less than 20% of the continent receives reliable rainfalls that support the growth of agricultural crops and improved pastures (Williams and Raupach, 1983). Approximately one-half of the continent's 7682 million hectares has a median annual rainfall of less than 300 mm, and 80% less than 600 mm. Over 75% of the continent, rainfall does not exceed evaporative loss in any month of the year (Taylor, 1983).

More than 30% of Australia's total agricultural production comes from the temperate semiarid zone on the cooler sides of the deserts extending from Geraldton, north of Perth, across southern Western Australia and South Australia and into the Wimmera, Mallee and northern districts of Victoria (Rovira, 1992). Thus, the majority of long-term agronomic studies (25+ years) in Australia are in this zone (Fig. 4.1) which is dominated by winter rainfall and Mediterranean climate. The soils range from deep, coarse-textured sands and sandy loams (Quartzipsamment), to duplex soils with coarse-textured sands over clay and fine-textured red–brown earths (Rhodoxeralf) and are derived from weathered, ancient rocks with no new exposures of fresh rock which occurred in the northern hemisphere due to glaciation. The widespread occurrence of inherent deficiencies of soil phosphorus and nitrogen have played a major part in the development of dryland farming practices in southern Australia, and led to the use of superphosphate and leguminous pastures, to increase N supply, and the development of rotations including several years of leys.

Fig. 4.1. Geographical location of long-term trials in Australia with respect to the semiarid climate zone.

The first recommendation to use superphosphate for wheat was made by Custance in South Australia in 1885 (Donald, 1964) and since then this practice has been commonplace throughout southern Australia (McClelland, 1968). Evidence on the value of legumes in the cropping sequence began to emerge in the 1940s (Hayman, 1943; Morrow *et al.*, 1948; Purchase *et al.*, 1949) and the use of subterranean clover subsequently revolutionized cereal growing methods in the country (Sims, 1953). Donald's (1963) review of Australian wheat yields for the period 1870–1960 (Fig. 4.2) is testimony to the fact that these inputs have played dominant roles in Australia's agronomic history.

The low and variable rainfall in semiarid areas and the lack of diversity of species that can be grown, limits options for increasing yields (Russell, 1977). For example, fallowing was one practice that was developed early in such areas but, although it may improve yields and reduce the risk of failure in marginal environments, fallowing also promotes a rapid decline in soil organic matter with conventional cultivation systems (Clarke and Russell, 1977). Up to the 1920s the value of fallow was based mainly on American data. Then, in the late 1920s and early 1930s many experiments were started in semiarid regions of southern Australia to examine the longer-term effects of fallowing and balance these against short-term yield advantages (Sims, 1977). Clarke and Russell (1977) suggested that these experiments, many continued over many years,

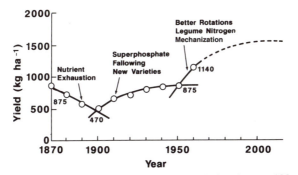

Fig. 4.2. The mean decennial yields of wheat in Australia during the past 100 years (Donald, 1964).

attempted to answer four key questions in relation to yields (predominantly wheat) and soil properties: (i) the effects of fallow, (ii) the effects of fertilizer, principally superphosphate, (iii) the effects of various lengths of pasture and its species composition and quality, and (iv) the effect of cereal crop rotations, including grain legumes. Interest in long-term comparisons of cultivation practices (conventional *vs* minimum or no-till), stubble handling (burning *vs* retention or mulching) and nitrogen applications has occurred only since the late 1970s and early 1980s (Carter and Steed, 1992; Chan *et al.*, 1992). To date, only data from these recent experiments have been analysed thoroughly, whereas those from on-going, long-term experiments in Australia have been somewhat neglected. For this reason only a selection of trials, mainly from locations in southern Australia, will be discussed in detail here. This list is probably not exhaustive.

Long-term Trials – Past and Present

Of the agronomic trials (Table 4.1) of at least 25 years duration undertaken in Australia during the past century 23 have been associated with state departments of agriculture. The exceptions are the Permanent Rotation (C1) and Top-dressing of Natural Pastures (P1) experiments at the Waite Agricultural Research Institute, which is part of the University of Adelaide in South Australia. Five of the 13 trials which still exist were initiated over 50 years ago, and the oldest, at the Rutherglen Research Institute in north-central Victoria, has been maintained for nearly 80 years. The emergence of south-western Western Australia as a major grain production area is reflected in the fact that a significant number of well-documented long-term trials commenced in this region in the late 1960s (Table 4.1). Of the eight trials with known termination dates since 1955, six were at centres (Mallee Research Station, Walpeup; Wimmera Research Station, Dooen and the Waite Agricultural Research Institute, Urrbrae), which still maintain other long-term experiments.

Table 4.1. Long-term agronomic trials in Australia.

Location	Type[a]	Commenced[b] (year)	Soil type[c]	Rain (mm)
Rutherglen (Vic)	C(RR1)	1913 (1973)	Natrixeralf	593
Rutherglen (Vic)	C(GM)	1913 n.a.	Natrixeralf	593
Rutherglen (Vic)	P	1914	Natrixeralf	593
Werribee (Vic)	C	1914 n.a.	Rhodoxeralf	541
Werribee (Vic)	C(GM)	1914 n.a.	Rhodoxeralf	541
Dooen (Vic)	C(LR1)	1917	Torrert	425
Dooen (Vic)	C(GM)	1917 n.a.	Torrert	425
Kybybolite (SA)	P	1919 (1958)	Natrustalf	549
Glen Innes (NSW)	C	1921	Natrustalf	796
Urrbrae (SA)	P(P1)	1925 (1955)[d]	Rhodoxeralf	627
Urrbrae (SA)	C(C1)	1926	Rhodoxeralf	627
Walpeup (Vic)	C(MR1)	1936 (1973)	Calciorthid	330
Walpeup (Vic)	C(MM1)	1940	Calciorthid	330
Rutherglen (Vic)	P	1946	Natrixeralf	593
Walpeup (Vic)	C(MC4)	1946 (1971)	Calciorthid	330
Walpeup (Vic)	C(MR2)	1946 (1975)	Calciorthid	330
Dooen (Vic)	C(LR2)	1946 (1981)	Pellustert	425
Walpeup (Vic)	C(MR3)	1951 (1980)	Calciorthid	330
Tamworth (NSW)	C	1966	Pellustert	690
Merredin (WA)	C(66M29)	1966	Pellustert	302
Newdegate (WA)	C(67N4)	1967	Natrustalf	390
Chapman (WA)	C(67C13)	1967	Palexeralf	412
Esperance (WA)	C(68E5)	1968	Paleaquult	496
Salmon Gums (WA)	C(68SG5)	1968	Calciorthid	337
Warwick (Qld)	C	1968	Pellustert	717

[a] P, continuous pasture trial; C, crop rotation trial, location specific identifier in parentheses.
[b] termination year in parentheses.
[c] Soil taxonomy approximation only (Soil Survey Staff, 1975).
[d] approximate termination date.

n.a.= not available.

The long-term effects of crop and pasture sequences on crop yield and soil properties, in particular C, N, P contents and pH, provide a common thread to the existing trials, with a shift away from long-term pasture productivity studies, a theme which dominated the establishment of some of the earliest trials in the early 1900s.

The long-term experiments at the Rothamsted Experimental Station at Harpenden in the UK played a significant role in the development of long-term agronomic experimentation in Australia. Crop rotation trials established at

Rutherglen, Werribee and Dooen (Longerenong Agricultural College) between 1913 and 1917 by the Victorian Department of Agriculture and the Permanent Rotation trial established in 1925 at the Waite Institute were all modelled on the Rothamsted experiments. The Rutherglen and Werribee rotations and associated Green Manurial (GM) trials established in the same year have been discontinued and data from them are difficult to locate except for brief summaries (Penman, 1949).

Yields and Soil Chemistry

Rutherglen Research Institute, Victorian Department of Agriculture

The Permanent Top-dressing Experiment, established in 1914, demonstrates the effect of superphosphate and lime on the stocking capacity of pastures growing on an acid, fine sandy loam with a medium to heavy clay B horizon. It has three unreplicated treatments on adjacent 1.5 ha plots. The control (U) has never received fertilizer, the other two plots get superphosphate (125 kg ha^{-1}) every second year. One (F) gets no lime, the other (F+L) has had nine applications of lime (1.25 t ha^{-1}) between 1914 and 1948. The stocking rates over the period 1914–86 were 7.6 dry sheep equivalents (DSE) ha^{-1} for the fertilizer treatments and 4.2 DSE ha^{-1} for the unfertilized control, with hay being removed (24.7, 21.3 and 4.1 t dry matter (DM) ha^{-1} for treatments F, F+L and U respectively) from alternate halves of each treatment between 1953 and 1975 (Ridley *et al.*, 1990a). Approximately 30% of the hay cut from the fertilized treatments was returned as feed whereas all that cut from the unfertilized control was returned together with an additional 4 t DM from treatment F.

An investigation of the effect of pasture composition on animal production was started in 1946 on two adjacent permanent pasture fields (each 1.5 ha) which had been sown to subterranean clover (*Trifolium subterranean*) and Wimmera rye grass (*Lolium rigidum*) in 1931, and which had had identical fertilizer and management histories. Both had been top-dressed annually with superphosphate (94 kg ha^{-1} between 1937 and 1953 and 125 kg ha^{-1} between 1954 and 1986). In 1946, one of the plots, the annual pasture, continued with ryegrass and clover whereas the other, the perennial pasture, was sown to a perennial grass, phalaris (*Phalaris tuberosa* cv. Australian) and subterranean clover. The pastures were stocked, on average, at 14.8 and 12.1 DSE ha^{-1} for the annual and permanent pastures, respectively. Hay was cut and removed during 1953–1975, 26 t DM ha^{-1} from the permanent pasture with 30% returned as stock feed, and 18.2 t DM ha^{-1} from the annual pasture with a 50% feed return (Ridley *et al.*, 1990b).

By 1986 the treatments in both these pasture experiments had affected soil organic carbon (OC) and pH (in CaCl$_2$) (Table 4.2). In the older trial, pasture improvement with superphosphate and lime had resulted not only in

Table 4.2. Soil organic C and pH at the Permanent Topdressing Experiments at Rutherglen in 1986.

Year commenced	Treatment	Depth (cm)	Organic C[a] (t ha⁻¹)	pH[b] in $CaCl_2$
1918	Unfertilized	0–10	28.5	4.67
		10–20	11.0	5.05
	Superphosphate total 4.5 t ha⁻¹	0–10	33.4	4.20
		10–20	10.2	4.51
	Superphosphate plus lime, total 11.25 t ha⁻¹	0–10	40.3	4.62
		10–20	13.4	5.27
1946	Annual pasture	0–10	40.0	4.21
		10–20	9.2	4.00
	Perennial pasture	0–10	38.9	4.27
		10–20	9.6	4.08

[a] Calculated from Ridley *et al.* (1990a, b).
[b] Calculated from graphical data in Ridley *et al.* (1990a, b).

more above ground biomass but, after 73 years, total C contents in the top 10 cm had increased by 17% and 41% (compared to the control) in the F and F+L treatments respectively. Soil pH was nearly 0.5 pH unit lower where P fertilizer alone had been added and the addition of lime nullified this effect.

In the more recent trial on pasture composition, the perennial pasture supported a slightly lower stocking rate and a marginally higher pH (compared to annual pasture) after 39 years. In 1986 the amounts of OC were similar in both treatments, and, in the top 10 cm, had increased, on average, by 46%, assuming an organic matter (OM) content of 4.67% in 1948, an OM-to-OC ratio of 1.9 (Ridley *et al.*, 1990a) and a bulk density of 1.1 Mg m⁻³.

Kybybolite Research Centre, South Australia

The practical problems of managing historically poor pastures in the southeast of South Australia were studied from 1919 to 1958 at Kybybolite in an experiment started by Professor A.J. Perkins. It had 13 plots (P1–P13), on an area of 21 ha, on an acid, grey clay loam with low levels of N, P and OM, overlying a clay subsoil (Perkins, 1928). The natural pasture, was grazed and various fertilizer strategies were tested between 1919 and 1937 (Cook, 1939). These consisted of treatments with super-, calcium rock-, or aluminium rock-phosphates, as either single or annual applications, some with lime or gypsum amendments. The treatments were interrupted in the later war years due to fertilizer shortages, but superphosphate continued to be supplied in the latter years allowing a broad delineation between low and high P treatments (202 and 484 kg ha^{-1} year^{-1} respectively).

Dry matter yields were not always recorded, but all plots treated with phosphatic fertilizers during 1926–1945 carried more than three times as many sheep as the untreated plot (Russell, 1960a). After 39 years, significant increases in OC (range 3.4–7.6 t ha^{-1} in the top 5 cm), were found in the fertilized plots (Fig. 4.3), together with decreases in bulk density, compared to the control. Russell (1962) estimated that OC levels would stabilize in the treated plots in just over 100 years. The effect of liming on the OM content of the Kybybolite soils is similar to that found by Richardson (1938) on grassland soils at Rothamsted. An additional 7.8 t C ha^{-1} had accrued during 39 years in the top 20 cm of the limed Kybybolite soils (cf. plots P3/P10 with P7). Soil pH (in water) was also one unit higher (6.6) in the limed plots.

Soils given treatments supplying superphosphate and adequate calcium (P3, P8, P9 and P10) showed, on average, a linear increase in soil N of 51.5 kg ha^{-1} year^{-1} in the top 15 cm over a 30-year period. Soil N contents in treatments receiving aluminium- and calcium-rock phosphates (P1 and P5 respectively) increased by an average of 33 kg N ha^{-1} year^{-1}. Russell (1960a) reported an annual increase in soil N of 0.5 kg ha^{-1} per kg superphosphate applied, with superphosphate applications above 67–100 kg ha^{-1} year^{-1} giving diminishing returns in terms of additional soil N accumulation.

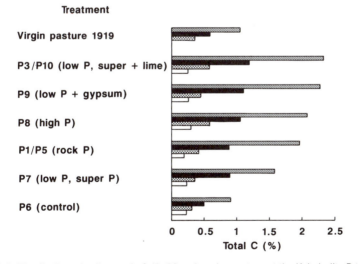

Fig. 4.3. Distribution of soil organic C (0–20 cm) under pasture at the Kybybolite P trial in 1958 (calculated from Russell, 1960a). Depth ▨ 0–5 cm; ■ 5–10 cm; ▨ 10–15 cm; ☐ 15–20 cm.

Waite Agricultural Research Institute, Urrbrae, South Australia

The Waite Agricultural Research Institute was established in 1925 as a result of a gift to the University of Adelaide by Peter Waite. The original endowment bequeathed an estate of 162 ha of cropping and grazing land on the Adelaide foothills and within 7 km of the city centre (University of Adelaide, 1934). The soil is a fine sandy loam; Red-brown earth (*A Handbook of Australian Soils*, Stace *et al.*, 1968). It is typical of the first, and still one of the most important, soils developed for wheat production in Australia (Williams, 1981). The Urrbrae series receives more rainfall than most other Red-brown earths in South Australia, and therefore has a higher OM content which reflects a longer growing season and higher annual organic matter input (Oades, 1981). Two long-term field trials were commenced in 1925 after the original vegetation had been cleared and the soil cultivated in late 1924.

The P1 top-dressing trial, 1925-55, investigated the influence of P and N on the yield, botanical composition and carrying capacity of natural pasture originally dominated by *Danthonia* spp. The experiment had five unreplicated treatments: a control, P (45 kg P_2O_5 ha^{-1} year^{-1}) as either basic slag, ground rock phosphate or superphosphate and superphosphate plus 14 kg N (as sodium nitrate). Each plot was 0.5 ha in area subdivided into four equal sections grazed in rotation (University of Adelaide, 1957). In the early years (1925-31), the superphosphate plots consistently outyielded the rock phosphate plots (38.2 and 28.3 kg ha^{-1} mean annual yield respectively) (Trumble and Fraser, 1932). Mean annual stocking capacity of the superphosphate plots was increased by 70% compared to only 21% in the rock phosphate plot. However, the benefit from superphosphate was gradually reversed over the next 5 years (Russell, 1960b) and overall between 1925 and 1955, the rock phosphate plots had outyielded the superphosphate only plots (Table 4.3). The overall success of rock phosphate was even more apparent in a logarithmic analysis (Russell, 1960b), and it was concluded that rock phosphates may have a role in pasture maintenance but are less effective in the early establishment phase.

The Permanent Rotation (C1) trial, started in 1925, compared 12 systems of crop rotation and in ten wheat is the principal crop (Table 4.4). Each course of each rotation is represented on 35 adjacent plots; there is no replication. Six of the rotations have continued in an unbroken sequence since 1925, but several of the original rotations (wheat-pea-fallow, wheat-oats-pea and wheat-barley-fallow) were discontinued after 1947, and replaced by rotations which included pasture legumes. Cereal straw is removed from the plots, and historically pasture phases have been grazed (except for 1992). Herbicides were not applied between 1978 and 1982 to study the influence of rotations on weed populations. Since 1973, initial cultivation for the fallow year has been with either a chisel plough, to reduce erosion, or a spring release cultivator rather than the mould board plough used in the early years. Relatively few modifi-

Table 4.3. Herbage yield and stock carrying capacity of grazed pastures on the top-dressing (P1) trial at the Waite Agricultural Research Institute, 1925–55.

Treatment[a]	Mean pasture yield 1925–55 (t ha⁻¹)	Mean stock capacity 1928–55 (sheep days ha⁻¹)
Nil	4.07	485
Basic slag	6.27	764
Rock-phosphate	5.86	716
Superphosphate	5.36	689
Superphosphate + 14 kg N ha⁻¹	6.04	764

[a] Applied each year at rates to supply 45 kg P_2O_5 ha⁻¹.

cations to the original experiment have been made since its inception, other than the rotation changes outlined above.

The wheat variety Seewari has been sown since 1935. A basal dressing of superphosphate of 209 kg ha⁻¹ year⁻¹ was applied between 1926 and 1943 but this was halved in 1944. The inclusion of a pasture phase in the rotation, whether oats for grazing, or Wimmera ryegrass and subterranean clover, has generally resulted in higher yields of wheat (1996 kg ha⁻¹) during the period 1952-1989 (Table 4.4) than those obtained from rotations without pasture (1355 kg ha⁻¹). Yields from all plots show considerable seasonal variation, particularly in the rotations where a fallow does not follow the pasture phase (University of Adelaide, 1991). Trends in rotation performance are therefore more easily identified if a moving average is used to compare grain yields (Fig. 4.4). All the rotation plots consistently yielded over 2 t ha⁻¹ of grain in the first 20 years of the trial, but experienced a sharp decline in yields between 1945 and 1955. Yields from the continuously cropped wheat plot levelled out at 776 kg ha⁻¹ after a rapid decline during the first 10 years of the experiment. An overall linear decline in wheat yields of 4.7 kg ha⁻¹ in the continuous wheat plot (Table 4.4) is small compared to the rotation plots but it must be put in the context that the mean yield for the plot is only 806 kg ha⁻¹ (1926-89) compared to 1433 kg ha⁻¹ in the next highest yielding treatment (wheat-oats-pea).

A limited number of measurements of soil OC in the C1 trial have been reported (Steward and Oades, 1972; Tisdall and Oades, 1980). A comprehensive analysis of historical soil samples by Keith (unpublished) and a preliminary 1993 sampling of selected plots by Grace (unpublished), form the basis of the OC data presented here. There is a positive linear relationship between the frequency of pasture in the rotation and soil OC in the top 10 cm in 1993 (Fig. 4.5). Although the change to pasture dominated rotations occurred in the late 1940s, the amount of OC in the top 10 cm continued to decline in crop-pasture rotations until at least 1973 (Fig. 4.6). Between 1973 and 1993, this

Table 4.4. Grain yields of wheat from the Permanent Rotation Trial (C1) at the Waite Agricultural Research Institute, 1925–89.

Rotation[a]	Mean grain yield (t ha⁻¹)		Linear trend 1925–89 (kg ha⁻¹ year⁻¹)
	1926–51	1952–89	
W	0.87	0.74	−7
WF	2.30	1.39	−17.2
WPF	2.66	1.97	−16.9
WBPe	2.26[b]	1.90	−7
WOF	2.34	1.80	−13.5
WOPF	3.00	2.47	−14.8
WPe	1.67	1.38	−4.9
W2P		2.04	
W2PF		2.36	
2W4P year 1		1.82	
year 2		1.50	

[a] W, wheat; F, fallow; P, pasture; Pe, peas; B, barley; O, oats.
[b] 1926–47.

decline was reversed and OC increased, on average by 13.6%, equivalent to 21.9 t ha⁻¹ (assuming a bulk density of 1.3 Mg m⁻³). The average linear decrease in OC for plots with little or no pasture since 1925 (i.e. wheat, wheat–peas and wheat–oats–pasture–fallow) was 0.026% OC year⁻¹.

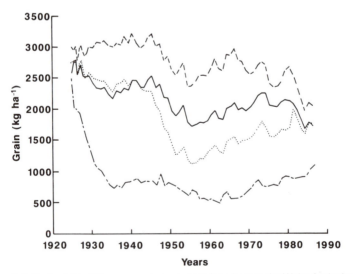

Fig. 4.4. Grain yields (10-year moving average) for wheat at the Waite Agricultural Research Institute C1 trial (1925–91) (W, wheat, F, fallow, P, pasture, O, oats). Continuous wheat, — – — – —; WF, ⋯⋯; 3-year rotation, ———; WOPF, – – – –.

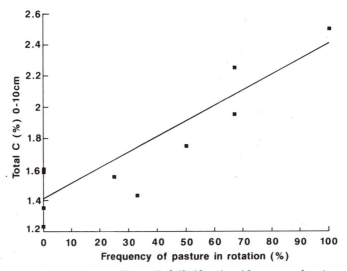

Fig. 4.5. Relationship between soil organic C (0–10 cm) and frequency of pasture phase in the rotation at the Waite Agricultural Research Institute C1 trial –1993 (Grace, unpublished). Total C = 1.41 + 0.01 × Pasture frequency ($P<0.01$).

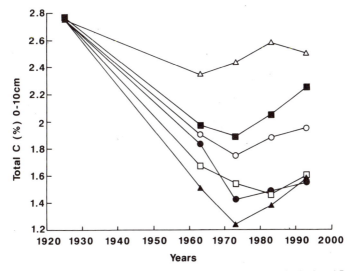

Fig. 4.6. Changes in soil organic C contents (0–10 cm) at the Waite Agricultural Research Institute C1 trial (1925–93). Rotations P (since 1951), △; 2W4P (since 1951), ■; W2P, ○; W, □; WOPF, ●; WPe, ▲. (W, wheat; F, fallow; P, pasture; O, oats; Pe, peas).

The Rothamsted C model (Jenkinson *et al.*, 1987) was modified to allow crop sequences to be simulated (Grace, unpublished), and the robustness of the modified model was tested for a different environment from that for which it was originally developed. All rate constants were left as in the original model. The annual plant residue C inputs for a particular crop were adjusted to give the best fit with observed data, but were left constant over time and not adjusted for seasonal variation. Simulated and observed OC contents (0–23 cm) for wheat, wheat–fallow and the 2 wheat–4 pasture rotations (previously a 3-year cereal–fallow rotation) are shown in Fig. 4.7. To achieve the simulated results, annual C inputs equivalent to 2.3 and 1.7 t ha^{-1} were required for the continuous wheat and wheat–fallow treatments respectively. In the wheat-pasture rotation, C inputs of 2.3 and 4.0 t ha^{-1} year^{-1} were required in the wheat and pasture phases respectively. The pasture plots in the 2 wheat–4 pasture rotation, produced 8897 kg ha^{-1} year^{-1} of aboveground DM (average 1950-1989), equivalent to 4003 kg OC ha^{-1} year^{-1} (assuming 45% OC t DM^{-1}). This was nearly all removed with grazing, but it is not unrealistic to apply a root/shoot ratio of 1 (or higher) to pastures which exactly equates to the 4 t C ha^{-1} year^{-1} required in the simulation. Applying similar assumptions to the continuous wheat and wheat–fallow plots, annual C inputs are estimated as 1.3 and 2.3 t ha^{-1} respectively. An annual shortfall of over 1 t C ha^{-1} to be input into the continuous wheat plot, would suggest that even with a refinement of the root-to-shoot assumptions in this calculation, more precise

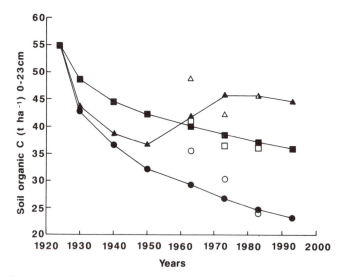

Fig. 4.7. Soil organic C contents (0–23 cm) for selected plots at the Waite Agricultural Research Institute C1 trial (1925–93). Rotations: Continuous wheat, ■; Wheat–fallow, ●; 2 wheat–4 pasture, ▲ (since 1951), solid symbols, as simulated by the Rothamsted C model (version 26.2), open symbols observed values.

measurements of other model input data, e.g. the size of the inert organic pool, is required for accurate simulations.

The effect of crop-rotation on both soil N dynamics and pH in the C1 trial is difficult to assess because of lack of detailed historical data. There is a linear relationship between pasture frequency and total soil N content of the top 10 cm for selected plots sampled in March 1993 (Grace, unpublished). Total soil N declined under both continuous wheat (by approximately 62% between 1925 and 1993) and in soils with fallow; the soil under continuous pasture accumulated N at 8.3 kg ha^{-1} year^{-1} since 1963 (Table 4.5). Under the original rotations which commenced in 1925 on soil with an initial pH 6.3 (0–23 cm), there had been little change in pH when sampled in 1963 and 1973. Under the newer rotations established in 1947, in which clover pasture makes up a substantial part of the rotation, there has been a decrease in pH in all cases, with a mean pH of 6.15 and 5.95 in 1963 and 1973 respectively (Donald, unpublished).

The Mallee Research Station, Walpeup, Victoria

This Research Station was established by the Department of Agriculture in Victoria in 1932, in response to the increasing amount of soil erosion in the Mallee region since settlement began in the 1890s. The disastrous effects of this erosion are still evident today, with subsoils being exposed in many places. Long-term erosion trials were focused on possible benefits from leguminous pastures, especially with barrel medic (*Medicago truncatula*), and trash fallowing in place of burning crop residues. The soils are sand to sandy loam, low in total N (<0.1%) and slightly alkaline (Newell, 1961).

The Permanent Fertilizer Experiment (MM1), started in 1940, investigates the effect of 0, 34, 67, 101 and 135 kg ha^{-1} superphosphate applied to wheat

Table 4.5. Changes in total soil N (0–10 cm) in selected plots from the Permanent Rotation (C1) trial at the Waite Agricultural Research Institute, 1925–93.

	Soil total N (kg ha^{-1})			Linear trend 1963–93 (kg ha^{-1} year^{-1})
Rotation*	1925[a]	1963[a]	1993[b]	
W	3974	1840	1495	-11.5
WF	3974	1280	1235	–1.5
WPF	4078	1760	1333	–14.2
W2P	4288	1920	2080	+5.3
2W4P	3869	2080	2340	+8.7
P	3974	2480	2730	+8.3

[a] Calculated from Greenland (1971) assuming a bulk density of 1.3 Mg m^{-3}.
[b] Grace (unpublished).
* Abbreviations as Table 4.4.

in a wheat–fallow rotation. In 1986 all plots were split to test a volunteer pasture legume in the fallow year; in practice one half plot was sown to barrel medic. There are two replicates, with each plot being 84 m², but the experiment is on two fields, both on fine sandy loam, approximately 1 km apart. The mean grain yield for the fertilizer treatments (1940–90) was 1621 kg ha⁻¹, approximately 40% higher than the unfertilized plots. In 1958 the trial recorded its highest yields, with a treatment average of 3284 kg ha⁻¹. Since then, yields have been decreasing at the rate of 18.8 kg ha⁻¹ each year. Total soil N has increased from 0.03 to 0.05% (1987 value).

Like most semiarid sites in southern Australia, wheat production is dependent on rainfall in the growing season. When yields over the first 26 years of this trial were adjusted to a constant mean monthly rainfall, McClelland (1968) reported that after 10 cycles of the rotation, 56% of the yield response to 34 kg ha⁻¹ superphosphate was attributable to residual fertilizer P. This residual fertilizer effect was equivalent to 71% in the 67 kg ha⁻¹ superphosphate treatment and yields effectively plateaued at this level. He concluded that above 60 kg ha⁻¹, the additional P was either unavailable, or not required because of lack of rain.

The Soil Fertility Experiment (MR3), 1951–80, investigated the effect of pasture type (barrel medic or grass ley) and its duration on subsequent wheat yield and quality. There were 16 treatments, each with three replicates, on plots approximately 350 m². Treatments consisted of an initial pasture phase of 2, 4 or 6 years, then cropping to fallow–wheat. From 1961, all treatments reverted to fallow–wheat, except for treatments representing 2 year-pasture-fallow–wheat. Continuous fallow–wheat was regarded as the control, with two adjacent sites, each with a complete set of treatments, being used so that wheat was grown each year. From 1955 to 1957 all pasture plots were top-dressed with 34 kg ha⁻¹ superphosphate, then from 1957 to 1964 they were split with one-half receiving 100 kg ha⁻¹ superphosphate. The wheat variety Insignia was grown throughout and sown with 100 kg ha⁻¹ superphosphate. Following the same length of the pasture phase, the yield of the first crop after barrel medic always exceeded both that of the control and the grass-dominated pasture (Table 4.6). The trial also demonstrated that, on these light soils, yields following a 2-year legume-based pasture exceeded those after 4- or 6-year grass pasture.

The Permanent Rotation Experiment (MR1), 1936-1973, evaluated the long-term effects of several different crop rotations on the yield of cereals. The 52 plots (four replicates of 13) tested a two-year (fallow–wheat), a three-year (fallow–wheat then barley or wheat or oats or pasture), or a four-year (fallow–wheat-2 pasture) rotation. Each plot was split into subplots of equal size, the number depending on the number of years in the rotation, so that wheat was grown each year. The minimum subplot size (four-year rotation) was 0.12 ha. The soil was a sandy loam, and had only been cropped for 3 years prior to 1936. The wheat varieties Ranee 4H (1936–56) and Insignia were sown with super-

Table 4.6. Effect of pasture phase duration on wheat yields, grain and protein content, of the first wheat crop grown after legume[a] or grass dominated pasture at the Soil Fertility Experiment (MR3) at Walpeup between 1957 and 1962 (Castleman, unpublished).

Rotation	Grain yield (t ha⁻¹)	Protein (%)
Fallow-wheat	1.45	8.6
2 years grass	1.66	9.1
4 years grass	1.76	9.7
2 years legume	1.82	9.1
4 years legume	1.92	12.2
6 years legume	1.90	13.1

[a] Barrel medic (*Medicago truncatula*).

phosphate at either 100 kg ha⁻¹ (after a fallow year) or 50 kg ha⁻¹ (after a cropping year). Elliott and Jardine (1972) reported that 73% of the variation in yield between 1940 and 1968 was attributable to climatic fluctuation. During this time, two replicates were eliminated through disease, and complete crop failures occurred in 1944, 1950 and 1967. Average yields, linear trends and total soil N are shown in Table 4.7. Even though all yields were beginning to plateau after 1958, the wheat–fallow treatment was the only system to show a marked decline in yields after that year. Grain yield on the wheat–fallow plots was only 4.4% below the trial average in 1958, but by 1968 the difference had increased to 43%.

Two other long-term trials began in 1946 at this site. The Fallow Modification Experiment (MC4), 1946–71, investigated the effect of different fallowing methods (e.g. mulch, retained and burnt) on wheat yield, nitrate accumulation, moisture conservation, fertilizer requirements and time of sowing. The Supplementary Rotation Experiment (MR2), 1946–75, compared fallow–wheat and fallow–wheat–2 pasture rotations on soil fertility and the incidence of soil drift. Data for both experiments are archived at Mallee Research Station.

Wimmera Research Station, Dooen, Victoria

The Longerenong Permanent Rotation No.1 (LR1) trial was started in 1917 by the Victorian Department of Agriculture at Longerenong Agricultural College on a grey self-mulching loam. This is one of three trials started by the state government between 1913 and 1917, modelled on the Rothamsted experiments and is the only one still in existence. The seven rotations tested have each phase represented every year, but without replication. There are two blocks each of 19 plots, each block being 812 m², the second block started in 1986, when the original plots were split for cereal cyst-resistant cultivars.

Table 4.7. The effect of rotation on grain yield and total soil N (0-15 cm) at MR1 trial at Walpeup.

	Rotation[a]					
	WF	WWF	WBF	WOF	WPF	W2PF
Grain yield[b] (t ha⁻¹)	1.51	1.56	1.58	1.70	1.71	1.77
Linear trend[b] (kg ha⁻¹ year⁻¹)	+11.0	+26.0	+28.4	+32.6	+31.2	+37.6
Total soil N[c] (%)	0.068	0.082	0.070	0.065	0.080	0.083

[a] W, wheat; F, fallow; P, pasture; O, oats; B, barley.
[b] Mean 1940–68.
[c] 1970 sample from Elliot and Jardine (1972).

Yields in Table 4.8 are for the northern plots, which are still sown with the same wheat (Ghurka), oats (Swan) and barley (Weeak) varieties used since 1917. Double superphosphate (38% P_2O_5) is applied to all grain crops. Grazing oats represent the pasture phase of this trial.

Grain yield from the continuous wheat plot averaged 0.60 t ha⁻¹ between 1918 and 1988, and there was a marginal annual increase in yield during this time. The inclusion of a fallow after wheat, increased yields threefold, but this system had the greatest annual decline in yield over time, 12.6 kg ha⁻¹ year⁻¹. The highest yields have consistently been on two plots which included a pas-

Table 4.8. Wheat yields, grain in the Longerenong Permanent Rotation (LR1) trial at Dooen (1918–88) and the annual linear trend.

	Main grain yield (t ha⁻¹)				Linear trend (kg ha⁻¹)
Rotation[a]	1938-40	1958-67	1978-87	1918-88	1918-88
W	0.51	0.72	0.54	0.60	+0.5
WF	2.24	1.81	1.49	1.89	−12.6
WPF	2.41	2.05	2.10	2.38	−2.0
WBPe	1.97	0.89	1.17	1.43	−4.3
WOPe	2.00	0.87	1.36	1.49	−4.3
WOF	2.48	1.92	1.89	2.16	−8.5
WOPF	2.38	1.88	2.18	2.37	+5.4

[a] W, wheat; F, fallow; P, pasture; Pe, peas; B, barley; O, oats.

ture phase in the rotation, with a 70-year average of 2.37 t ha⁻¹. Total soil N declined at approximately the same rate in all treatments (0.00053% year⁻¹) between 1949 and 1987. Penman (1949) reported that initially the soil had 0.11 % N, therefore, assuming a bulk density of 1.3 Mg m⁻³, losses of total soil N (0-15 cm) between 1917 and 1987 ranged from 214 kg ha⁻¹ in the wheat–barley–peas treatment to 1072 kg ha⁻¹ in the fallow–wheat rotation, and a loss of 946 kg ha⁻¹ in the high-yielding pasture treatments.

The Longerenong Permanent Rotation No.2 (LR2), 1946-81, compared crop residue management (burnt versus retained) and the inclusion of pasture in a fallow–wheat rotation. During 35 years, yields averaged 2573, 2479 and 2702 kg ha⁻¹ for the residue burnt, residue retained and pasture-based rotations respectively. There was a positive linear trend in grain yield (average 7.8 kg ha⁻¹ year⁻¹) for all treatments. Soil OC levels in the top 15 cm of the pasture-based rotation increased by 89 kg ha⁻¹ year⁻¹ (assuming a bulk density of 1.3 Mg m⁻³) between 1950 and 1967, and declined by 94 and 142 kg ha⁻¹ year⁻¹ in the residue retained and residue burnt rotations respectively.

The green manuring trial at Dooen (Table 4.1) was similar to those at Ruth-erglen and Werribee. The date of termination is unknown and there is little information except that reported by Penman (1949).

South-western Western Australia

The youngest group of long-term agronomic trials in Australia are situated in the south-western corner of Western Australia. There are five trials, each similar in design and each include cereal rotations with pasture (subterranean clover) phases of 1-4 years and grain legumes (lupins) but not all treatments are replicated. These trials are generally recognized for their fundamental rather than applied aspects because not all the management practices tested are realistic (Rowland and Perry, 1991). Summary yields (Table 4.9) and OC data (Fig. 4.8) are given only for selected rotations for the Chapman, Merredin and Newdegate trials. Trends in yields are consistent with those from experiments which have been in progress for similar or longer periods in south-eastern Australia, with the extended pasture phase rotation consistently giving the best response in terms of grain yield and % N. Continuous wheat gives the lowest yields (and declining) and, in all instances, results in the lowest soil organic C. Data from the Chapman site indicates that although wheat yields have been high, these soils may not be able to maintain current production levels for much longer considering the large linear declines in yield in the short rotations and the overall low soil OC content.

Table 4.9. Grain yield and grain N of wheat from long-term rotation trials at Merredin, Newdegate and Chapman (Rowland, unpublished).

Rotation[a]	Mean yield (kg ha^{-1})		Linear trend (kg ha^{-1} year^{-1}) 1967–90	Mean grain N (%) 1981–90
	1971–80	1981–90		
Merredin (66M29)				
2W4P	894	1119	+14.2	2.75
WP	858	1038	+9.7	2.63
W	661	931	+9.6	2.62
Newdegate (67N4)[b]				
2WP	1255	1268	+12.2	1.96
WP	1145	1176	+3.5	1.89
W	418	321	−30.8	1.95
Chapman (67C13)				
2WP	1610	1881	+2.8	2.12
WP	1617	1637	−23.2	2.03
W	1120	884	−45.7	1.88

[a] W, wheat; F, fallow; P, pasture.
[b] Linear trend calculated from 1968.

Northern New South Wales and southern Queensland

Relatively little information is available about long-term trials in these areas, particularly in Queensland, however, a brief description of three on-going trials is included because they are distinct in that they are either on the fringe, or in, a higher rainfall subtropical environment.

The Hermitage Fallow Management Trial near Warwick was started in 1968 by the Queensland Department of Primary Industry to investigate the influence of crop residue management and tillage on fallow water storage and winter cereal yields. The experiment is on a deep, alluvial black clay and has a randomized block design of 12 treatments in four replications. The treatments are a factorial combination of conventional and zero tillage during the fallow periods, crop residues being retained or burned after harvest, and three rates of urea application (0, 23 and 69 kg ha^{-1} year^{-1}). Plot size is about 400 m^2. The wheat and barley cultivars have varied but grain yields (wheat and barley) for the first 12 crops (1968–1979) were similar for each of the fallowing systems (average 2736 kg ha^{-1}). The lack of response in the zero-till, residue-retained treatments was due to a combination of disease and root lesion nematode infestation (Marley and Littler, 1989). After 22 years, soil total N (0–10 cm) had declined, on average, at 25 kg ha^{-1} year^{-1} under all treatments (Dalal, 1992), but

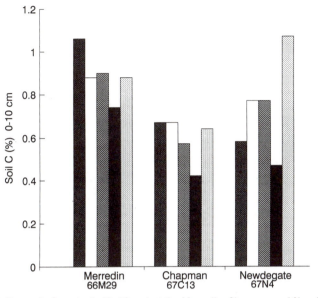

Fig. 4.8. Soil organic C contents (0–10 cm) at the Merredin, Chapman and Newdegate rotation trials (1970–90). ▨ % C in 1970. In 1990 % C in rotation 2W4P ☐, WP ▨, W ■, P ▨.

the zero-till, residue-retained and high-fertilizer treatments contained more total soil N than the tilled, residue-removed, low-fertilizer treatments.

The legume/cereal rotation experiment at Tamworth was started in 1966 to investigate the effects of three pasture legume treatments, a grain legume and long fallow on subsequent cereal yields and soil fertility. This experiment is on two soils, a self-mulching black earth and a hard-setting red clay; at the two sites plot sizes are 344 m² and 418 m², respectively. The design is a Latin square of six main treatments split for N on the cereals. The three pasture legume treatments were designed to determine the optimum duration of lucerne ley (phase 1), the optimum methods of lucerne establishment (phase 2) and to compare two annual legumes with lucerne (phase 3). Treatments have been evaluated by growing wheat in 1970–79 (phase 1), grain sorghum in 1983–87 (phase 2) and wheat in 1991–93 (phase 3). Major conclusions include: the beneficial effect of 3 year-lucerne leys may last up to 9 years on subsequent wheat yields, however there must be a sufficient fallow period (6–8 months) after the ley to rehydrate the profile; soil nitrate levels and cereal yields are usually much lower after grain legumes than after pasture legumes and long fallow, with the pasture legume rotation being capable of maintaining or increasing cereal yield and soil fertility (Holford, 1980, 1981, 1989, 1990).

The clover/maize rotation experiment at Glen Innes, started in 1921, investigates the effects of clover on yields of maize and oats and soil fertility. There are seven treatments of which four include clover. The rotations are sequences of two, three and four crops and plot size (720 m²) is sufficient to allow each crop in a rotation to be grown each year. To date, the major conclusions are that the size and sustainability of maize yields were always increased when clover was included in the rotation. The inclusion of autumn oats in the rotation increases maize productivity such that the clover effect is not significant (Holford, unpublished).

Soil Properties

Bulk density, aggregation and water infiltration

Soil structure and its stability are of prime importance in the relatively fragile soils used for cereal production in the semiarid environment of southern and southeastern Australia. Crops are grown during the winter after establishment following rain in May–June. Conventionally seedbeds were prepared following the first significant rain and sowing began immediately after follow-up rains. Yields depended in the first instance of rainfall from May to September, the 'growing season rainfall'. Other limitations to yield are efficient storage and use of the rainfall, the length of the growing season (early establishment and late finishing rains), adequate supplies of essential nutrients, lack of disease and, of course, optimum management of the whole system (French and Schultz, 1984a, b).

The importance of the management of the water supply and the maintenance of a stable soil structure under cropping regimes has been the subject of research projects in the 1950s which utilized long-term trials at the Waite Institute and elsewhere. Millington (1959) established relationships between rainfall after seeding, the bulk density of the tilled layer, the establishment of wheat plants and grain yield. Bulk densities greater than 1.5 Mg m⁻³ decreased the number of plants established. In a rotation containing three years of pasture and one year of wheat, bulk density remained low even in very wet seasons, but in soils under a two-year pasture–fallow–wheat rotation and especially the alternate fallow–wheat rotation, bulk densities in the month after seeding exceeded 1.5 Mg m⁻³. This resulted in poor establishment and in very poor yields. Maximum bulk densities occurred in those years with substantial rainfall in the month after seeding.

Changes in bulk density indicate a collapse of soil structure, changes in porosity and soil aggregation. Early work on the fallow–wheat and fallow–wheat–pea–pasture rotations compared with a soil which had not been cultivated (virgin) revealed changes in porosity and also dispersion of finer particles in the soil surface exposed to rainfall from the time of seedbed preparation until the crop was sufficiently established for the leaf canopy to protect the soil. The finer particles were repacked in surface layers to form a crust (McIntyre, 1955). The data (Table 4.10) indicate drastic changes in the percentage of water-stable aggregates in soils subjected to the cropping regimes compared to the virgin soil.

The dramatic impact of management systems on soil aggregation with depth was studied at the Waite Institute (Greacen, 1958). There were few water-stable aggregates (>2 mm) under continuous cultivation down to 150 mm (Fig. 4.9). Water-stable aggregation improved with time under pasture, even after only three years, to 25 mm depth, but many years would be required to return the soil structure at depth to its original state. Greacen (1958) also differentiated water-stable aggregation from the dispersive effects of cultivation and rainfall on surface layers of soil resulting in crusting which acted as a throttle on infiltration of rainfall. Three years of pasture renovated the top 25 mm of soil but soil at depth in the tilled layer became impermeable under artificial rainfall. Soil from layers below the depth of tillage was very stable under rainfall which illustrates the deleterious effects of exposure of soil to the weather. The decline in both hydraulic conductivity and sorptivity under wheat and the improvement in both under pasture at the Waite Institute was shown by Greacen (1981). There was much seasonal fluctuation in the measurements of both but the trends during the rotation were clearly expressed.

By the mid-1970s the major impact of tillage and pastures on the structure of the Red–brown earths was established and although yields per hectare were increasing it was obvious that summer fallows were very detrimental to the amount of soil organic matter. Related to the loss of organic matter was severe structural degradation which led to poor infiltration rates and decreased water storage; surface run-off led to erosion and poor plant establishment to small yields. Data from the long-term trials showed that a benign rotation should include about two years of pasture for each year of cereal. Such a ratio of pastures to cereals has rarely been maintained in commercial practice.

Table 4.10. Physical properties of cropped soils compared to those from a virgin site (McIntyre, 1955).

Treatment	Bulk density (Mg m^{-3})	Total porosity (%)	Air-filled porosity (%)	Water-stable aggregates (%)
Virgin	—	—	—	58
Fallow–wheat–peas–pasture	1.36	49	19	19
Fallow–wheat	1.46	45	13	10
Crust wet years	1.64	38	1	0

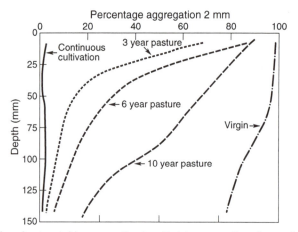

Fig. 4.9. Profiles of water stable aggregation in a Red–brown earth under continuous cultivation and after different periods under pasture (from Graecen, 1958).

Mechanisms of structural improvements under pasture

It has long been considered that there is a connection between soil structural stability and the amount of soil organic matter. In the 1960s and 1970s world-wide research attempted to identify the organic chemicals responsible for sta-bilizing soil aggregates and soils from various long-term trials, including those at the Waite Institute, were used to understand the mechanisms of stabiliza-tion of soil structure during the growth of pastures. Greenland *et al.* (1962) used permeability measurements of columns of aggregates from a wide range of rotations on long-term trials to assess the role of periodate oxidizable materials (polysaccharides) on structural stability. The data showed a relation between permeability of beds of 1–2 mm soil aggregates from the different rotations to 0.05 M-NaCl. The integrated measure of soil structure so obtained was strongly correlated with the organic carbon content of the soils which increased with the number of years of pasture in the rotation. After treatment with sodium periodate the permeabilities decreased indicating the role of per-iodate oxidizable materials (polysaccharide) in stabilizing soil structure. The data showed that structural stability stable to the treatment with sodium per-iodate increased linearly with years of pasture in the rotation, whereas the stabilizing role of polysaccharides was dominant in the soils that were cropped regularly.

After demonstrating the efficacy of ryegrass root systems in stabilizing soil aggregates in pots in the glasshouse (Tisdall and Oades, 1980), structural stab-ility in soils from the long-term trials was shown also to be dependent on root systems and associated vesicular arbuscular mycorrhiza (VAM). There was a good correlation between water stable aggregates (2 mm) and the organic car-bon content of the soil, and similar correlations with root length and with hyphal length. Roots and hyphae as biological entities are resistant to oxidation by periodate and their growth in soil controls aggregate stability in the Red-brown earths. However, the major effects were on larger aggregates several mm in diameter and it is this macroaggregation (>0.25 mm) which was changed dramatically by different cropping regimes (Fig. 4.10). Aggregates <0.25 mm were not so dramatically affected by management systems but, eventually, regular cultivation and exposure of soil to heavy rainfall around seeding time leads to clay dispersion. This is well illustrated in comparisons of the aggregate hierarchy which exists in the Red-brown earths as a legacy of the prolonged growth of roots, rhizospheres and VAM. It is now evident that plants can be rated on their ability to stabilize soil structure in the order grasses > cereals > dicotyledonous species based on the length of roots and associated VAM. These principles have been introduced into intensive management systems particularly for Red-brown earths under irrigation.

Soil organic chemistry

Long-term trials, like those at the Waite Institute, offer a range of opportunities to study changes in the chemistry of organic materials as organic matter accumulates or is lost (Oades, 1981). In summary, this work has shown a remarkable similarity in the chemistry of organic materials present under extremes of management from alternate wheat–fallow to permanent pasture. Although the organic matter content differs by a factor of three, the proportion present as carbohydrate was constant in soils from all rotations (Oades, 1967), and comparing the extremes of fallow–wheat and permanent pasture showed that the extractable polysaccharides had similar molecular weight ranges and monosaccharide compositions (Swincer *et al.*, 1968). Conclusions from these

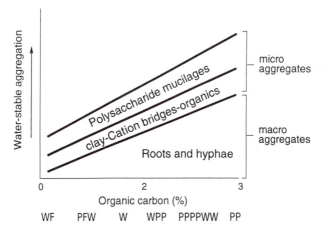

Fig. 4.10. The relationship of aggregation in Red–brown earths to soil organic carbon content as controlled by rotations and tillage (from Tisdall and Oades, 1982) W, wheat; P, pasture; F, fallow.

wet chemical investigations were confirmed by application of solid state ^{13}C nuclear magnetic resonance studies which indicated gross similarities in the chemistry of organic materials in a virgin soil, soil under an old pasture and a fallow–wheat rotation (Oades *et al.*, 1988) (Fig. 4.11). Although there were differences in the chemistry of the organic inputs, e.g. between cereal straw, grass/legume and eucalypt litter, when utilized by the soil microbial biomass the products were indistinguishable. Thus, the chemistry of soil organic matter is not influenced by soil management, but is controlled by the indigenous soil fauna and flora. How then is the organic matter accreted under pasture accommodated in soil?

Initially organic inputs under pastures as plant debris can be sieved out or floated from the soil. During decomposition this organic material interacts with the inorganic soil matrix and the resulting organomineral association controls further turnover of the organic materials and influences soil physical properties. Some of the accreted organic materials accumulate in silt-sized particles. These changes take place slowly, for example, when a soil after 50 years of fallow–wheat was compared with one after 40 years pasture there was a larger concentration of C, N and P in particles 1–20 μm in the pasture soil (Turchenek and Oades, 1979). The pasture soil is becoming similar to a Mollisol which has been under prairie grasses for hundreds of years. In an agricultural context this pool of C, N and P can be exploited, but only to the extent that soil structural problems are not created. It could represent the 'slow' pool of the 'carbon-turnover modellers' and represents an approach to giving reality to the conceptual pools used in simulating the dynamics of carbon and nitrogen in soils.

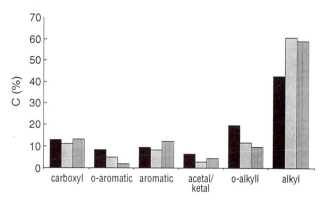

Fig. 4.11. Chemical composition of organic matter in a Red–brown earth in a virgin state and after decades of grass/legume pasture and an exploitive fallow–wheat rotation (from Oades *et al.*, 1988). ▮ virgin soil; ▨ wheat–fallow; ▥ permanent pasture.

Soil biology

Barley (1959) established the impact of intensively cultivated rotations on earthworm populations. The fallow–wheat rotation contained less than 10% of the population in a four course rotation sampled in wheat after fallow and two years of pasture, or in the permanent pasture plot.

Long-term trials were used by Wildermuth (1980) to study the suppression of 'take all' in some Australian soils. The suppression of 'take-all' caused by *Gaeumannomyces graminis tritici* (Ggt) was shown by pot experiments to be less in soils under both pasture and wheat and there was a greater decrease in the suppressive factor on some pasture soils demonstrating that the wheat plant was not necessary for the suppressive factor to operate.

Subsequently Chakraborty (1983) used the long-term plots at the Waite Institute to study the population dynamics of amoebae in soils shown to be suppressive and non-suppressive to wheat take-all and found that the numbers of mycophagous amoebae in the take-all suppressive soil under a pasture–pasture–wheat rotation were consistently larger than in a non-suppressive soil which grew wheat every year.

The Future of Long-term Trials in Australia

Long-term experiments in Australia are presently in a vulnerable position with respect to funding to support on-going maintenance and basic research. Overall reductions in government spending in the agricultural sector leaves long-term trials as one of the seemingly less profitable research options which can be eliminated in the wake of the promises of the genetic engineers. Even the most genetically superior plant cannot exist if its soil environment is devoid of structure, nutrition and water, therefore the development of sustainable management strategies can only be addressed in the long term. Although long-term trials from many locations in Australia may have suffered from lack of adequate reporting of even basic research findings, archives of information and historical soil samples still exist, which if properly audited, could address many of the problems agronomists and soil scientists are facing today. If only a small portion of the large amounts of money spent on short-term gains within the Australian grain industry was used to re-open these files it would allow research objectives to be narrowed to critical topics without the fear of expensive duplication. The fact that long-term trials also serve as a means of looking at climate change, a global topic very much in vogue at the present time, should be reason enough for government funding worldwide to be directed to a detailed assessment and the on-going maintenance of selected long-term trials.

Acknowledgements

The authors would like to thank Ian Rowland, Janine Carter and Murray Hannah for supplying valuable data in the preparation of this chapter. The assistance of Jan Ditchfield and Som Jarwal is also appreciated.

References

Barley, K.P. (1959) The influence of earthworms on soil fertility. 1. Earthworm populations found in agricultural land near Adelaide. *Australian Journal of Soil Research* 10, 171-178.

Carter, M.R. and Steed, G.R. (1992) The effects of direct-drilling and stubble retention on hydraulic properties at the surface of duplex soils in north-eastern Victoria. *Australian Journal of Soil Research* 30, 505-516.

Chakraborty, S. (1983) Population dynamics of amoebae in soils suppressive and non-suppressive to wheat take-all. *Soil Biology and Biochemistry* 15, 661-664.

Chan, K.Y., Roberts, W.P. and Heenan, D.P. (1992) Organic carbon and associated soil properties of a red earth after 10 years of rotation under different stubble and tillage practices. *Australian Journal of Soil Research* 30, 71-83.

Clarke, A.L. and Russell, J.S. (1977) Crop sequential practices. In: Russell, J.S. and Greacen, E.L. (eds) *Soil Factors in Crop Production in a Semi-arid Environment.* University of Queensland Press, St Lucia, pp. 279-300.

Cook, L.J. (1939) Further results secured in 'top dressing' poor south-eastern pasture lands with phosphatic fertilizers. *Journal of the Department of Agriculture of South Australia* 42, 791-808, 857-866.

Dalal, R.C. (1992) Long-term trends in total nitrogen of a Vertisol subjected to zero-tillage, nitrogen application and stubble retention. *Australian Journal of Soil Research* 30, 223-231.

Donald, C.M. (1963) Grass or crop in the land use of tomorrow. *Australian Journal of Science* 25, 386-395.

Donald, C.M. (1964) Phosphorus in Australian agriculture. *Journal of the Australian Institute of Agricultural Science* 30, 75-105.

Elliott, B.R. and Jardine, R. (1972) The influence of rotation systems on long-term trends in wheat yield. *Australian Journal of Agricultural Research* 23, 935-944.

French, R.J. and Schulz, J.E. (1984a) Water use efficiency of wheat in a Mediterranean-type environment. 1. Relation between yield, water use and climate. *Australian Journal of Agricultural Research* 35, 743-764.

French, R.J. and Schulz, J.E. (1984b) Water use efficiency of wheat in a Mediterranean-type environment. 2. Some limitations to efficiency. *Australian Journal of Agricultural Research* 35, 765-775.

Greacen, E.C. (1958) The soil structure profile under pastures. *Australian Journal of Agricultural Research* 9, 129-137.

Greacen, E.C. (1981) Physical properties and water relations. In: Oades, J.M., Lewis, D.G. and Norrish, K. (eds) *Red-brown Earths of Australia*, Waite Agricultural Research Institute/CSIRO (Australia) Division of Soils, Adelaide, pp. 83-96.

Greenland, D.J (1971) Changes in the nitrogen status and physical condition of soils under pastures, with special reference to the maintenance of the fertility of Australian soils used for growing wheat. *Soils and Fertilizers* 34, 237-251.

Greenland, D.J., Lindstrom, G.R. and Quirk, J.P. (1962) Organic materials which stabilize natural soil aggregates. *American Society of Soil Science Proceedings* 26, 366-371.

Hayman, R.H. (1943) Maintenance of soil fertility and production. *Journal of the Department of Agriculture of Victoria* 41, 329-335.

Holford, I.C.R. (1980) Effects of duration of grazed lucerne on long-term yields and nitrogen uptake of subsequent wheat. *Australian Journal of Agricultural Research* 31, 239-250.

Holford, I.C.R. (1981) Changes in nitrogen and organic carbon of wheat–growing soils after various periods of grazed lucerne, extended fallow and continuous wheat. *Australian Journal of Soil Research* 19, 239-249.

Holford, I.C.R. (1989) Yields and nitrogen uptake of grain sorghum in various rotations including lucerne, annual legume and long fallow. *Australian Journal of Agricultural Research* 40, 255-264.

Holford, I.C.R. (1990) Effects of eight year rotations of grain sorghum with lucerne, annual legume, wheat and long fallow on nitrogen and organic carbon in two contrasting soils. *Australian Journal of Soil Research* 28, 277-291.

Jenkinson, D.S., Hart, P.B.S., Rayner, J.H. and Parry, L.C. (1987) Modelling the turnover of organic matter in long-term experiments at Rothamsted. In: Cooley, J.H. (ed.) *Soil Organic Matter Dynamics and Soil Productivity*. Intecol Bulletin 15, pp. 1-8.

Marley, J.M. and Littler, J.W. (1989) Winter cereal production on the Darling Downs – an 11 year study of fallowing practices. *Australian Journal of Experimental Agriculture* 29, 807-827.

McClelland, V.F. (1968) Superphosphate of wheat: the cumulative effect of repeated applications on yield response. *Australian Journal of Agricultural Research* 19, 1-8.

McIntyre, D.S. (1955) Effect of soil structure on wheat germination in a Red-brown earth. *Australian Journal of Agricultural Research* 6, 797-803.

Millington, R.J. (1959) Establishment of wheat in relation to apparent density of the surface soil. *Australian Journal of Agricultural Research* 10, 487-494.

Morrow, J.A., Killeen, N.C. and Bath, J.G. (1948) Development of clover–ley farming at Rutherglen Research Station. *Journal of Department of Agriculture of Victoria* 46, 13-20.

Newell, J.W. (1961) *Soils of the Mallee Research Station*, Walpeup, Victoria. Technical Bulletin 13, Department of Agriculture, Victoria.

Oades, J.M. (1967) Carbohydrates in some Australian soils. *Australian Journal of Soil Research* 5, 103-115.

Oades, J.M. (1981) Organic matter in the Urrbrae soil. In: Oades, J.M., Lewis, D.G. and Norrish, K. (eds) *Red-brown Earths of Australia*, Waite Agricultural Research Institute/CSIRO (Australia) Division of Soils, Adelaide, pp. 63-81.

Oades, J.M., Waters, A.G., Vassallo, A.M., Wilson, M.A. and Jones, G.P. (1988) Influence of management on the composition of organic matter in a Red-brown earth as shown by ^{13}C nuclear magnetic resonance. *Australian Journal of Soil Research* 26, 289-299.

Penman, F. (1949) Effects on Victorian soils of various crop rotation systems with particular reference to changes in nitrogen and organic matter under cereal cultivation. British Commonwealth Scientific Official Conference. *Specialist Conference in Agriculture*, Australia.

Perkins, A.J. (1928) Some results secured in 'top dressing' poor south-eastern pasture land with phosphatic fertilizers. *Journal of the Department of Agriculture of South Australia* 31, 11-24, 1120-1135.

Purchase, H.F., Vincent, J.M. and Ward, L.M. (1949) The contribution of legumes to soil nitrogen economy in New South Wales. *Journal of the Australian Institute of Agricultural Science* 15, 112-117.

Richardson, H.L. (1938) The nitrogen cycle in grassland soils: with especial reference to the Rothamsted Park Grass Experiment. *Journal of Agricultural Science* 28, 73-121.

Ridley, A.M., Slattery, W.J., Helyar, K.R. and Cowling, A. (1990a) The importance of the carbon cycle to acidification of a grazed pasture. *Australian Journal of Experimental Agriculture* 30, 529-537.

Ridley, A.M., Slattery, W.J., Helyar, K.R. and Cowling, A. (1990b) Acidification under grazed annual and perennial grass based pastures. *Australian Journal of Experimental Agriculture* 30, 539-544.

Rovira, A.D. (1992) Dryland mediterranean farming systems in Australia. *Australian Journal of Experimental Agriculture* 32, 801-809.

Rowland, I. and Perry, M (1991) Rotations. In: Perry, M. and Hillman, B. (eds.) *The Wheat Book*. Department of Agriculture, Western Australia, pp. 81-86.

Russell, J.S. (1960a) Soil fertility changes in the long-term experimental plots at Kybybolite, South Australia. I. Changes in pH, total nitrogen, organic carbon, and bulk density. *Australian Journal of Agricultural Research* 11, 902-926.

Russell, J.S. (1960b) Soil fertility changes in the long-term experimental plots at Kybybolite, South Australia. I. Changes in phosphorus. *Australian Journal of Agricultural Research* 11, 927-947.

Russell, J.S. (1962) Changes in soil fertility under fertilized pastures. *Journal of the Department of Agriculture of South Australia* 65, 194-203.

Russell, J.S. (1977) Introduction. In: Russell, J.S. and Greacen, E.L. (eds) *Soil Factors in Crop Production in a Semi-arid Environment*. University of Queensland Press, St Lucia, pp. 3-5.

Sims, H.J. (1953) Some aspects of soil fertility in the cereal areas of Victoria. *Journal of the Australian Institute of Agricultural Science* 19, 89-98.

Sims, H.J. (1977) Cultivation and fallowing practices. In: Russell, J.S. and Greacen, E.L. (eds) *Soil Factors in Crop Production in a Semi-arid Environment*. University of Queensland Press, St Lucia, pp. 243-261.

Soil Survey Staff (1975) *Soil Taxonomy Handbook* No. 436, USDA, Washington.

Stace, H.C.T., Hubble, G.D., Brewer, R., Northcote, K.H., Sleeman, J.R., Mulcahy, M.J. and Hallsworth, E.G. (1968) *A Handbook of Australian Soils*. Rellim, Glenside.

Steward, D.G. and Oades, J.M. (1972) The determination of organic phosphorus in soils. *Journal of Soil Science* 23, 38-49.

Swincer, G.D., Oades, J.M. and Greenland, D.J. (1968) Studies on soil polysaccharides in soils under pasture and under a fallow-wheat rotation. *Australian Journal of Soil Research* 6, 225-235.

Taylor, J.K. (1983) The Australian environment. In: Division of Soils, CSIRO (ed.) *Soils: An Australian Viewpoint*. CSIRO, Melbourne, pp. 1–2.

Tisdall, J.M. and Oades, J.M. (1980) The effect of crop rotation on aggregation in a Red-brown earth. *Australian Journal of Soil Research* 18, 423–433.

Tisdall, J.M. and Oades, J.M. (1982) Organic matter and water-stable aggregates in soils. *Journal of Soil Science* 33, 141–163.

Trumble, H.C. and Fraser, K.M. (1932) The effect of top-dressing with artificial fertilizers on the annual yield, botanical composition and carrying capacity of a natural pasture over seven years. *Journal of the Department of Agriculture of South Australia* 35, 1341–1353.

Turchenek, L.W. and Oades, J.M. (1979) Fractionation of organomineral complexes by sedimentation and density techniques. *Geoderma* 21, 311–343.

University of Adelaide (1934) *Report of the Waite Agricultural Research Institute 1925-1932*, Hassell Press, Adelaide.

University of Adelaide (1957) *Report of the Waite Agricultural Research Institute 1954-55*, Griffin Press, Adelaide.

University of Adelaide (1991) *Biennial Report of the Waite Agricultural Research Institute 1988-1989*, Griffin Press, Adelaide.

Wildermuth, G.B. (1980) Suppression of take-all by some Australian soils. *Australian Journal of Agricultural Research* 31, 251–258.

Williams, C.H. (1981) Chemical properties. In: Oades, J.M., Lewis, D.G. and Norrish K. (eds) *Red-brown Earths of Australia*. Waite Agricultural Research Institute /CSIRO (Australia) Division of Soils, Adelaide, pp. 47–62.

Williams, C.H. and Raupach, M. (1983) Plant nutrients in Australian soils. In: Division of Soils, CSIRO (ed.) *Soils: An Australian Viewpoint*. CSIRO, Melbourne, pp. 777–793.

5 Long-term Experimentation in Forestry and Site Change

JULIAN EVANS
Chief Research Officer (S), Forestry Authority Research Division, Alice Holt Lodge, Farnham, Surrey GU10 4LH, UK.

Introduction

In one sense nearly all field experiments in forestry are long-term by lasting more than one year or one season. But the great contrast with arable cropping is asking the right questions about one crop which lasts 150 years, as many of our oak and beech do, rather than about 150 crops each lasting one year. The Forestry Commission's oldest permanent sample plot has been regularly measured since 1920 and several, still active, silvicultural experiments date back to the 1930s or earlier. However, the study of forest/site interaction began very much earlier. The relationship between species and site type has been observed throughout recorded history and when, in Britain, there were the first stirrings of concern about our impoverished forest resources, Evelyn, writing in 1664, plainly warned of the danger of not growing the right tree on ground it was suited to. Successive generations of textbooks in the 18th and 19th centuries refined and reinforced this fundamental silvicultural principle: for optimum growth there was an optimum match of tree species and site.

The 19th century saw a second development. The pace of plantation forestry gathered momentum in Germany and other European countries from after the Napoleonic wars and, by the 1860s, these 'new forests' were raising doubts about their sustainability. Indeed, German foresters were also worrying about the over-dependence on conifers and that there was almost no broadleaved forest left (Jones, 1965). Uniform plantations inevitably neglected variation in microsite. Would the imposed matching of tree species and site, underpinned at the time only by limited observations and anecdote, expose them to extra risk owing to stress and ill-health? The converse was also queried. Would these new plantations begin to modify soil processes and, themselves, alter site fertility?

By the end of the 19th century such concerns led authors to warn against monocultures, to advocate mixtures and to raise doubts whether the same ground could yield two timber crops of the same species (Grigor, 1868; Brown and Nisbet, 1894). Evidence of declining productivity of successive rotations was meagre, inconclusive and more qualitative than quantitative. A disenchantment with pure plantations grew with the implication of their causing soil deterioration, and many advocated a return to the 'natural forest conditions of mixed all-age crops'. There was a preoccupation with crop health, the apparent vigour of the stand as evinced by its appearance being the primary indicator of good silviculture. Superimposed on this was the confounding influence of fungal attack of a replanted crop being facilitated by the presence of old stumps acting as a source of infection.

Just after the First World War, Weidemann (1923) undertook a detailed investigation into the state of the planted forests of pure spruce growing in middle and lower Saxony. This was to study the phenomenon of checked growth which was then present in 8.3% of the forest area. The main conclusions of his investigation were an unequivocal statement that the growth of the spruce had declined considerably in the second and third generations and that the deterioration was primarily due to the adverse effects on the soil of clear-felling and replanting of pure stands. Ironically, only a few years earlier Leyendecker-Hilders (1910) was advocating the introduction of the conifers spruce, pine, fir or larch to arrest the decrease in timber yield of forests of pure beech. However, the effect of Weidemann's findings led several authors to stress the dangers of conifer monoculture (Troup, 1928).

Nineteenth century forestry in Europe, and its UK advocates like Simpson (1900) who enthusiastically dubbed the uniformity of species, sizes, age and orderliness of management the 'New Forestry', had pointed up the need to answer at least three questions in objective, i.e. experimental, ways:

1.   As new forests were planted, how could species be matched with site?
2.   Could site conditions be modified to suit particular species or ameliorate detrimental changes?
3.   Was intensive forestry, exemplified by monoculture, causing fundamental site change?

None of the above could be quickly answered.

Species and Site

Species choice was aided in the widely replicated but undesigned approach of numerous arboreta and collections of trees established by the wealthy during the great exploration era of 19th century Victorian travellers and botanists. This great unplanned experiment; this spreading of plant material around the world, conclusively demonstrated that, for Britain, Sitka spruce (*Picea sitchen-*

sis Bong. Carr.) was well fitted to our cool maritime climate and that *Pinus radiata* (D. Don), despite its highly restricted natural range in California's Monterey peninsula, was exceptionally well suited to the Mediterranean climatic regions of southern hemisphere countries.

Many such successes, as well as numerous failures, can be cited, but these early introductions were more by luck than judgement. As plantations expanded in temperate regions between the wars (Savill and Evans, 1986) and in the tropics since the 1950s (Evans, 1992) evaluation of species potential became increasingly systematic. And this was not only confined to species level, but also to taxa representing adequately the range of a species natural adaptability most notably its provenances. Across the world and none more so than in Britain, large-scale trials of potentially suitable species were laid down; one of the best accounts will be found in MacDonald *et al.* (1957).

The Forestry Commission's first species and provenance trials in the 1920s, while laid out on contrasting sites, were usually unreplicated. By the mid 1930s this had all changed. The work of Rothamsted's famous statisticians, R.A. Fisher and F. Yates, had impressed forestry researchers and randomized block and Latin square designs soon became familiar (Jeffers, 1952), not to mention intrusive in the upland landscape, especially when testing species at high altitude (Fig. 5.1). Indeed, the experiment in Fig. 5.1 (Beddgelert 12 photographed in 1945) was planted in 1929 and its Latin square design is plainly evident. By 1932 the Forestry Commission had laid down 98 experiments using such statistical designs (MacDonald, 1933). 'By 1935 the layouts of field trials in a form capable of statistical analysis was a commonplace in forest research, and the bad old methods of single, unreplicated field plots was a thing of the past' (Gillebaud, 1948). Of course, trials continued to be laid down on a representative range of site types. Today, this continues with study of GEI (genotype:environment interaction) where even the best clones or geneticist's selections for one site may not be the optimum for another.

In such long-term experiments to evaluate species potential, it was soon realized that design and plot size were crucial to obtaining worthwhile information. Small treatment plots or even single trees or lines of trees would reveal how a species fared on a site, but would tell you nothing about timber potential of a stand (*vide* forest) grown under such conditions. This can only be done by using permanent sample plots. In addition the buffer zone requirements to avoid between treatment interactions were very different for plots of small trees than large ones! In Britain the minimum acceptable plot area in which a permanent sample plot can be satisfactorily established is 0.1 ha. Add to that the buffer zone, multiply by the replications, and then by the number of species and provenances, and much ground (that most variable of all commodities) is occupied. Well-replicated trials designed to yield wood production information became very large experiments indeed. Nonetheless, continuous collection of data over time from such trials allows performance of individual trees to evolve to the stand-scale parameters of such interest to foresters.

Fig. 5.1. Species trial at high altitude showing Latin square design. Beddgelert Experiment 12, planted 1929, photographed in 1945. Copyright Forestry Commission c.229.

The timelessness of forestry intrudes in another way for species trials. Susceptibility to damage, biotic and abiotic, varies with tree age. Despite good quality of design and spatial replication a once-only establishment of such trials is dangerous – the year in question may be very atypical. Replication over time to expose the different species and provenances to varying weather patterns and climatic phenomena is also required to achieve confidence in a result.

Site Amelioration

The converse to choosing a tree species for a particular set of site characteristics is to modify the site itself to suit the tree, but there are clearly limits. With the exception of irrigation in a few subtropical countries, improvements to soil physical and chemical status have been the principal strategies. From the point of view of this chapter the question raised is how long do such ameliorative practices persist? There is only one opportunity in each forest rotation, i.e. once every 50 or 100 years, to cultivate or dig drainage ditches and, even with fertilizer, although additions can theoretically be applied annually, the economics of timber production would wholly preclude such intensive treatment. Indeed, it is only in forest nurseries where agricultural levels of input are reached, and for that foresters owe much to another Rothamsted scientist, Miss Blanche Benzian and her sustained work from the late 1940s to the 1960s which the Forestry Commission published in 1965 as a major two volume bulletin (number 37).

Site preparation

Long-term experiments have sought to answer questions about the importance of site preparation. It is now clear that, in Britain, for example, a single cultivation of inhospitable upland peats and peaty gleys changes a site's forestry potential from almost nil to reasonable in many situations. The improved surface soil aeration and temporary control of rank weed growth provides sufficient opportunity for a tree crop to become established. Much experimentation has examined the scale or degree of cultivation needed, but typically for all such research, conclusive answers must await canopy closure or mid rotation, i.e. 20–40 years. And in this topic, well researched in the 1940s, 1950s and 1960s, there was an end-of-rotation surprise. Some practices, especially single mouldboard ploughing left deep furrows which severely affected lateral root development and rendered tree crops on exposed sites unstable in strong winds. This only became clear 30 or 40 years after introduction of these otherwise successful establishment practices.

Fertilizer application

Use of fertilizers in forestry was slower in coming, the view generally being held that trees did not make large demands on a site for nutrients, at least compared with many agricultural crops. Although experiments with manure were certainly carried on more than 100 years ago much attention was paid to liming treatments to accelerate litter breakdown and relatively little to correction of inherent deficiencies. In Britain since the 1930s evidence accumulated that lack of phosphate severely limited tree growth on many upland peaty-gleys and lowland podzols. Wareham experiment no. 25, laid down in Dorset in 1934, still shows the effect of phosphate applications; today trees in plots given P are 25 m tall, those in adjacent untreated plots remain stunted and rarely over 5 m. Despite its early date of establishment, this experiment was a factorial with six replications designed to compare treatments of no fertilizer, liming and phosphate. It may be the oldest replicated, randomized trial still in existence. (Thanks to the stimulus of Dr E.M. Crowther of Rothamsted the powerful tool that factorial experiments represent became increasingly important in forestry research from the late 1930s.) Demonstration of marked contrast between P and nil-P took several years and only became really clear 15–20 years after laying down the experiment. Indeed local forest management did not accept the evidence of widespread phosphate deficiency at Wareham until the 1960s! Such slow uptake of research results is not unfamiliar. A celebrated example involved the Forestry Commission Chairman in 1950, Lord Robinson, who asserted that 'an ounce of patience was worth three ounces of phosphate' in a most unthrifty, and we now know acutely phosphate-deficient, forest on Wilsey Down in Devon. He could not have been more wrong, but so too perhaps were the researchers in how to communicate effectively the results of their experiments. Experimental contrasts need to be pronounced to be persuasive to non-researchers: well-designed experiments maintained over many years are required.

Proving the fact of response is one role of an experiment, but sometimes not the most important. Proving or demonstrating the nature, scale and persistence of response can be more important. Forestry literature abounds with reports of dose–response relationships, of transient responses, and of no responses at all despite foliar nutrient improvement, better leaf colorations, etc. In forestry you cannot jump to early conclusions as a more recent experiment also exemplifies.

In 1970, again at Wareham in Dorset, a large experiment was established to compare the performance of all tree species with the capability of outgrowing Corsican pine (*Pinus nigra* var. *maritima*) on the site in question. Maximum amelioration of the site was to be achieved by intensive weed control and annual fertilizer application if required. Many species grew at rates that far exceeded the height growth profiles in the yield models for top quality stands. Indeed, at one stage a Yield Class of 48 (a potential maximum mean annual increment of 48 m^3 ha^{-1}) was being predicted for Sitka spruce or *twice* the most productive Yield Class incorporated in the Forestry Commission's standard models (Fig. 5.2). Later measurements did not confirm this remarkable predicted yield and while the stand is clearly growing exceptionally it is exhibiting more of a subtropical growth pattern of fast early growth and early culmination of maximum productivity rather than inherently massively

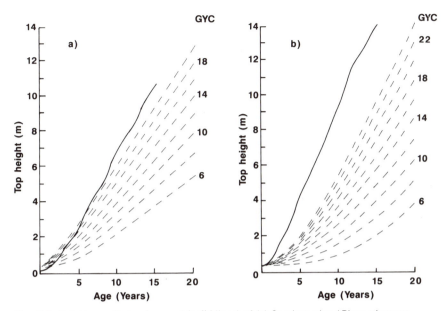

Fig. 5.2. Height growth development (solid lines) of (a) Corsican pine (*Pinus nigra* var. *maritima*) and (b) Sitka spruce (*Picea sitchensis*) in the Wareham 156 maximum amelioration trial compared with the range of height growth profiles (dashed lines) published by the Forestry Commission. The numbers next to each curve indicate the General Yield Class (GYC), i.e. the potential maximum mean annual increment in wood production as $m^3 ha^{-1} year^{-1}$.

increased yield in absolute terms. Only secure, long-term, well-recorded experiments can reveal such results – the value of the continuous dataset, and hence temper initial enthusiasm with reality!

In Britain, as noted earlier, many heathland and upland sites required addition of phosphate for satisfactory establishment of the first crop. It is now clear that in most instances no re-application is required when restocking and beginning the second rotation (Taylor, 1990). Research findings for one rotation may not apply to the next.

Forest/Site Interaction

The concern of foresters about the effects of trees on a site was noted earlier. The work of Weidemann (1923) appeared to confirm the worst fears, at least for coniferous monoculture. Since that time the problem has only come to the fore in a large way on three occasions, in South Australia, Swaziland and Central China, and to a lesser extent in several isolated examples. Yield decline with successive tree crops has not proved to be the rule, but rather an exception (Evans, 1990).

Yield decline in successive crops

In South Australia in the 1960s and 1970s young second rotation *Pinus radiata* showed an average yield drop of 25–30% on most sites compared with the first rotation. Fundamental to this decline, which occurred throughout the state, was not the effects of the trees themselves but the destruction of organic matter in slash disposal at harvesting and the near stagnation of second rotation growth due to weed infestation, notably by grasses. Today, with good silviculture and genetically improved planting stock, most second rotation pine is superior to the previous crop.

The South Australian experience led to comparable fears in the exotic pine plantations of the Usutu Forest, Swaziland. Twenty-five years growth monitoring, over three successive rotations of *Pinus patula*, has shown for the most part increasing productivity with each rotation (Fig. 5.3a) and yield decline confined to small areas of phosphate-deficient soils derived from gabbro (Fig. 5.3b). Again grass competition has been a critical factor because the second and third rotation crops were established in largely weed-free conditions unlike the first rotation which was planted into the inhospitable southern African grass veld. Research by Morris (1993) suggests that this satisfactory productivity position may not last indefinitely owing to increasing litter

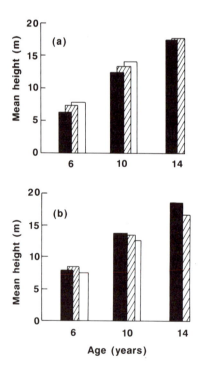

Fig. 5.3. Mean height at ages 6, 10 and 14 years, of three rotations of *Pinus patula* on (a) granite derived or (b) gabbro-derived soils in the Usutu Forest, Swaziland. Data are the mean of (a) 26 or (b) 9 plots and solid bars show first rotation; hatched bars, second rotation; open bars, third rotation. Adapted from Evans (1992).

accumulations with each rotation (Fig. 5.4) and consequent interruptions of nutrient cycling.

In China, soil degradation under Chinese fir (*Cunninghamia lanceolata*) is an accepted fact (Li and Chen, 1992). Yield declines of 10% and 40% in second and third rotation crops are regularly observed. Numerous detrimental changes have been recorded in soil parameters. However these changes, which are genuine, may well not arise from the fact that successive crops of Chinese fir are grown in monoculture, but rather from the harmful practices associated with crop harvesting and re-establishment. Every part of the tree including bark, leaves and twigs is removed from the site in whole tree harvesting, any remaining debris is burnt and whole hillsides are diligently reterraced: all practices which, directly or indirectly, maximize nutrient loss, conspicuously fail to conserve organic matter, and provide ideal conditions for grass and bamboo invasion (Evans and Hunter, unpublished results). The UK Overseas Development Administration has just initiated a three-year research programme to lay down long-term experiments to investigate this contention and recommend silvicultural remedies.

Forests and soil change

Much research has examined the question, how does continuous forest growth affect soil development? The question is asked, not so much in the context of forest soils which have always been part of a woodland ecosystem, but where open ground, such as grassland, field or moorland is afforested. What changes occur as a result of tree cover and how do such changes affect successor crops? All of us are familiar with horticulture's equivalent of 'specific replant diseases' associated with citrus, apple, cherry and peach.

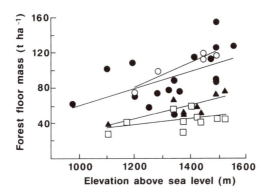

Fig. 5.4. Forest floor mass compared for first rotation (□); second rotation and litter burn (▲); second rotation and no litter burn (●); and third rotation (○) stands at various heights above sea level in Swaziland. Adapted from Morris (1993).

Most investigations have sought to answer the question of forest influence on soil by comparing adjacent areas of forested and unforested ground. The shortcomings of this approach, recognizing soil's high variability and historical influences determining land use, are well known, but the experimental alternative is necessarily long term. In Britain, and specifically as part of forestry research, two sets of long-term experiments have been laid down, both in the 1950s, to examine forest effects – at Gisburn Forest, Yorkshire, and the 'Ovington Plots' at Bedgebury National Pinetum, Kent, West Tofts in East Anglia and in the Forest of Dean. The most recent reports are Moffat and Boswell (1990) for Gisburn and Anderson (1987) for the Ovington Plots. From this work and research on Rothamsted's Geescroft and Broadbalk Wildernesses (Jenkinson, 1971; Johnston *et al.*, 1986; see also Southwood, Chapter 1; Johnston, Chapter 2) two main conclusions can be drawn: (i) organic matter accumulates; and (ii) soil acidity usually increases. (Flowing from these, of course, are numerous secondary changes of nitrogen economy, aeration, biological activity, nutrient availability, etc.) Throughout the world these two principal effects have been observed, as has their converse when deforestation occurs.

The acidification component of these results spawned great interest into 'mull' and 'mor' humus – under what conditions were the products of organic matter degradation causing decrease in pH? Broadleaved woodland was usually regarded as producing less acidifying and more rapidly decaying litter (mull) than coniferous forest. This generalization oversimplifies to the point of distortion and it is clear that the effect of any one species is strongly determined by site itself and stand management (Moffat, 1991). Experiments in the past have sought to harness this 'benefit', such as Professor Dimbleby's trials in the 1950s with birch as a 'soil improver' on the North York Moors, but today, all the pre-1970s records need to be reviewed, indeed re-examined, in the light of what we now know about acid aerosol inputs and the role trees play in scavenging such pollution (Fowler *et al.*, 1989). In particular, we need to know how much acidification arises from this effect as opposed to the 'humic acid' route?

This leads, finally, to note that as environments change and evolve so will the influence of the 'static' woodland cover. Tree crops rotate at long intervals on cycles longer than the presently predicted timescales of significant change, such as rising 'greenhouse gas' levels and the associated climatic phenomena. As a result, long-term experimentation in forestry must now have added to it a further dimension of change while the forest grows. New programmes to monitor this in the long term are now being implemented, most notable of which in the UK is the Environmental Change Network of which the Forestry Commission and Rothamsted, among others, are participants (see Tinker, Chapter 22).

Experimentation in Forestry Research: Some Lessons

Forestry research is, of its nature, necessarily long term in the large majority of cases. Forestry researchers have learnt several lessons.

1. Robust results derive from well-designed and well-replicated trials. Replication needs to be spatial within a site, across several contrasting sites and, in many instances, over time as well, i.e. in different years.

2. Experimental design for whole rotation studies must accommodate the factor of scale at the outset, i.e. the requirements of stand integrity and adequate buffer zone provision at maturity.

3. Scale and persistence of response are as important as the fact of it. Initial response to a treatment may differ radically from final response; it is dangerous to jump to early conclusions. Regular assessment and updating of measurements are essential.

4. Research findings in one rotation may not apply to the next.

5. Acceptance of research results by forest managers depends largely on the demonstration of a clear, unambiguous benefit; simple experimental designs facilitate this.

6. Commitment to long-term experiments is costly in time and resources, but in many cases unavoidable to achieve confidence in research results.

Much forestry research is necessarily long term, but only by commitment to such long-term research has a sound basis been laid for the success of so much of British silviculture.

Acknowledgements

I am grateful to David Mobbs, Dr A.J. Moffat and Dr P.H. Freer-Smith for contributions and helpful comments. Thanks are also due to George Gate and Mary Trusler for photographs and figure preparation, and to Hazel Payne for typing.

References

Anderson, M.A. (1987) The effects of forest plantations on some lowland soils. I. A second sampling of nutrient stocks. *Forestry* 60, 69-85.

Brown, J. and Nisbet, J. (1894) *The Forester*. Vol.1, Blackwood and Sons Ltd.

Evans, J. (1990) Long-term productivity of forest plantations – Status in 1990. *Proceedings of the 19th World Congress, International Union of Forestry Research Organisations*, Montreal. Vol.1, pp. 165-180.

Evans, J. (1992) *Plantation Forestry in the Tropics*. 2nd Edn, Clarendon Press, Oxford.

Fowler, D., Cape, J.N. and Unsworth, M.H. (1989) Deposition of atmospheric pollutants on forests. *Philosophical Transactions of Royal Society of London B* 324, 247-265.

Gillebaud, W.H. (1948) Some recent developments in forest research. *Forestry* 22, 145-157.

Grigor, J. (1868) *Arboriculture*. Emmonston and Douglas, Edinburgh.

Jeffers, J.N.R. (1952) The use of statistical methods in forest research. *Proceedings of the Sixth British Commonwealth Conference*, Canada.

Jenkinson, D.S. (1971) The accumulation of organic matter in soil left uncultivated. *Rothamsted Report for 1970*, Part 2, 113-137.

Johnston, A.E., Goulding, K.W.T. and Poulton, P.R. (1986) Soil acidification during more than 100 years under permanent grassland and woodland at Rothamsted. *Soil Use and Management* 2, 3-10.

Jones, E.W. (1965) Pure conifers in central Europe - a review of some old and new work. *Journal of Oxford University Forestry Society* 13, 3-15.

Leyendecker-Hilders (1910) The restoration of beechwoods with conifers. *Allgemeine Forst und Jagdzeitung* 86, 384-386.

Li, Y. and Chen, D. (1992) Fertility degradation and growth response in Chinese fir plantations. *Proceedings of the 2nd International Symposium on Forest Soils*, Ciudad, Guyana - Venezuela, pp. 22-29.

MacDonald, J. (1933) Experimental methods as applied to silvicultural problems in Great Britain. *Comptes rendus du Congrès Forestier de Nancy*, 1932.

MacDonald, J., Wood, R.F., Edwards, M.V. and Aldhous, J.L. (1957). Exotic forest trees in Great Britain. *Forestry Commission Bulletin No. 30*. HMSO, London.

Moffat, A.J. (1991). Forestry and soil protection in the U.K. *Soil Use and Management* 7, 145-151.

Moffat, A.J. and Boswell, R.C. (1990) Effect of tree species and species mixtures on soil properties at Gisburn Forest, Yorkshire. *Soil Use and Management* 6, 46-51.

Morris, A.R. (1993) Forest floor accumulation, nutrition and productivity of *Pinus patula* in the Usutu Forest, Swaziland. *Symposium on Nutrient Uptake and Cycling in Forest Ecosystems*, Halmstad, Sweden, IUFRO, June 1993.

Savill, P.S. and Evans, J. (1986) *Plantation Silviculture in Temperate Regions - with Special Reference to the British Isles*. Clarendon Press, Oxford.

Simpson, J. (1900) *The New Forestry*. Pawson & Brailsford, Sheffield.

Taylor, C.M.A. (1990) Nutrition of Sitka spruce on upland restock sites. *Forestry Commission Research Information Note 164*. HMSO, London.

Troup, R.S. (1928) *Silvicultural Systems*. Clarendon Press, Oxford.

Weidemann, E. (1923) *Zuwachsruckgang und Wuchstockungen der Fichte in den mittleren und den unteren Hohenlagen der Sachsischen Staatsforsten*. Tharandt. (Seen in Translation No. 302 U.S. Department of Agriculture, C.P. Blumenthal, 1936.)

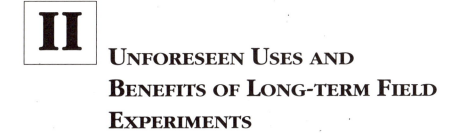

II

Unforeseen Uses and Benefits of Long-term Field Experiments

6 Quantification of Nutrient Cycles Using Long-term Experiments

D.S. POWLSON
Institute of Arable Crops Research, Rothamsted Experimental Station, Harpenden, Herts, AL5 2JQ, UK.

Introduction

Long-term field experiments provide ideal opportunities to study nutrient cycling. First, meaningful nutrient balances can only be calculated by comparing inputs and outputs over long periods: small changes in nutrient balances, whether positive or negative, are difficult to measure accurately over short periods because of year-to-year variation in crop growth and because the changes must be measured against the large quantities of nutrients usually present in soil. Such calculations provide an excellent starting point for more detailed studies of specific agroecosystems and can highlight areas of uncertainty. Second, the impact of different agricultural managements on the quantity and forms of nutrients (and pollutants; see Jones *et al.*, Chapter 9) in soil can be studied readily if soil and plant samples have been archived and are available for analysis using current techniques. Third, short-term studies on specific aspects of nutrient cycling and loss can be superimposed on long-term experimental sites, thus capitalizing on the well-documented history of the site and providing information on the sustainability and environmental impacts of different agricultural practices. This chapter gives examples of all three, drawing on recent work based mainly, though not exclusively, on the long-term Classical sites at Rothamsted (described in Johnston, Chapter 2). It also draws attention to the value of establishing new experiments with a view to continuing them for long periods so that sites and data will be available for addressing new questions in the future.

Table 6.1. Decline in readily soluble P and K in soil of the Exhaustion Land experiment at Rothamsted. The data are for plots which received manure between 1856 and 1901 but none thereafter and have been cropped continuously to spring barley (from Johnston and Poulton, 1977).

Nutrient measured	Year				
	1903	1951	1958	1965	1974
P soluble in 0.5 M-NaHCO$_3$ mg kg^{-1}	69.8	25.5	17.6	14.5	8.9
K exchangeable to 1 M-NH$_4$OAc mg kg^{-1}	454	150	154	158	112

Long-term Phosphate and Potassium Balances

Table 6.1 shows the rate of decline of plant-available P and K in the soil of plots within the Exhaustion Land experiment at Rothamsted, which have been cropped with spring barley for over 70 years but without any inputs of P or K fertilizers. Soil P and K contents were initially high because of previous manure applications but because of continued removal by crops they progressively declined and by 1974 bicarbonate-extractable P had decreased to 13% of its original value and ammonium acetate-extractable K to 25%. This type of information is necessary when considering the sustainability of reduced input agricultural systems and in calculating the inputs of nutrients needed to maintain a given level of productivity.

Table 6.2 shows a comparison of inputs and outputs of P for three treatments of the Exhaustion Land experiment, and the changes in soil total P content for the period 1856–1903. The data indicate that any losses of P that

Table 6.2. Inputs and outputs of P, and changes in soil P content, in three treatments of the Exhaustion Land experiment (from Johnston and Poulton, 1977).

Treatment[a]	Total P in soil (kg ha^{-1})			P balance[b] (kg ha^{-1})	Difference between P balance and change in soil P as % of total soil P
	1856	1903	Change		
None	1600	1606	+6	−80	5
FYM	1600	2612	+1012	+1031	1
PK	1600	2685	+1085	+1217	5

[a] Treatment applied annually, FYM, 35 t ha^{-1} fresh farmyard manure, P, 33 kg ha^{-1} as superphosphate, K, 90 kg ha^{-1} as potassium sulphate.
[b] P balance, P applied *minus* P removed in crops.

occurred must have been small; the P balance (defined as total P applied *less* total P removed in crop over the period) is almost equal to the change in total P in the soil. Phosphorus balances calculated for several of the Rothamsted long-term experiments show up to 85% of the P added over the last 100 years or so, as farmyard manure (FYM) or superphosphate, has been retained in the soil. Almost all the remainder has been removed in crops with losses too small to detect. Table 6.3 shows the distribution of total P in the soil profile within two long-term experiments in which the treatments have continued for well over 100 years. When P was added as superphosphate to the arable site there was little movement of phosphate down the profile and values in the subsoil were similar to those in the plot receiving no P. However, where a large quantity of organic matter was present (either because P was added as FYM or because of the high organic matter content of the grassland soil) there was rather more vertical movement. This may indicate that organic forms of P are more prone to movement within the soil than inorganic P and this possibility is currently being investigated. Although evidence from such experiments indicates very small losses of P this does not mean that they are unimportant, as even small increases in P concentration in surface waters may contribute to eutrophication. For the average winter drainage at Rothamsted an annual loss of 3 kg P ha^{-1} would be equivalent to about 1 mg P l^{-1} in drainage. This compares with a suggested limit of 1–2 mg P l^{-1} for effluent from sewage treatment works to waterways at risk to eutrophication.

Some studies on the forms of P in soil resulting from long-continued treatments suggest that the presence of organic matter is highly significant in determining P speciation and, hence, potential for movement. Table 6.4 shows the

Table 6.3. Total P (mg kg^{-1}), in soil at different depths where superphosphate or farmyard manure were applied to surface soils at pH 6.5 (from Johnston, 1976; Johnston and Poulton, 1993).

| | Cropping and fertilizer treatment[a] | | | | |
| | Arable crops (Barnfield) | | | Permanent grass (Park Grass) | |
Soil depth (cm)	None	P	FYM	None	P
0–23	780	1295	1375	575	1425
23–30	465	525	650	555	785
30–46	415	450	525	500	600
below 46	400	395	440	—	—

[a] Treatments applied annually: P, 33 kg ha^{-1} as superphosphate, FYM 35 t ha^{-1} fresh farmyard manure.

Table 6.4. Total, 0.5 M-NaHCO₃-soluble P and 0.01 M-CaCl₂-soluble P in surface soils from treatments of the Barnfield experiment receiving superphosphate and FYM either separately or together, measurements made in 1958 (from Johnston, 1976; Johnston and Poulton, 1993).

Treatment[a]	P pool measured[b]		
	Total (mg kg⁻¹)	NaHCO₃ (mg kg⁻¹)	CaCl₂ (μM)
Control	670	18	0.5
Superphosphate (P)	1215 (545)	69 (51)	3.0 (2.5)
FYM	1265 (595)	86 (68)	12.8 (12.3)
FYM + P	1875 (1205)	145 (127)	22.3 (21.8)

[a] see Table 6.3 for rates of application of FYM and P.
[b] Figures in parentheses are increases over the control.

quantities of total P in the topsoil of some treatments of the Barnfield Experiment and also the quantities extracted by 0.5 M-NaHCO₃ (Olsen's reagent) or 0.01 M-CaCl₂. Additions of superphosphate or FYM both increased soil total P by about 2-fold and 0.5 M-NaHCO₃-extractable P by 4–5-fold. Where *both* superphosphate and FYM were applied together the increases were approximately the sum of those produced by the two treatments individually. In the case of 0.01 M-CaCl₂ extractable P, considered to be representative of the concentration of phosphate in the soil solution, the increase in the FYM treatment was about five times that with superphosphate. In the treatment receiving both superphosphate and FYM the total increase was much greater than the sum of the individual increases (Table 6.4). This suggests that the organic matter derived from FYM was increasing the proportion of P in this readily extractable form.

Recently a linear relationship has been observed between 0.5 M-NaHCO₃-extractable P in soil and inorganic P concentration in the water draining from the Broadbalk wheat experiment (G. Heckrath and P.C. Brookes, personal communication). Again, concentrations were not large (up to 2 mg l⁻¹) for soils containing about 70 mg kg⁻¹ 0.5 M-NaHCO₃-extractable P, but in view of the environmental concerns regarding P in surface waters it is necessary to investigate such losses thoroughly. Phosphate in water reaching the Broadbalk drains appears to be mainly inorganic and in solution, rather than inorganic P adsorbed on colloidal particles, but in other situations this is not the case. Where water leaving agricultural land has less contact with soil, for example because of surface runoff or by-pass flow, P sorbed on colloids or particles makes a major contribution (Rekolainen, 1989; Catt *et al.*, 1994).

Nitrogen Cycle Studies

Inputs and outputs of N

From the time of their inception, the long-term experiments at Rothamsted have been used to probe aspects of the nitrogen cycle. Johnston (Chapter 2) describes some of the early work on N and the controversy with Liebig which ensued. The site continues to be of great value for addressing current topics involving the nitrogen cycle. Figure 6.1 shows how the total N content of soil has changed in selected plots of the Broadbalk wheat experiment. These data can only be obtained because soil samples have always been taken from the same depth (0–23 cm, the current depth of ploughing) and have been archived and so are available for analysis by current methods. The results show that the total N content of soil in the plot receiving no N fertilizer since 1843 has been almost constant for well over 100 years, indicating that inputs and outputs of N are in balance. As the crop on this plot removes, on average, about 30 kg N ha^{-1} year^{-1} there must be inputs of N of at least this magnitude from external sources. Jenkinson (1977) constructed N balances for selected treatments and concluded that N dissolved in rain plus that in dry deposition were the most likely inputs.

Powlson *et al.* (1986) reported results from a study in which ^{15}N-labelled fertilizer was applied to winter wheat in microplots within the main plots of the Broadbalk experiment. Labelled N was applied at the usual time in spring (mid-April) and its distribution in crop and soil was measured at harvest in

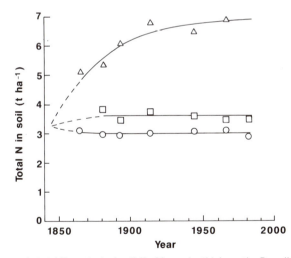

Fig. 6.1. Changes in total N content of soil (0–23 cm depth) from the Broadbalk experiment. Soils taken from plots which have been unmanured (○) or given 35 t ha^{-1} FYM (△) annually since 1843 or 144 kg N ha^{-1} (□) with PK fertilizers since 1852 (from Jenkinson, 1990).

August. Experiments were conducted in four separate years. Recovery in the above-ground crop ranged from, 51% to 68% with 15-20% retained in soil, almost entirely in organic forms. An average of 20% (range 9-27%) was unaccounted for in crop and soil at the time of harvest. Fig. 6.2 (taken from Jenkinson and Parry, 1989) is a nitrogen cycle diagram for one plot of the

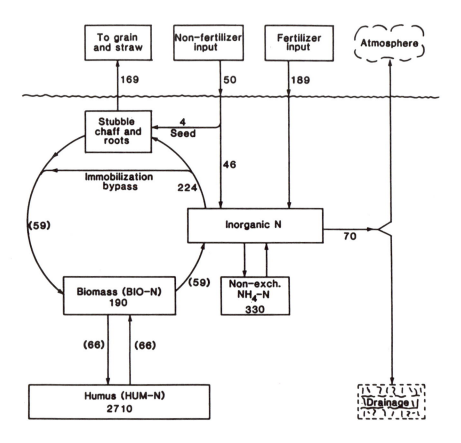

N cycle under continuous winter wheat

Fig. 6.2. Nitrogen cycle diagram for the continuous wheat plot of the Broadbalk experiment receiving 192 kg N ha^{-1} as fertilizer. Average application of ^{15}N-labelled fertilizer for the 4 years of microplot experiments (1980–83) was 189 kg N ha^{-1}. All transformations are assumed to take place in the plough layer (0–23 cm). Figures *within* boxes are kg N ha^{-1} (means for 1980–83); figures *between* boxes are kg N ha^{-1} year^{-1}. Figures in parentheses are calculated from a mathematical model; other figures between boxes are means of the 1980 and 1981 ^{15}N experiments (from Jenkinson and Parry, 1989).

experiment calculated using a combination of short-term data derived from the ^{15}N experiment and information from the long-term N balance. The total annual loss of N (i.e. labelled plus unlabelled) was estimated from the loss of labelled N during the growing season and by taking account of separate experiments in which over-winter losses of ^{15}N were measured (see Powlson *et al.*, 1986 for details). A key simplifying factor in constructing this nitrogen cycle is the fact that total soil nitrogen content is at an equilibrium value so inputs and outputs must be in balance (Fig. 6.2). Although the direct losses of fertilizer N during the year of application were usually quite small (often 15% or less), the total loss of N from the soil–plant system over the course of the year is equivalent to 29% of the total N inputs. This is because much of the N present in soil as nitrate during winter, when most leaching occurs, is derived from the mineralization of soil organic nitrogen.

Measurements of the fate of ^{15}N residues in soil over a number of years are extremely valuable in constructing mathematical models of soil N dynamics. Data from Broadbalk (Shen *et al.*, 1989; Jenkinson and Parry, 1989) and other long-term sites on which ^{15}N studies have been superimposed (Ladd and Amato, 1986; Hart *et al.*, 1993) have been used in this way; this is discussed in detail by Jenkinson *et al.* (Chapter 7).

One of the interesting features of Fig. 6.2 is the large calculated input of N from external sources other than fertilizer amounting to about 50 kg N ha^{-1} year^{-1}, considerably greater than the minimal value of 30 kg N^{-1} year^{-1} calculated from the long-term N balance alone. N inputs in rain have been measured over many years, but recently Goulding (1990) reported measurements of N in dry deposition. Dry deposition of N in gaseous, aerosol and particulate forms now account for a larger proportion of the input than that in rain; ammonia, oxides of nitrogen and nitric acid vapour being major sources of dry deposition (Table 6.5). Goulding (1990) estimated that 50–60 kg N ha^{-1} year^{-1} enters cereal-growing systems at Rothamsted. These are partly offset by gaseous losses giving a net annual input of about 40 kg N ha^{-1}, in good agreement with the value estimated in Fig. 6.2. Using a mathematical model for the nitrogen cycle of the Broadbalk experiment Whitmore and Goulding (1992) estimated the fate of this input. The calculations indicate that 51% of the atmospheric input is taken up by the wheat crop on the plot receiving 192 kg N ha^{-1} as fertilizer, but 29% is leached and the remaining 20% is returned to the atmosphere in gaseous form. The proportion leached represents 29% of the estimated total leaching from the system – a large proportion considering that atmospheric inputs are only 17% of total N inputs.

Long-term impacts of inorganic N fertilizer

There is much evidence that direct losses of N from fertilizer applied to cereals make only a small contribution to the nitrate leaching in the year of application (Macdonald *et al.*, 1989; Chaney, 1990; Addiscott *et al.*, 1991). However, a

D.S. Powlson

Table 6.5. Nitrogen deposited[a] to arable land from the atmosphere at Rothamsted.

	N deposited (kg ha^{-1} year^{-1})[b]	
Source	Bare soil	Winter cereal
Wet		
Ammonium-N	7.9	7.9
Nitrate-N	6.3	6.3
Dry		
Ammonia	6.4	9.1
Nitrogen dioxide	6.4	7.6 (15.2)[c]
Nitric acid	4.3	10.1
Particulate nitrate	1.7	2.3
Total	33.0	43.3 (50.9)[c]

[a] There may be other inputs, e.g. peracetyl nitrate (PAN) and nitrous acid (HONO) but the amounts are not known.
[b] Direct measurements of gross inputs. Data updated from Goulding (1990).
[c] Larger value for NO$_2$ deposition calculated using newer estimate of deposition velocity.

knowledge of the longer-term effects of N fertilizer is required in order to assess the overall environmental impacts of agricultural activities. Figure 6.1 shows that the Broadbalk plot receiving 144 kg N ha^{-1} each year since the start of the experiment now contains 15% more total N than the plot receiving no N fertilizer. This mainly arises because the larger crops grown where N is applied leave larger residues of stubble and roots, containing more N, than the unfertilized crop. However, there is evidence that the impact on readily mineralizable forms of organic N is considerably greater. For example, when ^{15}N-labelled fertilizer was applied, crop uptake of unlabelled N from soil by the crop receiving 144 kg N ha^{-1} was about 60 kg N ha^{-1} compared to a total uptake of 30 kg N ha^{-1} by the unfertilized crop (Powlson et al., 1986). Shen et al. (1989) also found a greater mineralization rate in soil from this plot when it was incubated under laboratory conditions. This effect has also been observed, to a greater or lesser extent, in many other long-term experiments throughout the world; this subject has recently been reviewed by Glendining and Powlson (1994). At present, nitrate derived from mineralization of the extra organic N in soil resulting from long-continued fertilizer N applications is not explicitly considered in systems for advising farmers on fertilizer N applications to crops. It is also likely to lead to greater nitrate leaching as part of the additional mineralization will occur at periods when crop growth is minimal and leaching risk is high. There is some preliminary evidence to support this (Glendining et *al.*, 1992; Sylvester-Bradley and Chambers, 1992; K.W.T Goulding and C.P. Webster, personal communication).

A change made to Broadbalk in 1968 has proved valuable in studying the medium-term impacts of inorganic N fertilizer. A plot which had previously received 48 kg N ha^{-1} was changed to 192 kg N ha^{-1}. Within 20 years there was a measurable, though small, increase in the total N content of the soil so that it is now about equal to that of the plot that has received 144 kg N ha^{-1} since 1852. Also, mineralizable N in this soil now equals that in the 144 kg N ha^{-1} treatment (Powlson *et al.* 1986; Glendining *et al.*, 1992).

N cycle studies at the Brimstone experiment

This experiment is sited on a cracking clay soil near Faringdon, Oxfordshire, and is operated jointly with the ADAS Soil and Water Research Centre. Although it is only one-tenth the age of Broadbalk, it has provided extremely valuable insights into N transformations during the last 15 years. In the first ten-year period the impacts of drainage and cultivation treatments on the losses of nitrate were studied. When autumn-sown crops were grown, over-winter leaching losses were usually in the range of 20–50 kg N ha^{-1}. Winter losses averaged 21% more from plots receiving conventional tillage than those that were direct drilled (Goss *et al.*, 1993). In only one year did leaching in spring after N fertilizer was applied exceed 10 kg N ha^{-1}. More recently nitrate leaching has been measured in a number of different crop rotations (Catt *et al.*, 1992). Leaching was always greatest when soil was bare during the winter; any form of crop cover during the winter decreasing leaching to some extent, although on this soil winter cover crops such as mustard were less effective than winter wheat because of poor establishment. At other sites cover crops have been very effective (Christian *et al.*, 1992; Schröder *et al.*, 1992). Various treatments tested at Brimstone decreased nitrate leaching in the year they were applied by causing retention of N in the crop–soil system in organic forms; these treatments included straw incorporation, a grass ley, reduced tillage and cover crops. In all cases, however, some of the retained N was mineralized later leading to increased nitrate leaching in subsequent years. These results emphasize the importance of continuing experiments on nutrient dynamics for a number of years (though not necessarily 150 years in all cases!) in order to reveal the true impact of a particular management strategy. It is noteworthy that nitrate concentrations in many of the drainflow samples collected over the course of the Brimstone experiment have exceeded the 50 mg nitrate l^{-1} (the European Union's limit for nitrate in drinking water), even in treatments involving good agricultural practices to minimize leaching. This is a warning that it is likely to be extremely difficult to decrease nitrate leaching to the extent required under current legislation on water quality.

Gaseous fluxes associated with the N cycle

Soil can act as both source and sink for several gases which have significant environmental effects, including the greenhouse gases, nitrous oxide and methane. Much very useful work can be done using short-term experiments to examine the effect of different agricultural practices on the fluxes of these gases and the controlling mechanisms. Long-term experimental sites provide the opportunity of examining the impact of a range of agricultural systems when continued for some years. One example is shown in Fig. 6.3, which shows total losses of N by denitrification during autumn on two treatments of the Hoosfield barley experiment (Goulding and Webster, 1989). In one treatment, soil organic matter content has been increased to 3.1% C as a result of FYM applications since 1852. In the other, which has received inorganic fertilizers only, soil organic C is 0.9%. The additional organic matter in the FYM treatment is, of course, beneficial in many respects (e.g. Johnston, 1991). However, Fig. 6.3 shows that denitrification losses were much greater in the high organic matter soil, at least over the period studied, amounting to 30 kg N ha^{-1} compared to only 5 kg N ha^{-1} in the low organic matter soil: part of the N will have been lost as N$_2$O. By making more extensive measurements on soils receiving contrasting treatments, such as these, it should be possible to build up a picture of relative importance of soils under different managements as sources and sinks of N$_2$O. In contrast to the data in Fig. 6.3, denitrification rates in spring, following N fertilizer application, were almost equal in soils of high and low organic matter content resulting from long periods in either a continuous arable or a ley–arable rotation (Goulding *et al.*, 1993).

In a recent study, plots within the Broadbalk experiment proved to be invaluable in investigating the impact of N inputs on methane oxidation (Hütsch *et al.*, 1993). Aerobic soil acts as a sink for methane because soil bacteria oxidize methane to carbon dioxide which is a much less potent greenhouse gas. The recent studies on Broadbalk confirmed earlier indications that this oxidation is slower in soils with a history of N fertilizer applications. When soil cores were exposed to methane under laboratory conditions, the rate of disappearance due to oxidation was much less in soil from a plot that had received inorganic N at 144 kg N ha^{-1} since 1843 than in that from a plot receiv-

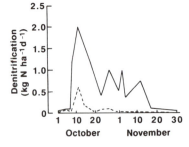

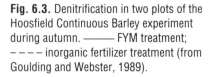

Fig. 6.3. Denitrification in two plots of the Hoosfield Continuous Barley experiment during autumn. ———— FYM treatment; – – – – inorganic fertilizer treatment (from Goulding and Webster, 1989).

ing no N (Fig. 6.4). This did not appear to be a short-term blocking by ammonium of the enzyme(s) involved because the effect occurred in soils taken one year after the last fertilizer N application. Interestingly, the methane oxidation rate in the FYM treatment, which has received more than 200 kg N ha⁻¹ as a mixture of organic N and ammonium N, was almost the same as in the no-N control. As yet the mechanisms involved are not clear but there are indications that a change in the balance of the soil microbial population between ammonium oxidizers and methane oxidizers may be involved. In the case of FYM treatment it may be that the total microbial population is increased so much that a decrease in the proportion of methane oxidizers does not affect the overall rate of methane oxidation. Whatever the mechanism, the result is of considerable environmental importance as it shows an unexpected impact of agricultural practice.

Organic inputs and N transformations

In the coming decades, a major issue in designing sustainable agricultural systems will be the management of soil organic matter and the rational use of organic inputs such as animal manures, food industry wastes, sewage sludge and crop residues. Maintaining or increasing soil organic matter content is of great benefit in terms of recycling plant nutrients, minimizing the need for inorganic fertilizers and improving soil physical structure. However, inputs of organic matter often lead to large losses of nutrients, especially N, with the consequent problems of pollution. Jarvis and Pain (1990) have reviewed

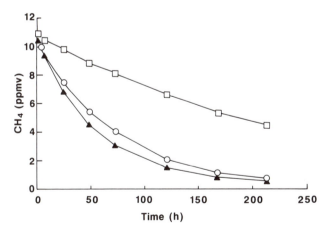

Fig. 6.4. Rates of methane oxidation by soil cores exposed to 10 ppm methane and incubated at 25°C. Soils taken from section 1 (continuous wheat) of the Broadbalk experiment where plots have been unmanured (○) or given 35 t ha⁻¹ FYM (▲) annually since 1843 or 144 kg N ha⁻¹ (□) with PK fertilizers since 1852 (from Hütsch *et al.*, 1993).

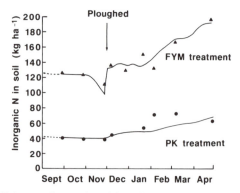

Fig. 6.5. Inorganic N (ammonium + nitrate) in soil to a depth of 110 cm in two plots of the Hoosfield continuous barley experiment as measured (▲, FYM treatment; ●, PK treatment) or simulated (lines) (from Powlson *et al.*, 1989). Annual treatments were PK, 33 kg P ha⁻¹ and 90 kg K ha⁻¹; FYM, 35 t ha⁻¹.

recent work on the losses of N that can occur soon after animal wastes are applied to soil; these include ammonia volatilization, denitrification and nitrate leaching. It is extremely important that strategies should be developed to minimize losses of nitrogen occurring soon after manure applications. However, the longer-term impact of different management practices must also be addressed and long-term experiments can be valuable in such studies.

Figure 6.5 shows the quantities of mineral N (mainly nitrate) present in the soil, to a depth of 110 cm, on two contrasting plots of the Hoosfield Barley experiment at Rothamsted during the winter of 1986–87. The plot which receives FYM every year contained at least 80 kg ha⁻¹ *more* N as nitrate than the plot receiving inorganic fertilizer throughout the period. This was not surprising as the past additions of FYM had increased the organic nitrogen content of the soil almost threefold so greater mineralization of N is inevitable. Part of the extra nitrate formed is taken up by crops and part is present in soils during the winter period when, in this experiment, no crop is present and leaching is at a maximum. Nitrate leaching could not be measured directly in this experiment, but Powlson *et al.* (1989) were able to simulate soil nitrate content using a mathematical model of mineralization and nitrate leaching. This predicted that over-winter nitrate leaching on the FYM plot was 124 kg N ha⁻¹ compared to 25 kg N ha⁻¹ from the inorganic fertilizer plot. The calculated concentrations of NO_3-N in leachate were in the range 40–50 mg l⁻¹ for the FYM plot compared to 4–20 mg l⁻¹ in the inorganic fertilizer plot. The FYM plot on the experiment may not be typical of most soils that receive organic manures because of the very large quantity applied over the years but the result illustrates the difficulty of utilizing organic inputs in an environmentally acceptable way in arable systems. Indeed, any farming system that tends to increase soil organic matter content will increase the problem of untimely nitrate production.

There are similar difficulties in minimizing nitrate leakage from crop rotations involving leys of grass or grass–legume mixtures. The benefits of N being released in the year following the breaking of the ley are well known and are illustrated by the yields from the ley–arable experiments at Rothamsted and Woburn (e.g. Johnston, 1986). However, large losses of nitrate have been observed because mineralization of the N at this stage far outstrips crop uptake in the autumn period (Johnston *et al.*, 1994). Increased nitrate production is not necessarily restricted to the first year after breaking the ley; Macdonald *et al.* (1989) found inorganic N contents of about 50 kg ha⁻¹ in soil to a depth of 50 cm one year after ploughing at Woburn compared to less than 20 kg ha⁻¹ in continuous arable treatments. Of course these losses of nitrate must be set against the small losses that occur during the ley phase provided it receives only small dressings of N fertilizer and is either cut or only grazed at a low intensity. There is a dearth of experimental data on this topic and experimental sites designed to be maintained for at least ten years are needed in order to provide adequate information. Learning to manage crop nutrients in ley–arable rotations is an important challenge for the future.

The incorporation of cereal straw has both short- and long-term impacts on nitrogen dynamics. In the short term, the decomposition of straw, having a wide C:N ratio, leads to some immobilization of N so that rather less nitrate is present in soil in winter and at risk to leaching. The maximum quantity of N likely to be immobilized can be estimated (Johnston and Jenkinson, 1989) although recent work suggests that this may be less than previously thought (Ocio *et al.*, 1991). In the longer term it would be expected that a proportion of the immobilized N would be remineralized leading to a greater quantity of nitrate in soil than where straw has not been incorporated. This was observed in a study of soils from two experimental sites in Denmark in which spring barley straw had either been burned or incorporated for 18 years (Powlson *et al.*, 1987). The increase in soil total N content as a result of continuous straw incorporation was barely measurable but the quantity of N held in the soil microbial biomass was increased by 40–50% compared to the burned treatment (Table 6.6). When the soils were incubated for 60 days at 25°C in the laboratory, 40–50% more N was mineralized in the soils from the straw-treated plots. A similar effect was observed in the Woburn Organic Manuring experiment. After 6 years of straw incorporation mineralizable N, as assessed in laboratory incubations of up to 8 weeks, was about 30% greater than in soil not receiving straw; in this experiment a small increase in soil total N content was also detectable (Mattingly *et al.*, 1974).

New experiments on straw incorporation were established at Rothamsted, Woburn and other sites in the UK in 1985 and incorporation also began on one section of Broadbalk in 1986. In the coming years these sites will

Table 6.6. Effects of 18 years of incorporating spring barley straw on some properties of the soil at two sites in Denmark (from Powlson *et al.*, 1987)[a].

Site	Straw treatment	Soil total N content (%)	Soil biomass N content (μg g^{-1})	N mineralized[b] (μg g^{-1})
Studsgaard	Burned	0.115	10.5	19.7
	Incorporated	0.127	15.8	29.6
	% increase	10	50[c]	50[c]
Rønhave	Burned	0.119	18.8	23.4
	Incorporated	0.129	17.3	32.3
	% increase	8	46[c]	43[c]

[a] Soils were sampled to 25 cm except for the burned treatment at Studsgaard which was sampled to 23.4 cm to obtain equal weights of soil in the two treatments because of a change in bulk density.
[b] Measured after incubation in the laboratory for 60-days at 25°C.
[c] Indicates that the increase is significant at the 5% level; increases in total N content not significant.

become a valuable resource for investigating the longer-term impact of straw incorporation on soil nitrogen dynamics and many other processes.

Conclusions

Long-term experiments have significant limitations. For example, many of the oldest agricultural experiments in the world, including those at Rothamsted, do not contain replicate plots. This was the case for about half the experiments older than 40 years reviewed by Glendining and Powlson (1994). Consequently, differences in soil properties or crop growth that now exist between plots cannot always be assigned unequivocally to the effect of the treatment. In theory at least, some current differences could reflect soil differences that pre-dated the start of the experiment. In many cases this is inconceivable but it does mean that the significance of small differences in observations between plots must be interpreted carefully. Very often the consistency of differences between treatments, or of a trend with time, provide reliable indications of a real effect and thus overcome this limitation. In the Broadbalk and Park Grass experiments some degree of replication can be obtained as there are some plots with treatment histories that differ only slightly from others and in ways that are considered unlikely to influence current soil properties. Of course such proxy replicates must be used with caution. Statistical approaches relevant to long-term unreplicated experiments are discussed by Barnett (Chapter 10).

If soil samples are taken from all the plots of a replicated long-term experiment at the start, before the treatments are imposed, differences observed at this time can enhance the value of later measurements. An example is a study of a site in Australia in which measurements of soil organic C, N and microbial biomass were made five years after different tillage treatments and sorghum residue managements were imposed (Saffigna *et al.*, 1989). By taking account of pre-existing spatial variability of soil organic C and N within the site it was possible to decrease the standard errors of later measurements. This made it easier to detect trends in organic matter and biomass content caused by the treatments.

Another limitation concerns the linked factors of plot size and soil movement. If plots are small, and not bounded by some form of barrier, the spread of soil during cultivation will eventually destroy the integrity of the plot (see Leigh *et al.*, Chapter 14). This issue became apparent during studies of heavy metals introduced to soil from sewage sludge at the Woburn Market Garden experiment. Initial calculations of metal balances suggested substantial losses of metals but McGrath and Lane (1989) showed that this was erroneous and resulted from soil movement. They used a diffusion-type model, developed from that previously used by Sibbesen and Andersen (1985), to account for the soil movement that occurred over a 40-year period. After making this correction it was clear that there had been virtually no loss of metals and their half-lives in soil were mostly greater than 10,000 years. Soil movement resulting from cultivation also has to be considered when ^{15}N plots are established and monitored over several years. Hart *et al.* (1993) measured ^{15}N enrichment in plants along transects spanning the 2 m × 2 m ^{15}N microplots within the Broadbalk plots five years after ^{15}N application and were able to make approximate corrections for the spread of ^{15}N.

Despite these limitations, long-term experiments are the only means of obtaining information on the physical, chemical and biological consequences of agricultural activities and the resulting loss, accumulation or transformation of nutrients. They also provide data on which to base rational judgements on the biophysical aspects of sustainability. A recent study has also attempted to use them to make deductions on the economic sustainability of different agricultural systems and to assign an economic value to environmental impacts (Barnett, Chapter 10).

Naturally, caution and intelligence must be exercised in extrapolating results from long-term experiments to larger areas. The maintenance of crop growth and soil fertility within well-managed plots on an experimental farm is not necessarily proof that the system is applicable more widely, but there are no other means of obtaining advance information on the sustainability or suitability of a system.

Long-term experiments can provide valuable data and experimental material for new research on subjects of vital current interest, often far removed from those envisaged by their founders. One example is the effect of nitrogen use on methane oxidation in soil (Fig. 6.4). Another is the unexpected effects of low concentrations of heavy metals on soil microbial processes discovered in the course of research on the Woburn Market Garden experiment (Brookes and McGrath, 1984). Yet another is the build-up of organic and inorganic pollutants in soils as a result of anthropogenic activity (Jones *et al.*, Chapter 9). Powlson and Johnston (1994), Johnston and Powlson (1994) and Leigh *et al.* (Chapter 14) give other examples. Research on nutrient cycles and related matters over the past few decades has benefited greatly from the foresight of Lawes, Gilbert and others throughout the world who established long-term experiments. The present generation has a responsibility both to continue the present experiments and establish new long-term sites so that these are available for future research. They are an invaluable resource for use in testing hypotheses and they frequently reveal significant issues that would otherwise have been missed. The use of data and observations from long-term experiments is, therefore, one of the vital components in the formulation of future research agendas.

References

Addiscott, T.M., Whitmore, A.P. and Powlson, D.S. (1991) *Farming, Fertilizers and the Nitrate Problem*. CAB International, Wallingford, UK.

Brookes, P.C. and McGrath, S.P. (1984) The effects of metal toxicity on the soil microbial biomass. *Journal of Soil Science* 35, 341–346.

Catt, J.A., Christian, D.G., Goss, M.J., Harris, G.L. and Howse, K.R. (1992) Strategies to reduce nitrate leaching by crop rotation, minimal cultivation and straw incorporation in the Brimstone Farm Experiment, Oxfordshire. *Aspects of Applied Biology* 30, 255–262.

Catt, J.A., Quinton, J.N., Rickson, R.J. and Styles, P. (1994) Nutrient loss and crop yields in the Woburn Erosion Reference Experiment. Proceedings of the First European Society for Soil Conservation Conference, Silsoe, April 1992, *Catena* (in press).

Chaney, K. (1990) Effect of nitrogen fertilizer rates on soil nitrate nitrogen content after harvesting winter wheat. *Journal of Agricultural Science, Cambridge* 114, 171–176.

Christian, D.G., Goodlass, G. and Powlson, D.S. (1992) Nitrogen uptake by cover crops. *Aspects of Applied Biology* 30, 291–300.

Glendining, M.J. and Powlson, D.S. (1994) The effects of long-continued applications of inorganic nitrogen fertilizer on soil organic nitrogen – a review. *Advances in Soil Science* (in press).

Glendining, M.J., Poulton, P.R. and Powlson, D.S. (1992) The relationship between inorganic N in soil and the rate of fertilizer N applied to the Broadbalk Wheat Experiment. *Aspects of Applied Biology* 30, 95–102.

Goss, M.J., Howse, K.R., Lane, P.W., Christian, D.G. and Harris, G.L. (1993) Losses of nitrate-nitrogen in water draining from under autumn-sown crops established by direct drilling or mouldboard ploughing. *Journal of Soil Science* 44, 35–48.

Goulding, K.W.T. (1990) Nitrogen deposition to land from the atmosphere. *Soil Use and Management* 6, 61-63.

Goulding, K.W.T. and Webster, C.P. (1989) Denitrification losses of nitrogen from arable soils as affected by old and new organic matter from leys and farmyard manure. In: Hansen, J.A. and Henriksen, K. (eds) *Nitrogen in Organic Wastes Applied to Soils*. Academic Press, London, pp. 223-234.

Goulding, K.W.T., Webster, C.P., Powlson, D.S. and Poulton, P.R. (1993) Denitrification loss of nitrogen fertilizer applied to winter wheat following ley and arable rotations as estimated by acetylene inhibition and ^{15}N balance. *Journal of Soil Science* 44, 63–72.

Hart, P.B.S., Powlson, D.S., Poulton, P.R., Johnston, A.E. and Jenkinson, D.S. (1993) The availability of the nitrogen in the crop residues of winter wheat to subsequent crops. *Journal of Agricultural Science, Cambridge* 121, 355-362.

Hütsch, B.W., Webster, C.P. and Powlson, D.S. (1993) Long-term effects of nitrogen fertilization on methane oxidation in soil of the Broadbalk Wheat Experiment. *Soil Biology and Biochemistry* 25, 1307-1315.

Jarvis, S.C. and Pain, B.F. (1990) Ammonia volatilization from agricultural land. *Proceedings of the Fertilizer Society* 298, 35pp.

Jenkinson, D.S. (1977) The nitrogen economy of the Broadbalk experiments. I. Nitrogen balance in the experiments. *Rothamsted Experimental Station Report for 1976, Part 2*, pp. 103-109.

Jenkinson, D.S. (1990) The turnover of organic carbon and nitrogen in soil. *Philosophical Transactions of the Royal Society, London B* 329, 361-368.

Jenkinson, D.S. and Parry, L.C. (1989) The nitrogen cycle in the Broadbalk Wheat Experiment: a model for the turnover of nitrogen through the soil microbial biomass. *Soil Biology and Biochemistry* 21, 535-541.

Johnston, A.E. (1976) Additions and removals of nitrogen and phosphorus in long-term experiments at Rothamsted and Woburn and the effect of residues on total soil nitrogen and phosphorus. In: *Agriculture and Water Quality*, Ministry of Agriculture, Fisheries and Food Technical Bulletin No. 32, pp. 111-144.

Johnston, A.E. (1986) Soil organic matter, effects on soils and crops. *Soil Use and Management* 2, 97-105.

Johnston, A.E. (1991) Soil fertility and soil organic matter. In: Wilson, W.S. (ed.) *Advances in Soil Organic Matter Research: the Impact on Agriculture and the Environment*, Royal Society of Chemistry, Cambridge, UK, pp. 299-314.

Johnston, A.E. and Jenkinson, D.S. (1989) The nitrogen cycle in UK arable agriculture. *Proceedings of the Fertilizer Society* No. 286, pp.1-24.

Johnston, A.E. and Powlson, D.S. (1994) The setting up, conduct and applicability of long-term, continuing field experiments in agricultural research. In: Greenland, D.J. and Szabolcs, I. (eds) *Soil Resilience and Sustainable Lane Use*, CAB International, Wallingford, UK, pp. 395-421.

Johnston, A.E. and Poulton, P.R. (1977) Yields on the Exhaustion Land and changes in the NPK content of the soils due to cropping and manuring, 1852-1975. *Rothamsted Annual Report for 1976, Part 2*, pp. 53-85.

Johnston, A.E. and Poulton, P.R. (1993) The role of phosphorus in crop production and soil fertility: 150 years of field experiments at Rothamsted, United Kingdom. *Proceedings of an International Workshop, Phosphate Fertilizers and the Environment*, Tampa, Florida, USA, pp. 45-63.

Johnston, A.E., McEwen, J., Lane, P.W., Hewitt, M.V., Poulton, P.R. and Yeoman, D.P. (1994) Effects of one to six year old ryegrass-clover leys on soil nitrogen and on subsequent yields and fertilizer nitrogen requirements of the arable sequence winter wheat, potatoes, winter wheat, winter beans (*Vicia faba*) grown on a sandy loam soil. *Journal of Agricultural Science, Cambridge* 122, 73-89.

Ladd, J.N. and Amato, M. (1986) The fate of nitrogen from legume and fertilizer sources in soils successively cropped with winter wheat under field conditions. *Soil Biology and Biochemistry* 18, 417-425.

Macdonald, A.J., Powlson, D.S., Poulton, P.R. and Jenkinson, D.S. (1989) Unused fertilizer nitrogen in arable soils - its contribution to nitrate leaching. *Journal of the Science of Food and Agriculture* 46, 407-419.

Mattingly, G.E.G., Chater, M. and Poulton, P.R. (1974) The Woburn Organic Manuring Experiment. II. Soil analyses 1964-72 with special reference to changes in carbon and nitrogen. *Rothamsted Experimental Station Report for 1973, Part 2*, pp. 134-151.

McGrath, S.P. and Lane, P.W. (1989) An explanation of the apparent losses of metals in a long-term field experiment with sewage sludge. *Environmental Pollution* 60, 235-256.

Ocio, J.A., Brookes, P.C. and Jenkinson, D.S. (1991) Field incorporation of straw and its effects on soil microbial biomass and soil inorganic N. *Soil Biology and Biochemistry* 23, 171-176.

Powlson, D.S. and Johnston, A.E. (1994) Long-term field experiments: their importance in understanding sustainable land use. In: Greenland, D.J. and Szabolcs, I. (eds) *Soil Resilience and Sustainable Land Use*, CAB International, Wallingford, UK, pp. 367-394.

Powlson, D.S., Pruden, G., Johnston, A.E. and Jenkinson, D.S. (1986) The nitrogen cycle in the Broadbalk Wheat Experiment: recovery and losses of ^{15}N-labelled fertilizers applied in spring and inputs of nitrogen from the atmosphere. *Journal of Agricultural Science, Cambridge* 107, 591-609.

Powlson, D.S., Brookes, P.C. and Christensen, B.T. (1987) Measurement of soil microbial biomass provides an early indication of changes in total soil organic matter due to straw incorporation. *Soil Biology and Biochemistry* 19, 159-164.

Powlson, D.S., Poulton, P.R., Addiscott, T.M. and McCann, D.S. (1989) Leaching of nitrate from soils receiving organic or inorganic fertilizers continuously for 135 years. In: Hansen, J.A. and Henriksen, K. (eds) *Nitrogen in Organic Wastes Applied to Soils*, Academic Press, London, pp. 334-345.

Rekolainen, S. (1989) Effect of snow and soil frost melting on the concentration of suspended solids and phosphorus in two rural watersheds in Western Finland. *Aquatic Sciences* 51, 211-223.

Saffigna, P.G., Powlson, D.S., Brookes, P.C. and Thomas, G.A. (1989) Influence of sorghum residues and tillage on soil organic matter and soil microbial biomass in an Australian vertisol. *Soil Biology and Biochemistry* 21, 759-765.

Schröder, J., de Groot, W.J.M. and van Dijk, W. (1992) Nitrogen losses from continuous maize as affected by cover crops. *Aspects of Applied Biology* 30, 317-326.

Shen, S.M., Hart, P.B.S., Powlson, D.S. and Jenkinson, D.S. (1989) The nitrogen cycle of the Broadbalk Wheat Experiment: ^{15}N-labelled fertilizer residues in the soil and in the soil microbial biomass. *Soil Biology and Biochemistry* 21, 529–533.

Sibbesen, E. and Andersen, C.E. (1985) Soil movement in long-term field experiments as a result of cultivations. II. How to estimate the two-dimensional movement of substances accumulating in the soil. *Experimental Agriculture* 21, 109–117.

Sylvester-Bradley, R. and Chambers, B.J. (1992) The implications of restricting use of fertilizer nitrogen for the productivity of arable crops, their profitability and potential pollution by nitrate. *Nitrate and Farming Systems: Aspects of Applied Biology* 30, 85–94.

Whitmore, A.P. and Goulding, K.W.T. (1992) The Nitrogen Cycle. *AFRC Institute of Arable Crops Research, Annual Report for 1991*, p. 36.

7

How the Rothamsted Classical Experiments Have Been Used to Develop and Test Models for the Turnover of Carbon and Nitrogen in Soil

D.S. Jenkinson[1], N.J. Bradbury and K. Coleman
Institute of Arable Crops Research, Rothamsted Experimental Station, Harpenden, Herts AL5 2JQ; and at [1]Department of Soil Science, University of Reading, London Road, Reading, Berks RG1 5AQ, UK.

Introduction

Models of the turnover of organic carbon and nitrogen in soil have three main functions: (i) to bring disparate data into a comprehensible (never comprehensive) whole; (ii) to formulate hypotheses in a way that can be tested and (iii), within strictly specified limits, to predict future events from past experience. They have been used for tasks such as predicting the effect of management on the amount of organic matter in soil, or for calculating the likely effects of climate change on the global stock of soil organic carbon – a stock (Ajtay *et al.*, 1979; Post *et al.* 1982; Buringh, 1984; Schlesinger, 1984; Eswaran *et al.*, 1993) twice as big as that present in the atmosphere as CO_2-C. Similarly, nitrogen models are now being developed to predict how much nitrogen will be supplied by the soil to a crop growing in a particular field, thus allowing the additional quantity of nitrogen needed as fertilizer to attain a (specified) target yield to be calculated.

Soil organic carbon and nitrogen – the two are usually closely associated – respond slowly to external change. Thus when land at Rothamsted that had been arable for many centuries was put down to grass, it took 25 years to move halfway from the organic N level of arable soil to that of old grassland (Jenkinson, 1988). The long-term experiments at Rothamsted and Woburn provide unique material for examining such slow changes and for developing and testing models to describe these changes.

Models for the Turnover of Organic Carbon in Surface Soils

Turnover can be modelled over time spans that range from hours to millennia. Here attention will be focused on the years-to-centuries time period, as this is the scale most relevant to the transformations undergone by the bulk of soil organic carbon. The simplest model for the decomposition of organic carbon in soil was proposed by Henin and Dupuis (1945): it is an elaboration of the single compartment model originally introduced for soil organic nitrogen by Jenny (1941) and further developed by Bartholomew and Kirkham (1960) and by Nye and Greenland (1960). In it, the annual input of plant carbon (A_c) decomposes very rapidly, forming a quantity of humus carbon (fA_c) that is biologically identical to that already in the soil. The factor f is known as the isohumic coefficient; it is commonly about 0.3 for residues from agricultural crops but much larger for materials such as peat (Janssen, 1984). For this model:

$$C = fA_c/k + (C_o - fA_c/k)e^{-kt} \tag{7.1}$$

where C is the organic carbon content of the soil, t is time in years and k is the fraction of this carbon decomposing each year. The quantity of fresh plant carbon present in the soil at any particular time is assumed to be negligible compared with the amount of 'humus' carbon. Equation (7.1) closely fits changes in soil organic carbon over the 10–100-year period (Jenkinson and Johnston, 1977), giving values of A_c that are far more realistic than if the whole of the plant input is assumed to decompose in the same way as the carbon already in the soil.

Many multicompartmental models for the decomposition of organic matter in or on the soil have been described over the past 20 years (Russell, 1975; Jenkinson and Rayner, 1977; Van Veen and Paul, 1981; Voroney *et al.*, 1981; Parton *et al.*, 1983, 1987, 1988; O'Brien, 1984; Van der Linden *et al.*, 1987; Jenkinson *et al.*, 1987; Woodmansee, 1988; Bouwman, 1989; Hsieh, 1989; Melillo *et al.*, 1989; Stewart and Cole, 1989; Wolf *et al.*, 1989; Kovda *et al.*, 1990; Esser, 1990; Aber *et al.*, 1990; Verberne *et al.*, 1990; Hunt *et al.*, 1991; Paustian *et al.*, 1992). In most of them the material in a compartment is assumed to decay by first-order kinetics, as in equation (7.1), so that the rate of decomposition in a particular compartment is deemed to be a feature of the organic matter it contains, and is never retarded by a lack of competent organisms. In the majority of these long-term models the rate constants for the various compartments are multiplied by one or more 'rate modifiers' that alter the rate constants to speed decomposition as the temperature increases or to decrease it as the soil dries out, etc. The rate modifiers for temperature, water content, etc. are usually multiplied together (Hunt, 1977; Van Veen and Paul, 1981; Jenkinson *et al.*, 1987; Van der Linden *et al.*, 1987; Parton *et al.*, 1988, 1994a), although this is by no means the only way of adjusting such models for environmental constraints (Frissel and Van Veen, 1978; Woodmansee, 1978). Some of the models specify that decomposition is faster when soils are tilled (Parton *et al.*, 1988), but others that it is slowed when the soil is covered with vegetation (Jenkinson *et al.*, 1987). Other things being equal, heavy clay soils contain more organic matter than light-textured sandy soils and any mechanistic model must take this into account. This can be done by using a rate modifier to decrease rate constants as clay content increases (Parton *et al.*, 1987, 1994a) or by assuming that any decomposition product (including the microbial biomass) that is not stabilized by clay is decomposed very quickly (Van Veen and Paul, 1981; Verberne *et al.*, 1990) or by decreasing the proportion of microbial substrate converted into CO_2, relative to that remaining in the soil, as clay content increases (Jenkinson *et al.*, 1987).

The composition of the incoming plant debris has a marked effect on its rate of decomposition: this has been modelled by partitioning the incoming litter into resistant and readily decomposable material from its lignin and nitrogen contents (Verberne *et al.*, 1990; Parton *et al.*, 1994a) or by postulating that the debris from different vegetation types has characteristic (and different) proportions of resistant and decomposable organic matter (Jenkinson, 1990).

A good carbon turnover model should be able to account for changes in the radiocarbon content of soil organic matter (whether this radiocarbon is naturally produced in the upper atmosphere or comes from the thermonuclear tests of the early 1960s), as well as for changes in total organic carbon. Indeed, radiocarbon measurements provide a critical test of the workings of any carbon turnover model (Harkness *et al.*, 1991; Jenkinson *et al.*, 1992; Hsieh, 1992, 1993; Trumbore, 1993).

With the development of more sophisticated models, there has been a move to model soil organic carbon turnover on the regional (Parton *et al.*, 1987, 1989a, b; Burke *et al.*, 1989, 1990; Cole *et al.*, 1990; Schimel *et al.*, 1990; Ågren *et al.*, 1991; Carter *et al.*, 1993; Woomer, 1993) and global scales (Esser, 1990; Jenkinson *et al.*, 1991; Kirschbaum, 1993; Parton *et al.*, 1994b) a move greatly facilitated by the increasing power and accessibility of modern computers.

Multicompartmental 'box' models are not the only way of handling the turnover of organic matter in soil: Bosatta and Ågren (1985) introduced the idea that decomposition is a continuum, organic matter moving down a 'quality' scale as it decays, fresh decomposable organic matter having a high quality, with the most resistant material present in the system being of zero quality. In this approach the problem of selecting compartments and fitting rate constants to them is replaced by the problem of establishing the decay pattern for each incoming cohort of plant material as it enters the soil and decomposes (Ågren and Bosatta, 1987; Bosatta and Ågren, 1991). The mathematics of this approach is more complex than that of compartmental models and it has not, as yet, been applied to long runs of data from agricultural soils.

The Rothamsted Model for the Turnover of Organic Matter in Soil

The current Rothamsted model (ROTHC-26.2: Jenkinson *et al.*, 1992) is a simplified descendant of earlier versions (Jenkinson and Rayner, 1977; Hart, 1984; Jenkinson *et al.*, 1987; see Jenkinson (1990) for a detailed description). This five-compartment model (Fig. 7.1) has two input compartments: decomposable plant material (DPM) and resistant plant material (RPM), both of which decompose, by first-order processes which have characteristic (and different) rates, to CO_2 (lost from the system), microbial biomass (BIO) and humified organic matter (HUM). Both microbial biomass and humified organic matter decompose at their characteristic rates by first-order processes to give more CO_2, biomass and humified matter. The soil is also assumed to contain a small organic compartment (IOM) that is inert to biological attack. The model allows for the effects of temperature, soil moisture content, plant cover (decomposition being faster in bare soil than under vegetation, Jenkinson, 1977a) and soil clay content on the decomposition processes. It works in monthly intervals, using mean monthly air temperatures. Soil moisture is updated monthly from the rainfall and evaporation of the preceding month. The amount of moisture held in a soil layer between field capacity and wilting point is calculated from the clay content.

Here we are concerned with how this model is fitted, tuned and tested, mainly against data from long-term experiments at Rothamsted. The fitting,

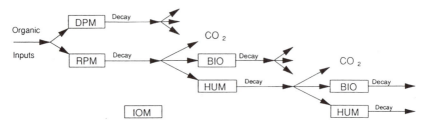

Fig. 7.1. Structure of the current version of the Rothamsted model for the turnover of organic carbon in soil – ROTHC-26.2. Incoming organic carbon is partitioned between decomposable plant material (DPM) and resistant plant material (RPM). During decomposition, microbial biomass (BIO), humus (HUM) and CO_2 are formed, in proportions α, β and $(1-\alpha-\beta)$, respectively. IOM is inert organic carbon. (From Jenkinson, 1990.)

tuning and testing of complex multicompartmental models is often an inces-
tuous affair, the model being tuned using the same (or similar) data that are
used for its validation. What we have attempted to do is to tune the model to
Rothamsted data on the *total* organic carbon content of topsoils and then use
it, *without further tuning*, to predict both soil microbial biomass and the
effect of bomb radiocarbon on soil radiocarbon age – two independent
measurements.

The model was first tuned to measurements of the decomposition of ryegrass in bare and cropped soil over a 10-year period at Rothamsted and to
measurements on the decomposition of the same plant material in bare soil
over a 2.5-year period in Ibadan, Nigeria (Fig. 7.2). The second stage in the
tuning was to adjust the parameters in the model so that the output fitted the
changes in soil organic C in three of the plots on the Broadbalk continuous
wheat experiment, started in 1843. The size of the inert organic carbon compartment (IOM) was set to give the measured radiocarbon age of the (unfertilized) soil in 1881 (1400 years). Figure 7.3 shows the data points and the model
outputs, as adjusted to the data. Figure 7.4, for the continuous spring barley
experiment on a field (Hoosfield) adjacent to Broadbalk, shows the same comparison, but for a different crop; data from this particular experiment were not
used in tuning.

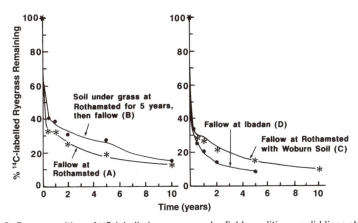

Fig. 7.2. Decomposition of ^{14}C-labelled ryegrass under field conditions: solid lines show
the model output. Curve A is fitted to decay in bare Rothamsted soil (23.4% clay),
incubated at Rothamsted (mean annual temperature 9.3°C). Curve B is fitted to decay in
grassed Rothamsted soil, incubated at Rothamsted, the soil being under grass for the first
5 years and bare for the next 5 years. The model retainment factor (i.e. the factor used to
modify the rate constants for the effects of plant cover) is set at 0.6 for the first 5 years and
1.0 for the next 5 years. Curve C shows decomposition of ^{14}C labelled ryegrass in Woburn
soil (10.7% clay), but incubated in the field at Rothamsted. Curve D shows decomposition
under field conditions at Ibadan, Nigeria (Egbeda soil, 10% clay), where the mean annual
temperature was 26°C (Ayanaba and Jenkinson, 1990). (From Jenkinson *et al.*, 1987.)

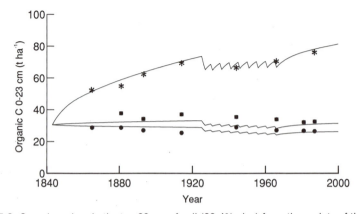

Fig. 7.3. Organic carbon in the top 23 cm of soil (23.4% clay) from three plots of the Broadbalk Continuous Wheat Experiment; the treatments are, unmanured (●), NPK fertilizers (■), and FYM (*) The solid lines show the model output: the data points are calculated from % N in soil (as given by Jenkinson, 1977b, apart from the 1981 results, which are from Powlson *et al.*, 1986) using a C:N ratio of 9.71 for the unmanured plot, 9.91 for the plot receiving inorganic fertilizers and 10.28 for the plot receiving farmyard manure (FYM). P.R. Poulton supplied the unpublished data for 1987. The weight of soil in the 0–23 cm layer was taken as 2.91 M kg ha^{-1} throughout the experiment in the unmanured and inorganically fertilized plots. Bulk density decreased on the FYM plot during the experiment: the organic C contents given in the Figures for this plot have been adjusted to allow for soil expansion, using the equivalent depth concept, as described by Jenkinson (1977b). The FYM plot receives 35 t FYM ha^{-1} annually, containing about 3 t C, and the NPK Mg plot 144 kg N, 35 kg P, 90 kg K and 12 kg Mg ha^{-1}, all annually, apart from fallow years. The FYM, applied in early autumn, was assumed to contain DPM, RPM and HUM (but no biomass) in the proportion 0.49, 0.49 and 0.02, respectively. In fallow years (two initially, followed by a rotation of one-year's fallow in five between 1926 and 1966), decomposition was assumed to proceed as usual with no fresh FYM or plant debris entering the soil. The IOM compartment contained 3.0 t C ha^{-1} (see Jenkinson *et al.*, 1992). The carbon inputs used, all in t C ha^{-1} year^{-1} were: unmanured plot 1.3 (fitted); NPKMg plot 1.7 (fitted); FYM plot 1.7 (plant debris, fitted) + 3.0 FYM (set from its C content). The incoming plant residues were assumed to contain DPM and RPM in the proportion 0.59 and 0.41 respectively.

Data from three medium-term field experiments, two at Rothamsted and one at Woburn Experimental Farm, on a lighter soil than at Rothamsted, are shown in Fig. 7.5, which also gives the model outputs for specified annual inputs of organic carbon. It should be noted that *all* the model parameters in Figs 7.4 and 7.5 are exactly as used to generate the model outputs in Figs 7.2 and 7.3; the only driving variable is the annual input of organic carbon.

Two small areas of old arable land on Rothamsted Farm were fenced off and allowed to revert to deciduous woodland a little over 100 years ago – Broadbalk and Geescroft 'Wildernesses'. Figure 7.6 shows how soil organic

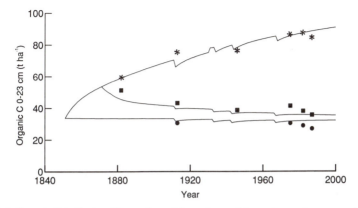

Fig. 7.4. Organic C in the top 23 cm of a soil from three plots under continuous spring barley (Hoosfield); the treatments are, unmanured (●), FYM (∗) and FYM 1852–71 none since (■). The solid lines show the model output: the data points, adjusted for changes in bulk density, are from Jenkinson and Johnston (1977), except for 1982 and 1987 (unpublished results, supplied by P.R. Poulton). The FYM plot receives 35 t FYM annually; the FYM residues plot received 35 t FYM annually between 1852 and 1871 and nothing since. The FYM, applied in February between 1852 and 1930 and in November after 1932, was assumed to contain DPM, RPM, and HUM (but no biomass) in the proportions 0.49, 0.49 and 0.02, respectively. The IOM compartment contained 3.8 t C ha⁻¹. The C inputs used, all in t C ha⁻¹ year⁻¹, were: unmanured plot, 1.76 (fitted); FYM plot, 2.4 (plant debris, fitted) + 3.0 FYM (set from its C content); FYM residues plot, as FYM plot during 1852–71, 2.4 during 1872–76, thereafter, 1.76. Note that the fitted inputs of plant debris are greater than on the corresponding plots in Broadbalk: this is presumably because weeds have much more opportunity to develop during the period between harvest and ploughing in Hoosfield (ploughed in early spring before 1930 and in winter after 1932) than on Broadbalk (nearly always ploughed in early autumn). The incoming plant material was deemed to have the same DPM/RPM proportions as on Broadbalk.

carbon has accumulated in Broadbalk Wilderness, together with the model outputs for various specified inputs. All the model parameters *except* the ratio of decomposable plant material (DPM) to resistant plant material (RPM) were as used for Fig. 7.3; the proportion of resistant material was (abruptly) increased 25 years after reversion started, to allow for the change from herbaceous vegetation to the less decomposable organic returns from woody vegetation.

Figure 7.6 also shows the measured and modelled values for the radiocarbon content of the soils from Broadbalk Wilderness. The increased radiocarbon content of soil organic matter caused by bomb-derived radiocarbon from the thermonuclear tests of the early 1960s is closely mirrored by the model as set up and parameterized for total organic carbon. This provides an independent test of its workings.

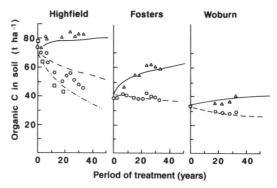

Fig. 7.5. Organic C in the top 23 cm of soil from two ley–arable experiments at Rothamsted (Highfield and Fosters) and one at Woburn. Data, corrected for changes in bulk density, are from Johnston (1973); later points unpublished. The fitted C inputs, all in t C ha⁻¹ year⁻¹, were: Highfield grassland, first 12 years (grazed) 5, thereafter 4; Highfield arable crop rotation 1.4; Highfield bare fallow 0; Fosters grassland, first 12 years (grazed) 5, thereafter 4; Fosters arable crop rotation 1.4. The Woburn experiment (on a soil containing 10.7% clay) does not have a continuous grassland treatment: the ley rotation has 3 years grass (input 4), followed by 2 years arable (input 1.3), giving a mean annual input of 2.9; the arable crop rotation has an input of 1.1. (From Jenkinson *et al.*, 1987.)

Another independent test of the model is provided by the correctness of its prediction for soil microbial biomass (Table 7.1). Agreement between bio-

Table 7.1. Biomass C content of soils, as measured and as modelled.

Site	Sampling date	Measured biomass (t C ha⁻¹) By fumigation/ incubation	By fumigation/ extraction	Modelled biomass[a]
Broadbalk Wilderness	June 1985[b]	1.57	1.33	1.63 (3.5)
Broadbalk wheat, unmanured	March 1989[c]	0.47	0.47	0.52 (1.3)
Broadbalk wheat, NPK	March 1989[c]	0.76	0.62	0.67 (1.7)
Broadbalk wheat, FYM	July 1973	1.29	—	1.66 (4.7)
Highfield grassland	March 1982[d]	2.27	—	2.15 (4.0)
Highfield arable rotation	March 1982[d]	0.82	—	1.16 (1.4)
Woburn arable rotation	March 1982[d]	0.32	—	0.49 (1.1)

[a] Figures in parentheses are C inputs to the model, in t C ha⁻¹ year⁻¹.
[b] From Vance *et al.* (1987).
[c] Measured by Jinshui Wu (personal communication).
[d] From Brookes *et al.* (1984).

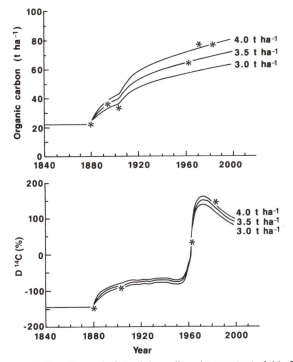

Fig. 7.6. The accumulation of organic C (and the radiocarbon content of this C, expressed as D ^{14}C‰, where D ^{14}C‰ is the normalized radiocarbon content relative to 0.95 of the NBS standard oxalic acid) in topsoil (to a depth that contains the same weight of ignited 'fine soil' as when sampled to a depth of 22.9 cm in 1985) from the wooded section of Broadbalk Wilderness. The lines show the model predictions for annual inputs of 3.0, 3.5 and 4.0 t ha^{-1} year^{-1}. (From Jenkinson *et al.*, 1992.)

mass as measured and as modelled is tolerable, although by no means perfect, particularly with the soil receiving farmyard manure on Broadbalk.

The Effects of Climate Change on the Global Stock of Soil Organic Carbon

Climate change can influence the stock of soil organic matter in two ways: by altering plant production, thus altering the annual return of plant debris to the soil and by changing the rate at which this input decays in or on the soil. There can be little doubt that warming will increase decomposition rates and thus, *if inputs are unchanged*, cause a decline in the world stock of organic matter, the extra carbon dioxide thus released contributing to the atmospheric burden and hence to carbon dioxide-caused warming (Schleser, 1981; Lashof, 1989; Buol *et al.*, 1990; Franz, 1990; Kohlmaier *et al.*, 1990; Schimel *et al.*,

1990; Tinker and Ineson, 1990; Jenkinson *et al.*, 1991; Gifford, 1992; Kirschbaum, 1993). A similar positive feedback will be caused by an increase in rainfall (wetlands excepted) in situations where decomposition is restricted by
drought. Jenkinson *et al.* (1991), using the Rothamsted model, calculated that
an annual increase of 0.03°C (the increase considered most likely by the Intergovernmental Panel on Climatic Change; Houghton *et al.*, 1990) would cause a
net decrease of 61×10^{15} g C in the global stock of soil organic carbon by the
year 2050, assuming that inputs remain unchanged. Although this decrease is
only 4% of the global stock of some 1500×10^{15} g C, it is 19% of the carbon
dioxide carbon that will be released by 2050 if present usage of fossil fuel (5.4
$\times 10^{15}$ g C year^{-1}) continues unabated.

In reality, inputs will also change as the climate changes. Anthropogenic
emissions of carbon dioxide and of chemically combined nitrogen (Thornley
et al., 1991; Kirschbaum, 1993) may well act as fertilizers, increasing plant
growth directly, quite apart from the indirect effects brought about by geo-
graphic shifts in vegetation zones, by changes in sea level, etc. (Prentice and
Fung, 1990). Models are now being constructed to examine the effects of
increasing concentration of greenhouse gases on both the production and
decomposition of organic matter (Gifford, 1992; Kirschbaum, 1993).

Soil Organic Matter and Net Primary Production

Net primary production is hard to measure, yet is a necessary input to models
for carbon cycling, whether in the field, biome or global scale. Two methods
are commonly used: gravimetric assessment of annual plant production, or
micrometeorological techniques, based on measurement of CO_2 flux across
the plant/atmosphere interface. Both methods have disadvantages: the
measurement of annual belowground production is not easy with perennial
plants (trees in particular) and the micrometeorological methods are subject
to the problems that arise when measurements made over a few days or weeks
are extrapolated to an annual basis. A new method for estimating NPP has been
proposed (Jenkinson *et al.*, 1992), based on measurements made on soil
organic matter.

The amount of organic carbon in a soil depends on two factors: (i) the
quantity of organic debris entering the soil each year and (ii) the rate at which
this organic debris decomposes in or on the soil. When steady-state conditions
have been attained, the annual input of organic matter is equal to NPP, less any
organic matter burnt or otherwise removed.

If you have a model for the turnover of organic matter in soil that can
predict the steady-state content of soil organic carbon for a given annual input,
then it should also be possible to use the model in *inverse* mode, i.e. to calculate the annual input of organic matter needed to maintain a known soil
organic carbon content. Table 7.2 shows such NPP calculations for three

Table 7.2. Net primary production (t C ha^{-1} year^{-1}) in five sites at Rothamsted. (From Jenkinson *et al.*, 1992.)

Site	Land use	Return of organic C to soil[a]	Removal in harvest[b]	Accumulating in standing crop[b,c]	NPP
Broadbalk Wilderness	Regenerating woodland	3.5	0	1.27	4.8
Geescroft Wilderness	Regenerating woodland	2.5	0	0.84	3.3
Park Grass, unmanured	Permanent grassland	3.0	0.97[d]	0	4.0
Broadbalk wheat, unmanured	Continuous arable	1.3	0.93[e]	0	2.2
Broadbalk wheat, NPK	Continuous arable	1.7	3.51[e]	0	5.2

[a] As modelled.
[b] Assuming that plant dry matter contains 40% C.
[c] Mean annual accumulation in trees, 1883–1969 (Jenkinson, 1971).
[d] Mean annual removal in two cuts of hay, 1970–85 (from *Numerical Results of the Field Experiments*, Rothamsted).
[e] Mean annual removal in grain plus straw, continuous wheat, 1970–85 (from *Numerical Results of Field Experiments*, Rothamsted).

steady-state and two non-steady-state sites. This procedure gives a value for NPP that is integrated over a run of years, the exact weight given to say the last 10 or 100 years depending on the particular climate.

A problem with the new method is that many soils contain biologically inert organic matter – charcoal, clay-bound kerogen, etc. Such material is not in the turnover process and must be excluded when calculating annual inputs from soil organic carbon content. This was done from measurements of soil radiocarbon by Jenkinson *et al.* (1992), by Jenkinson and Coleman (1994) and by Tate *et al.* (1993), but other approaches, for example the chemical determination of inert carbon, are also possible. The new method of measuring NPP needs to be tested over a much wider range of land uses, soils and climates than yet examined. It should, however, provide a relatively simple way of investigating spatial variability of NPP on the local, regional and global scales and thus of directly relating NPP to soil and climatic conditions. It was used by Jenkinson *et al.* (1991) to estimate global NPP at 75.8×10^{15} g C year^{-1}, a little greater than the published estimate of 60×10^{15} g C year^{-1} (Houghton *et al.*, 1990).

The Rothamsted Model for the Turnover of Nitrogen in Soil (SUNDIAL)

This model (Bradbury *et al.*, 1993) was developed to describe the flow of nitrogen between crop and soil on the field scale. It is designed to be part of a system for predicting how much nitrogen will be supplied by the soil to a particular crop growing in a particular field in a particular year, from soil type, past weather and a history of the field. From this prediction, the quantity of fertilizer nitrogen required to attain a (specified) target yield can then be calculated.

The model has a compartmental structure (Fig. 7.7) and runs on a weekly time-step, in contrast to the monthly time-step of the carbon model. Nitrogen enters via atmospheric deposition and by application of fertilizer or organic manures, and is lost through denitrification, leaching, volatilization and removal in the crop at harvest. Organic nitrogen is contained within three of the model compartments – crop residues (including plant material dying off through the growing season), soil microbial biomass and humus. Inorganic nitrogen is held in two pools as NH_4^+ or NO_3^-. Nitrogen flows in and out of these inorganic pools as a result of mineralization, immobilization, nitrification, leaching, denitrification and plant uptake. The model requires a description of the soil and the meteorological records for the site – mean weekly air temperature, weekly rainfall and weekly evapotranspiration. The model also allows the addition of ^{15}N as labelled fertilizer, and follows its progress through crop and soil.

Our philosophy in constructing the model was to make sure it dealt with *all* the major processes affecting the behaviour of N in the cereal/soil system,

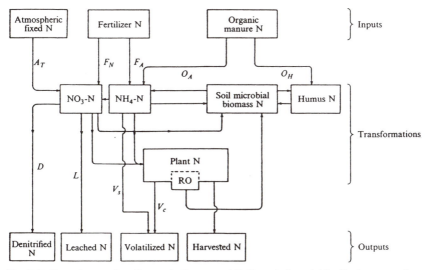

Fig. 7.7. Flow diagram for nitrogen in the current Rothamsted model for the turnover of nitrogen in soil: SUNDIAL. The portion of the plant nitrogen returned each year to the soil is termed RO: it may or may not include straw nitrogen. (From Bradbury *et al.*, 1993.)

even though each individual process is expressed in greatly simplified form. As far as possible, the model is modular in structure; if a particular module, for example that representing leaching, proves unsatisfactory at a later stage, it can be replaced by a more sophisticated module without rewriting the whole model. A central feature is that the model is designed to be used in a 'carry-forward' mode – one year's run providing the input for the next, and so on. It is constructed and tuned so that it does not, for example, allow soil organic N to build up to unrealistic levels, however long it is run. Needless to say, it has many ideas in common with other contemporary N models, particularly with SOILN (Bergstrom *et al.*, 1991), ANIMO (Rijtema and Kroes, 1991), DAISY (Hansen *et al.*, 1991) and NCSOIL (Molina *et al.*, 1983). Likewise, it has features in common with Thornley and Verberne's (1989) model of carbon and nitrogen flow through the plant/soil system, with studies of long-term changes in soil nitrogen dynamics (Wolf *et al.*, 1989), and with models designed to provide fertilizer recommendations (Neeteson *et al.*, 1987; Richter *et al.*, 1988).

The model parameters were set from experiments with ^{15}N-labelled fertilizers, superimposed on certain of the old Broadbalk plots (Hart *et al.*, 1993). The advantage of using these plots is that they are under steady-state conditions. Thus, over a run of years, inputs of combined N from fertilizers and the atmosphere must exactly balance N removed in grain and straw, leached as NO_3^-, volatilized as NH_3, or lost by denitrification as N_2 and N_2O. Steady-state conditions are particularly valuable in setting the parameters of models in that they constrain the range over which those parameters can vary. At the heart of the model is a carbon model similar to (but not identical to) that of ROTHC-26.2, as shown in Fig. 7.1. A carbon model is an essential component of a nitrogen model if processes such as nitrogen immobilization, mineralization and denitrification, all of which are carbon driven, are to be modelled in a realistic way.

Data from the plot (09) on the Broadbalk continuous wheat experiment that receives 192 kg N ha⁻¹ annually were used to set the model parameters. Microplots located on this plot received single applications of ¹⁵N-labelled fertilizer in 1980, given at (nominally) the same rate as the main plot (Powlson *et al.*, 1986). In subsequent years unlabelled fertilizer nitrogen was applied to the microplots at the customary rate. The pulse of labelled N was followed in crop and soil for four years.

After the model had been tuned to the plot 09 data, it was subjected to a series of statistically validated tests (Whitmore, 1991). The first (Fig. 7.8) was to see how accurately it could predict the behaviour of labelled nitrogen in the Broadbalk plots receiving less fertilizer (48, 96 and 144 kg N ha⁻¹ year⁻¹), all of

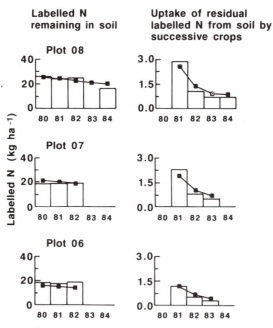

Fig. 7.8. Comparison of measured (histogram) and modelled (■) values for residual labelled nitrogen in the soil and in successive crops from Broadbalk, 1980–84. The results are for a single application of labelled fertilizer in 1980 to plot 06 (receiving an annual fertilizer application of 48 kg N ha⁻¹ year⁻¹), plot 07 (96 kg N ha⁻¹ year⁻¹) and plot 08 (144 kg N ha⁻¹ year⁻¹): for details of the experiment see Hart *et al.* (1993).

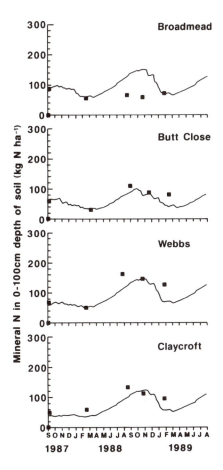

Fig. 7.9. Measured and modelled quantities of mineral nitrogen (nitrate plus ammonium) in the soil profile. Winter wheat had been grown on all four sites during the period October 1986–August 1987: the sites were then bare fallowed for the following two years. The Broadmead site was on a heavy clay soil, Butt Close on sandy loam, Webbs on chalky loam and Claycroft on silty clay loam.

which received a single pulse of ^{15}N-labelled fertilizer in 1980. The model was then tested on wheat grown on contrasting soils and in different years (Bradbury *et al.*, 1993); subsequently the tests were extended to cover crops other than winter wheat grown at four separate sites.

Another test of the model is its ability to predict the NO_3-N content of the soil profile throughout the year. Some measurements of NO_3-N in four contrasting soil types show encouraging, although by no means perfect, concordance between model and data (Fig. 7.9).

Conclusions

Well-run long-term experiments are of great value to model builders, particularly if they are concerned with slow changes in soil fertility and the sustainability of agricultural systems. Some of these benefits are obvious: known site histories, long runs of yield data, measurements of slow changes in the soil, etc. However, sometimes data from long-term experiments can be used to show that a particular system is under (or close to) steady-state conditions. Steady-state conditions are particularly valuable for setting the parameters of complex multicompartmental models, in that all the flows into and out of the system must balance. A pulse of tracer, moving through an undisturbed 'steady-state' system, provides the modeller with even more powerful tools for his or her work, whether the pulse enters the system inadvertently, as with ^{14}C from thermonuclear tests, or deliberately, as with the experiments using ^{15}N-labelled fertilizers on Broadbalk.

Lawes and Gilbert, far sighted as they both were, could not possibly have envisaged how their long-term experiments would one day be used to establish and validate complex computer-based models. How could they have foreseen, for example, that the soil samples they took so carefully from Broadbalk in 1881 would be radiocarbon dated a century later to establish the turnover time of soil organic carbon? Long-term field trials, properly run, documented and sampled, are a resource to be cherished and used by successive generations of scientists.

Acknowledgements

The Leverhulme Trust supported the work on the effects of climatic warming on the global stock of soil organic carbon; the Home-Grown Cereals Authority, the construction of SUNDIAL. We thank P.R. Poulton for access to recent (unpublished) measurements of organic carbon in soils from Broadbalk and Hoosfield; A.J. Macdonald and P.R. Poulton for unpublished data on the leaching of nitrate in different soils.

References

Aber, J.D., Melillo, J.M. and McCaugherty, C.A. (1990) Predicting long-term patterns of mass loss, nitrogen dynamics, and soil organic matter formation from initial fine litter chemistry in temperate forest ecosystems. *Canadian Journal of Botany* 68, 2201–2208.

Ågren, G.I. and Bosatta, E. (1987) Theoretical analysis of the long-term dynamics of carbon and nitrogen in soils. *Ecology* 68, 1181–1189.

Ågren, G.I., McMurtrie, R.E., Parton, W.J., Pastor, J. and Shugart, H.H. (1991) State-of-the-art models of production decomposition linkage in conifer and grassland ecosystems. *Ecological Applications* 1, 118–138.

Atjay, G.L., Ketner, P. and Duvigneaud, P. (1979) Terrestrial primary production and phytomass. In: Bolin, B., Degens, E., Kempe, S. and Ketner, P. (eds) *The Global Carbon Cycle. Scope 13*. Wiley, Chichester, pp. 129–182.

Ayanaba, A. and Jenkinson, D.S. (1990) Decomposition of carbon-14 labelled ryegrass and maize under tropical conditions. *Soil Science Society of America Journal* 54, 112-115.

Bartholomew, W.V. and Kirkham, D. (1960) Mathematical descriptions and interpretations of culture-induced soil nitrogen changes. *Transactions of the 7th International Congress of Soil Science* Vol 2, 471-477.

Bergstrom, L., Johnsson, H. and Tortenson, G. (1991) Simulation of soil nitrogen dynamics using the SOLIN model. In: Groot, J.J.R., de Willigen, P. and Verberne, E.L.J. (eds) *Nitrogen Turnover in the Soil/Crop System*. Kluwer Academic, Dordrecht, pp. 181-188.

Bosatta, E. and Ågren, G.I. (1985) Theoretical analysis of decomposition of heterogeneous substrates. *Soil Biology and Biochemistry* 17, 601-610.

Bosatta, E. and Ågren, G.I. (1991) Theoretical analysis of carbon and nutrient interactions in soils under energy-limited conditions. *Soil Science Society of America Journal* 55, 728-733.

Bouwman, A.F. (1989) Modelling soil organic matter decomposition and rainfall erosion in two tropical soils after forest clearing for permanent agriculture. *Land Degradation and Rehabilitation* 1, 125-140.

Bradbury, N.J., Whitmore, A.P., Hart, P.B.S. and Jenkinson, D.S. (1993) Modelling the fate of nitrogen in crop and soil in the years following application of ^{15}N-labelled fertilizer to winter wheat. *Journal of Agricultural Science, Cambridge* 121, 363-379.

Brookes, P.C., Powlson, D.S. and Jenkinson, D.S. (1984) Phosphorus in the soil microbial biomass. *Soil Biology and Biochemistry* 16, 169-175.

Buol, S.W., Sanchez, P.A. and Weed, S.B. (1990) Predicted impact of climatic warming on soil properties and use. In: *Impact of Carbon Dioxide, Trace Gases and Climate Change on Global Agriculture*. American Society of Agronomy Special Publication 53, Madison, pp. 71-82.

Buringh, P. (1984) Organic carbon in soils of the world. *SCOPE* 23, 91-109.

Burke, I.C., Yonker, C.M., Parton, W.J., Cole, C.V., Flach, K. and Schimel, D.S. (1989) Texture, climate and cultivation effects on soil organic matter content in USA grassland soils. *Soil Science Society of America Journal* 53, 800-805.

Burke, I.C., Schimel, D.S., Yonker, C.M., Parton, W.J., Joyce, L.A. and Lauenroth, W.K. (1990) Regional modelling of grassland biogeochemistry using GIS. *Landscape Ecology* 4, 45-54.

Carter, M.R., Parton, W.J., Rowland, I.C., Schultz, J.E. and Steed, G.R. (1993) Simulation of soil organic carbon and nitrogen changes in cereal and pasture systems of Southern Australia. *Australian Journal of Soil Research* 31, 481-491.

Cole, C.V., Burke, I.C., Parton, W.J., Schimel, D.S., Ojima, D.S. and Stewart, J.W.B. (1990) Analysis of historical changes in soil fertility and organic matter levels in the North American Great Plains. In: *Proceedings International Conference in Dryland Farming*. Texas Agricultural Experiment Station, Texas, pp.436-438.

Esser, G. (1990) Modelling global terrestrial sources and sinks of carbon dioxide with special reference to soil organic matter. In: Bouwman, A.F. (ed.) *Soils and the Greenhouse Effect*. Wiley, Chichester, pp. 225-261.

Eswaran, H., Van den Berg, E. and Reich, P. (1993) Organic carbon in soils of the world. *Soil Science Society of America Journal* 57, 192-194.

Franz, E.H. (1990) Potential influence of climatic change on soil organic matter and tropical agroforestry. In: Scharpenseel, H.W., Shomaker, M. and Ayoub, A. (eds) *Soils on a Warmer Earth*. Elsevier, Amsterdam, pp. 109-120.

Frissel, M.J. and Van Veen, J.A. (1978) Computer simulation modelling for nitrogen in irrigated croplands - a critique. In: Nielsen, D.R. and MacDonald, J.G. (eds) *Nitrogen in the Environment*. Academic Press, New York, pp. 145-162.

Gifford, R.M. (1992) Interaction of carbon dioxide with growth limiting environmental factors in vegetation productivity. *Advance in Bioclimatology* 1, 24-58.

Hansen, S., Jensen, H.E., Neilsen, N.E. and Svendsen, H. (1991) Simulation of nitrogen dynamics and biomass production in winter wheat using the Danish simulation model DAISY. In: Groot, J.J.R., de Willigen, P. and Verberne, E.L.J. (eds) *Nitrogen Turnover in the Soil/Crop System*. Kluwer Academic, Dordrecht, pp. 245-259.

Harkness, D.D., Harrison, A.F. and Bacon, P.J. (1991) The potential of bomb-^{14}C measurements for estimating soil organic matter turnover. In: Wilson, W.S. (ed.), *Advances in Soil Organic Matter Research: the Impact on Agriculture and the Environment*. Royal Society of Chemistry, Cambridge, pp. 239-251.

Hart, P.B.S. (1984) Effects of soil type and part cropping on the nitrogen supplying ability of arable soils. Unpublished PhD Thesis. University of Reading.

Hart, P.B.S., Powlson, D.S., Poulton, P.R., Johnston, A.E. and Jenkinson, D.S. (1993) The availability of the nitrogen in the crop residues of winter wheat to subsequent crops. *Journal of Agricultural Science, Cambridge* 121, 355-362.

Henin, S. and Dupuis, M. (1945) Essai de bilan de la matière organique du sol. *Annales Agronomiques* 15, 17-29.

Houghton, J.T., Jenkins, G.J. and Ephraums, J.J. (1990) *Climatic Change*. Cambridge University Press, Cambridge.

Hsieh, Y.-P. (1989) Dynamics of soil organic matter formation in cropland - conceptual analysis. *Science of the Total Environment* 82, 381-390.

Hsieh, Y.-P. (1992) Pool size and mean age of stable organic carbon in cropland. *Soil Science Society of America Journal* 56, 460-464.

Hsieh, Y.-P. (1993) Radiocarbon signatures of turnover rates in active soil organic pools. *Soil Science Society of America Journal* 57, 1020-1022.

Hunt, H.W. (1977) A simulation model for decomposition in grassland. *Ecology* 58, 469-484.

Hunt, H.W., Trlica, M.J., Redente, E.F., Moore, J.C., Detling, J.K., Kittel, T.G.F., Walter, D.E., Fowler, M.C., Klein, D.A. and Elliott, E.T. (1991) Simulation model for the effects of climate change on temperate grassland ecosystems. *Ecological Modelling* 53, 205-246.

Janssen, B.H. (1984) A simple method for calculating decomposition and accumulation of 'young' soil organic carbon. *Plant and Soil* 76, 297-304.

Jenkinson, D.S. (1971) The accumulation of organic matter in soil left uncultivated. *Rothamsted Experimental Station Report for 1970* Part 2, 113-137.

Jenkinson, D.S. (1977a) Studies on the decomposition of plant material in soil V. *Journal of Soil Science* 28, 424-434.

Jenkinson, D.S. (1977b) The nitrogen economy of the Broadbalk Experiments. I. Nitrogen balance in the experiments. *Rothamsted Experimental Station Report for 1976* Part 2, 103-109.

Jenkinson, D.S. (1988) Soil organic matter and its dynamics. In: Wild, A. (ed.) *Russell's Soil Conditions and Plant Growth*. Longman, London, pp. 564-607.

Jenkinson, D.S. (1990) The turnover of organic carbon and nitrogen in soil. *Philosophical Transactions of the Royal Society B* 329, 361–368.

Jenkinson, D.S. and Coleman, K. (1994) Calculating the annual input of organic matter to soil from measurements of total organic carbon and radiocarbon. *European Journal of Soil Science* 45, 167–174.

Jenkinson, D.S. and Johnston, A.E. (1977) Soil organic matter in the Hoosfield Continuous Barley Experiment. *Rothamsted Experimental Station Annual Report for 1976* Part 2, 87–101.

Jenkinson, D.S. and Rayner, J.H. (1977) The turnover of soil organic matter in some of the Rothamsted Classical Experiments. *Soil Science* 123, 298–305.

Jenkinson, D.S., Hart, P.B.S., Rayner, J.H. and Parry, L.C. (1987) Modelling the turnover of organic matter in long-term experiments at Rothamsted. *INTECOL Bulletin* 15, 1–8.

Jenkinson, D.S., Adams, D.E. and Wild, A. (1991) Global warming and soil organic matter. *Nature* 351, 304–306.

Jenkinson, D.S., Harkness, D.D., Vance, E.D., Adams, D.E. and Harrison, A.F. (1992) Calculating net primary production and annual input of organic matter to soil from the amount and radiocarbon content of soil organic matter. *Soil Biology and Biochemistry* 24, 295–308.

Jenny, H. (1941) *Factors of Soil Formation*. McGraw-Hill, New York.

Johnston, A.E. (1973) The effects of ley and arable cropping systems on the amounts of soil organic matter in the Rothamsted and Woburn Ley Arable Experiments. *Rothamsted Experimental Station Annual Report for 1972* Part 2, 131–159.

Kirschbaum, M.U.F. (1993) A modelling study of the effects of changes in atmospheric carbon dioxide concentration, temperature and atmospheric nitrogen input on soil organic carbon storage. *Tellus (Ser.13)* 45, 321–334.

Kohlmaier, G.H., Janecek, A. and Kindermann, J. (1990) Positive and negative feedback loops within the vegetation/soil system in response to a CO_2 greenhouse warming. In: Bouwman, A.F. (ed.) *Soils and the Greenhouse Effect*. Wiley, Chichester, pp. 415–422.

Kovda, V.A., Bugrovskii, V.V., Kerzhentsev, A.S. and Zelenskaya, N.N. (1990) Model of organic matter transformation in soil for the quantitative study of soil function in ecosystems. *Doklady Akademii Nauk SSSR* 312, 759–762.

Lashof, D.A. (1989) The dynamic greenhouse: feedback processes that may influence future concentrations of atmospheric trace gases and climatic change. *Climatic Change* 14, 213–242.

Melillo, J.M., Aber, J.D., Linkins, A.E., Ricca, A., Fry, B. and Nadelhoffer, K.J. (1989) Carbon and nitrogen dynamics along the decay continuum: plant litter to soil organic matter. *Plant and Soil* 115, 189–198.

Molina, J.A.E., Clapp, C.E., Schaffer, M.J., Chichester, F.W. and Larson, W.E. (1983) NCSOIL, a model of nitrogen and carbon transformation in soil: description, calibration and behaviour. *Soil Science Society of America Journal* 47, 85–91.

Neeteson, J.J., Greenwood, D.J. and Draycott, A. (1987) A dynamic model to predict yield and optimum nitrogen fertilizer application rate for potatoes. *Proceedings No. 262*. The Fertilizer Society, London.

Nye, P.H. and Greenland, D.J. (1960) The soil under shifting cultivation. *Technical Communication No. 51* CAB, Harpenden.

O'Brien, B.J. (1984) Soil organic carbon fluxes and turnover rates estimated from radiocarbon enrichments. *Soil Biology and Biochemistry* 16, 115-120.

Parton, W.J., Anderson, D.W., Cole, C.V. and Stewart, J.W.B. (1983) Simulation of soil organic matter formation and mineralisation in semi-arid agroecosystems. In: Lowrance, R.R. *et al.* (eds) *Nutrient Cycling in Agricultural Ecosystems.* University of Georgia Special Publications 23, pp. 533-550.

Parton, W.J., Schimel, D.S., Cole, C.V. and Ojima, D.S. (1987) Analysis of factors controlling soil organic matter levels in Great Plains Grasslands. *Soil Science Society of America Journal* 51, 1173-1179.

Parton, W.J., Stewart, J.W.B. and Cole, C.V. (1988) Dynamics of C, N, P and S in grassland soils: a model. *Biogeochemistry* 5, 109-131.

Parton, W.J., Cole, C.V., Stewart, J.W.B., Ojima, D.S. and Schimel, D.S. (1989a) Simulating regional patterns of soil C, N and P dynamics in the US Central Grassland Region. In: Clarholm, M. and Bergstrom, L. (eds) *Ecology of Arable Land.* Kluwer Academic Press, Dordrecht, pp. 99-108.

Parton, W.J., Sanford, R.L., Sanchez, P.A. and Stewart, J.W.B. (1989b) Modelling soil organic matter dynamics in tropical soils. In: Coleman, D.C., Oades, J.M. and Uehara, G. (eds) *Dynamics of Soil Organic Matter in Tropical Ecosystems NifTAL Project*, University of Hawaii Press, Honolulu, pp.153-171.

Parton, W.J., Schimel, D.S., Ojima, D.S. and Cole, C.V. (1994a) A general model for soil organic matter dynamics: sensitivity to litter chemistry, texture and management. *Soil Science Society of America Journal* (submitted).

Parton, W.J., Scurlock, J.M.O., Ojima, D.S., Gilmanov, T.G., Scholes, R.J., Schimel, D.S., Kirchner, T., Menaut, J-C., Seastedt, T., Garcia Moya, E., Kamnalrut, A. and Kinyamario, J.L. (1994b) Observations and modelling of biomass and soil organic matter dynamics for the grassland biome worldwide. *Global Biogeochemical Cycles* (submitted).

Paustian, K., Parton, W.J. and Persson, J. (1992) Modelling soil organic matter in organic-amended and nitrogen-fertilized long-term plots. *Soil Society of America Journal* 56, 476-488.

Post, W.M., Emanuel, W.R., Zinke, P.J. and Stangenberger, A.G. (1982) Soil carbon pools and world life zones. *Nature* 298, 156-159.

Powlson, D.S., Pruden, G., Johnston, A.E. and Jenkinson, D.S. (1986) The nitrogen cycle in the Broadbalk Wheat Experiment: recovery and losses of ^{15}N-labelled fertilizer applied in spring and inputs of nitrogen from the atmosphere. *Journal of Agricultural Science, Cambridge* 107, 591-609.

Prentice, K.C. and Fung, I.Y. (1990) The sensitivity of terrestrial carbon storage to climate change. *Nature* 346, 48-51.

Richter, J., Kersebaum, K.Chr. and Utermann, J. (1988) Modelling of the nitrogen regime in arable field soils for advisory purposes. In: Jenkinson, D.S. and Smith, K.A. (eds) *Nitrogen Efficiency in Agricultural Soils.* Elsevier Applied Science, London, pp. 371-383.

Rijtema, P.E. and Kroes, J.G. (1991) Some results of nitrogen simulations with the model ANIMO. In: Groot, J.J.R., de Willigen, P. and Verberne, E.L.J. (eds) *Nitrogen Turnover in the Soil/Crop System.* Kluwer Academic, Dordrecht, pp. 189-198.

Russell, J.S. (1975) A mathematical treatment of the effects of cropping system on soil organic nitrogen in two long-term sequential experiments. *Soil Science* 120, 37-44.

Schimel, D.S., Parton, W.J., Kittel, T.G.F., Ojima, D.S. and Cole, C.D. (1990) Grassland biogeochemistry links to atmospheric processes. *Climatic Change* 17, 13–26.

Schleser, G.H. (1981) The response of CO_2 evolution from soils to global temperature changes. *Naturf* A37, 287–319.

Schlesinger, W.H. (1984) Soil organic matter: a source of atmospheric CO_2. In: Woodwell, G.M. (ed.) *The Role of Terrestrial Vegetation in the Global Carbon Cycle. Scope 23*, Wiley, Chichester, pp. 111–127.

Stewart, J.W.B. and Cole, C.V. (1989) Influences of elemental interactions and pedogenic processes in organic matter dynamics. *Plant and Soil* 115, 199–209.

Tate, K.R., Roass, D.J., O'Brien, B.J. and Kelliher, F.M. (1993) Carbon storage and turnover and respiratory activity, in the litter and soil of an old-growth southern beech (*Nothofagus*) forest. *Soil Biology and Biochemistry* 25, 1601–1612.

Thornley, J.H.M. and Verberne, E.L.J. (1989) A model of nitrogen flows in grassland. *Plant, Cell and Environment* 12, 863–886.

Thornley, J.H.M., Fowler, D. and Cannell, M.G.R. (1991) Terrestrial carbon storage resulting from CO_2 and nitrogen fertilisation in temperate grasslands. *Plant, Cell and Environment* 14, 1007–1011.

Tinker, P.B. and Ineson, P. (1990) Soil organic matter in relation to climate change. In: Scharpenseel, H.W., Schomaker, M. and Ayoub, A. (eds) *Soils on a Warmer Earth*. Elsevier, Amsterdam, pp. 71–87.

Trumbore, S.E. (1993) Comparison of carbon dynamics in tropical and temperate soils using radiocarbon measurements. *Global Biogeochemical Cycles* 7, 275–290.

Vance, E.D., Brookes, P.C. and Jenkinson, D.S. (1987) An extraction method for measuring soil microbial biomass C. *Soil Biology and Biochemistry* 19, 703–707.

Van Der Linden, A.M.A., Van Veen, J.A. and Frissel, M.J. (1987) Modelling soil organic matter levels after long-term applications of crop residues, and farmyard and green manures. *Plant and Soil* 101, 21–28.

Van Veen, J.A. and Paul, E.A. (1981) Organic carbon dynamics in grassland soils. I Background information and computer simulation. *Canadian Journal of Soil Science* 61, 185–201.

Verberne, E.L.J., Hassink, J., De Willigen, P., Groot, J.J.R. and Van Veen, J.A. (1990) Modelling organic matter dynamics in different soils. *Netherlands Journal of Agricultural Science* 38, 221–238.

Voroney, R.P., Van Veen, J.A. and Paul, E.A. (1981) Organic C dynamics in grassland soils. 2. Model validation and simulation of the long-term effects of cultivation and rainfall erosion. *Canadian Journal of Soil Science* 61, 211–224.

Whitmore, A.P. (1991) A method for assessing the goodness of computer simulation of soil processes. *Journal of Soil Science* 42, 289–299.

Wolf, J., De Wit, C.T. and Van Keulen, H. (1989) Modelling long-term response to fertiliser and soil nitrogen. *Plant and Soil* 120, 11–22.

Woodmansee, R.G. (1978) Critique and analyses of the grassland ecosystem model ELM. In: Innis, G.S. (ed.) *Grassland Simulation Model: Ecological Studies*. Springer-Verlag, New York, pp. 258–281.

Woodmansee, R.G. (1988) Ecosystem processes and global change. In: Risser, P.G., Woodmansee, R.G. and Rosswall, T. (eds) *Scales and Global Change. Scope Report 35*. Wiley, New York, pp. 11–27.

D.S. Jenkinson et al.

Woomer, P.L. (1993) Modelling soil organic matter dynamics in tropical ecosystems: model adoption, uses and limitation. In: Mulongoy, K. and Merckx, R. (eds) *Soil Organic Matter Dynamics and Sustainability of Tropical Agriculture*. Wiley, Chichester, pp. 279-294.

8

Polyetic Epidemics by Plan or Contingency

J.C. ZADOKS
Wageningen Agricultural University, Department of Phytopathology, P.O.Box 8025, 6700 EE Wageningen, The Netherlands.

Introduction

Plant disease epidemiology studies disease in plant populations. Epidemics are often considered as explosive events. However, suddenness is but a small though spectacular part of a complex truth. Even the swiftly striking pathogen of the foliage may need years of preparation. Plant pathogens dwelling in the soil slowly build up their capacity for devastation and disaster.

Long-term experiments, by plan or contingency, may teach the dynamics of such epidemics and the ways to interfere with them. Thus, the topic of this volume, long-term approaches, is well chosen. It is a neglected but utterly relevant area in plant disease epidemiology. This chapter discusses polyetic epidemics somewhat haphazardly and hops from the arbitrary to the design, and from the individual interest to the public interest.

Polyetic Epidemics

Polyetic epidemics are epidemics which span several years (Zadoks and Schein, 1979). A classical example is the epidemic of a soil-borne disease around Saint Petersburg, clubroot in cabbage, which led to Woronin's famous publication of 1877 on *Plasmodiophora brassicae* (Woronin, 1934). Any neglect of rotational principles may induce polyetic epidemics, even in foliar diseases of annual crops such as yellow rust (*Puccinia striiformis*) of wheat (Zadoks, 1961). The dynamics of polyetic epidemics are fairly well described for nematodes, beginning with the work by Seinhorst (1966). The spatiotemporal dynamics of polyetic epidemics, including the sawtooth progress in time and the oilslick spread in space, have been described mathematically (van den Bosch and Zadoks, 1994).

Most epidemics on crops are made or at least induced by humans. Epidemics do occur on wild plants but they are seldom destructive to the degree that a plant population is wiped out. The one exception seems to be the case where a pioneer plant species appears massively and suddenly dies away. For the polyetic case the rust (*Melampsora biglowii*) on willows in Alaska stands as a model (Baxter and Wadsworth, 1939).

Experimentation with polyetic epidemics is not easy due to lack of steadfastness and money. Three types of experimentation can be distinguished, research-driven experimentation, 'unintentional' experimentation which is not the *contradictio in terminis* it sounds to be, and policy-driven experimentation. The scale of such experiments varies from 1 m² to a continent.

Research-driven experiments

Intentional experiments revealing polyetic epidemics are not rare but, at least in the impatient Netherlands with their disrespect for tradition, they usually do not last long. The oldest agricultural field experiment, Broadbalk, I visited as a young scientist in 1957 and I felt like entering a sanctuary. In The Netherlands, Ritzema Bosch (1921) experimented with continuous cropping on plots of about 10 m² to see if and when 'soil sickness' would appear. As this condition did not appear in any of his crops, he concluded that there is no soil sickness without a pathogen.

Severe problems with the potato cyst nematodes, then indicated as *Heterodera rostochiensis*, induced Oostenbrink (1950) to strongly advocate and even impose crop rotation. With his chequerboard rotation experiments to study the dynamics of nematodes and the effects of various rotations, Oostenbrink (1959) became the artist on the square metre. A few long-term experiments are on record in The Netherlands, which demonstrated the existence of '*Ophiobolus* decline' in wheat (Gerlagh, 1968), as reported a generation earlier by Glynne (1935) for barley.

'Unintentional' experimentation

Among the 'unintentional' experiments are a variety of human actions with epidemiological consequences. The unintentional introduction of *Phytophthora infestans* causing potato late blight and the intentional introduction but unintentional escape of *Peronospora tabacina* causing tobacco blue mould into Europe produced polyetic epidemics with a model spatiotemporal behaviour (Heesterbeek and Zadoks, 1987; van den Bosch and Zadoks, in press).

A more intricate situation developed along the Dutch coast, on the isle of Terschelling, where a high sandbank separated the rough North Sea from the quieter Dutch Shallows. The sandbank of about 20 km^2 was virtually free of vegetation. The 'experiment' consisted of building a sanddike of nearly 10 km along the sandplate in the late 1930s. Under the protection of the dike, vegetation appeared at the side of the Dutch Shallows with sea lavender (*Limonium vulgare*) as the dominant plant and sea lavender rust (*Uromyces limonii*) and mildew (*Erysiphe communis*) as its dominating diseases (van Leur, 1981).

The vegetation can be visualized as a pancake-flat forest of deciduous trees, each measuring some 4 m in diameter, with a closed canopy, consisting of trees which grew after the dike was built but interspersed with a few very much older trees, even a hundred years old, survivors from the original vegetation. One lesson to be drawn is that different biotrophic fungi may have different ecological functions. Whereas the rust usually does not kill but gradually suppresses reproduction, the mildew seems to be an end-of-succession killer preparing the ground for the next round of colonization.

Another opportunity for 'unintentional' experimentation offered itself when long-term field monitoring could be analysed statistically. Annual surveys of winter wheat fields over a period of ten years or more yielded interesting correlations between prevalence of a disease and mean monthly weather data (Daamen and Stol, 1992; Daamen *et al.*, 1992). The interpretation of such correlations is still questionable and their usefulness for managerial purposes doubtful.

Policy-driven experimentation

Policy-driven experiments are a relatively new phenomenon. Whereas the Broadbalk experiment was primarily crop oriented, the Boxworth experiment in the UK is one at farm level, looking also at the farm environment. In The Netherlands, a farming systems experiment began in 1979 (Zadoks, 1989a) with the dual purpose of comparing three farming systems and to develop one of them, 'integrated farming'. The experiment is still continuing.

Ideas about sustainability and durability had been developed by various groups long before these ideas were taken up by policy makers. The United Nations Food and Agriculture Organization officially launched the concept of sustainability at its Stockholm 1972 conference (FAO, 1992). It remained practically unnoticed but a 1984 report on integrated agriculture (van der Weijden *et al.*, 1984) caused shock waves in Dutch governmental circles. Continued political pressure induced the Dutch Ministry of Agriculture to initiate the experiment on farming systems, in which three adjoining farms under one management were given three different regimes, called Current Farming, Integrated Farming and Biological Farming.

The Current Farm was to reflect the type of high input farming current in the farm's neighbourhood. The Biological Farm was to be a mixed farm with some dairy cattle, to be run without artificial fertilizers and pesticides. The

Integrated Farm was to follow the new ideas, with reduced inputs of agro-chemicals, but with economic results at par with the Current Farm. The experiment had no replications. Development (Vereijken, 1989) was given precedence over sequential replication (Zadoks, 1989b). The explicit aim was to produce economically feasible farm technology integrating various policy objectives such as sustainability, profitability, nature conservation, landscape and rural job opportunity.

The job was difficult but in the end it could be shown that integrated farming was an ecologically and economically attractive proposition, at least on the best of Dutch soils with the best of scientific council. In integrated farming, some pests and diseases were alleviated due to the choice of more-resistant cultivars and/or reduced nitrogen application. Weed control was the major problem. The Farming Systems Experiment is being continued and extended.

Crop Protection Policy in The Netherlands

When epidemics, annual and polyetic, became ever more limited by the use of pesticides, simultaneously various lobbies became better informed on undesirable side-effects of pesticides. Gradually, an increasingly vociferous opposition to the use of pesticides developed to which the Dutch Ministry of Agriculture turned a deaf ear. A typical 'don't bother, we will look into the matter' document was offered to Parliament in 1984.

However, in 1989, a well-considered Multi-Year Crop Protection Plan was published (Anonymous, 1989), inspired by a Danish example (Anonymous, 1986). The Plan, approved by Dutch Cabinet in 1991 (Anonymous, 1991), was specific in setting targets in quality, quantity and time, and demanding regular reporting to Parliament. The first report (Anonymous, 1993) describes a clear reduction in the volume of pesticides being used.

In the wake of the new crop protection policy, and with the Farming Systems Experiment as a tug, there followed the establishment of more farming systems experiments for the development of integrated farming and horticulture (Vereijken, 1992), and the organization of the network of some 40 closely monitored 'innovation farms' (Wijnands *et al.*, 1992; Wijnands and Vereijken, 1992) and of hundreds of farmers' 'study groups' (Wijnands, 1992).

Meanwhile, in the second half of the 1980s, the notion of good agricultural practice (GAP) spread. It is a friendly notion, a little bit vague, and rather similar in effect to Alternative agriculture in the USA (Committee, 1989) in its effect. GAP seems more a matter of science-based protocols than of rigorous scientific calculation.

Scientific rigour is the characteristic of another and related notion, best technical means (BTM). BTM is defined as the minimization of inputs with simultaneous maximization of outputs (WRR, 1992), the 'minimax principle'. The outputs do not have to be expressed in kilograms or guilders, but may also be expressed in other dimensions such as nature conservation values or job opportunity. Between such outputs, expressed in very different units, a trade off may be negotiated according to the scenario desired, such as yield-oriented, environment-oriented and land use-oriented productions.

Epilogue

The circle can be closed. Epidemics, the annual and the polyetic, are of all times. They do exist in nature and they plagued humankind as soon as agriculture began (Orlob, 1964). Damaging epidemics, intensified by modern agriculture (Zadoks and Schein, 1979), are an inherent and characteristic side effect of the Green Revolution.

The strategy to stop epidemics by means of pesticides, either applied to the foliage or to the soil, failed at times for a variety of reasons. Long-term observation, by plan or contingency, reveals their polyetic behaviour and suggests approaches to non-chemical control. Quantitative analysis of long-term observations may teach many lessons on the behaviour of epidemics in time and space and thus help to manage the disease instead of trying to kill its causal agent. The mixed stand of sea lavender indicates that genetic diversity (Luo Yong and Zadoks, 1992) and partial resistance (Jacobs and Parlevliet, 1993) do help. Rotation experiments demonstrate the existence of 'disease decline' and indicate that biological control helps. By broadening our outlook and respecting the minimax principle, we will be able to make the Green Revolution look greener (Swaminathan, 1990).

Summary

Three types of long-term experimentation are discussed, research-driven experiments, 'unintentional' experiments and policy-driven experiments. A few examples from The Netherlands are indicated, and some lessons taken. They lead to the new crop protection policy of The Netherlands. The policy, of which on-farm experimentation is a part, seems successful in reducing the volume of pesticides applied annually.

References

Anonymous (1986) *Miljøministerens handlingsplan for nedsættelse af forbruget af bekæmpelsesmidler*. Miljøministeriet J.nr. D 86-27000-4, Copenhagen, December 1986, 13 pp.

Anonymous (1989) *Meerjarenplan Gewasbescherming. Beleidsvoornemen.* (Multi-year Crop Protection Plan.) Sdu Publishers, The Hague, 138+133 pp.

Anonymous (1991) *Meerjarenplan Gewasbescherming. Regeringsbeslissing.* (Multi-year Crop Protection Plan.) Sdu Publishers, The Hague, 298 pp.

Anonymous (1993) *Voortgangsrapportage Meerjarenplan Gewasbescherming 1992*. Ministerie van Landbouw, Den Haag, 11 pp.

Baxter, D.V. and Wadsworth, F.H. (1939) Forest and fungus succession in the Lower Yukon Valley. *Bulletin of the University of Michigan School of Forestry and Conservation*, Ann Arbor, no. 9, 52 pp.

Committee (1989) *Alternative Agriculture*. National Academy Press, Washington DC, 448 pp.

Daamen, R.A. and Stol, W. (1992) Surveys of cereal diseases and pests in the Netherlands. 5. Occurrence of Septoria spp. in winter wheat. *Netherlands Journal of Plant Pathology* 98, 369-376.

Daamen, R.A., Stubs, R.W. and Stol, W. (1992) Surveys of cereal diseases and pests in the Netherlands. 4. Occurrence of powdery mildew and rusts in winter wheat. *Netherlands Journal of Plant Pathology* 98, 301-312.

FAO (1992) *Sustainable Development and the Environment: FAO Policies and Actions, Stockholm 1972 - Rio 1992*. FAO, Rome, 88 pp.

Gerlagh, M. (1968) Introduction of *Ophiobolus graminis* into new polders and its decline. *Netherlands Journal of Plant Pathology* 74 Suppl. no. 2, 97 pp.

Glynne, M.D. (1935) Incidence of take-all on wheat and barley on experimental plots at Woburn. *Annals of Applied Biology* 22, 225-235.

Heesterbeek, J.A.P. and Zadoks, J.C. (1987) Modelling pandemics of quarantine pests and diseases; problem and perspectives. *Crop Protection* 6, 211-221.

Jacobs, Th. and Parlevliet, J.E. (eds) (1993) *Durability of Disease Resistance*. Kluwer Academic Publishers, Dordrecht, 375 pp.

Luo Yong and Zadoks, J.C. (1992) A decision model for variety mixtures to control yellow rust on winter wheat. *Agricultural Systems* 38, 17-32.

Oostenbrink, M. (1950) Het aardappelaaltje (*Heterodera rostochiensis* Wollenweber), een gevaarlijke parasiet voor de eenzijdige aardappel-cultuur. *Verslagen en Mededelingen van de Plantenziektenkundige Dienst No 115*, 230 pp.

Oostenbrink, M. (1959) Enkele eenvoudige proefveldschema's bij het aaltjesonderzoek. *Mededelingen van de Landbouwhogeschool te Gent* 24, 615-618.

Orlob, G.B. (1964) Vorstellungen ueber die Aetiologie in der Geschichte der Pflanzenkrankheiten. *Pflanzenschutz Nachrichten 'Bayer'* 17, 185-272.

Ritzema Bosch, J. (1921) Mijn proefveldje bij het Instituut voor Phytopathologie van 1906 tot 1920. *Tijdschrift over Plantenziekten* 27, 29-44.

Seinhorst, J.W. (1966) The relationship between population increase and population density in plant parasitic nematodes. I. Introduction and migratory nematodes. *Nematologica* 12, 157-169.

Swaminathan, M.S. (1990) The Green Revolution and small-farm agriculture. *CIMMYT 1990 Annual Report*, 12-15.

van den Bosch, F. and Zadoks, J.C. (1994) Continental expansion of plant disease: a survey of some recent results (in press).

Van der Weijden, W.J., van der Wal, H., der, de Graaf, H.J., van Brussel, N.A. and ter Keurs, W.J. (1984) *Bouwstenen voor een geïntegreerde landbouw*. Wetenschappelijke Raad voor het Regeringsbeleid. WRR Voorstudies en achtergronden V44, Den Haag, Staatsuitgeverij, 196 pp.

van Leur, J.G.A. (1981) Onderzoek naar het evenwicht tussen waard en pathogenen in een natuurlijke situatie. *Vakblad Biologen* 61, 466-471.

Vereijken, P. (1989) The DFS farming systems experiment. In: Zadoks, J.C. (ed) *Development of Farming Systems. Evaluation of the Five-year Period 1980-1984.* Pudoc, Wageningen, pp. 1-8.

Vereijken, P. (1992) A methodic way to more sustainable farming systems. *Netherlands Journal of Agricultural Sciences* 40, 209-223.

Wijnands, F.G. (1992) Evaluation and introduction of integrated arable farming in practice. *Netherlands Journal of Agricultural Sciences* 40, 239-249.

Wijnands, F.G. and Vereijken, P. (1992) Region-wise development of prototypes of integrated arable farming and outdoor horticulture. In: van Lenteren, J.C., Minks, A.K. and de Ponti, O.M.B. (eds) *Biological Control and Integrated Crop Protection: Towards Environmentally Safer Agriculture.* Pudoc, Wageningen, pp. 125-138.

Wijnands, F.G., Janssens, S.R.M., van Asperen, P. and van Bon, K.B. (1992) *Innovatiebedrijven geïntegreerde akkerbouw. Opzet en eerste resultaten.* PAGV Verslag 144. PAGV, Lelystad, 88 pp.

Woronin, M. (1934) *Plasmodiophora brassicae*, the cause of cabbage hernia. (Translation Ch. Chupp) *Phytopathological Classics* 4, 32 pp.

WRR (1992) Netherlands Scientific Council for Government Policy. *Ground for Choices. Four Perspectives for the Rural Areas in the European Community.* Sdu Publishers, The Hague, 144 pp.

Zadoks, J.C. (1961) Yellow rust on wheat, studies in epidemiology and physiologic specialization. *Tijdschrift over Plantenziekten (Netherlands Journal of Plant Pathology)* 67, 69-256.

Zadoks J.C. (ed) (1989a) *Development of Farming Systems. Evaluation of the Five-year Period 1980-1984.* Pudoc, Wageningen, 90 pp.

Zadoks, J.C. (1989b) Comments on the research methodology for DFS. In: Zadoks, J.C. (ed) *Development of Farming Systems. Evaluation of the Five-year Period 1980-1984.* Pudoc, Wageningen, pp. 73-82.

Zadoks, J.C. and Schein, R.D. (1979) *Epidemiology and Plant Disease Management.* Oxford University Press, New York, 427 pp.

Historical Monitoring of Organic Contaminants in Soils

K.C. JONES[1], A.E. JOHNSTON[2] AND S.P. McGRATH[2]
*[1]Environmental Science Division, Institute of
Environmental and Biological Sciences, Lancaster
University, Lancaster, LA1 4YQ, UK; [2]Rothamsted
Experimental Station, Harpenden, Hertfordshire
AL5 2JQ, UK.*

Introduction

One of the most exciting unforeseen benefits of the Classical Experiments established by Lawes and Gilbert at Rothamsted in the 1840s and 1850s has been the opportunity to study contaminant trends over time, by comparing the chemical composition of samples from the archived collection with contemporary material. Public and scientific interest in Man's impact on the environment has increased since the 1960s, and is being reflected now in the growing financial support from international agencies and Governments for programmes on 'Global Change'. In the mid-1980s we began our collaboration to look at long-term changes in soil and crop composition on the Classical Experiments, which were extended in the late-1980s to include the sewage sludge-amended plots at Woburn. As a result several papers have been published on trends in metals in soils and herbage, focussing on cadmium, lead and selenium (Jones *et al.*, 1987a, b, 1991a, 1992a; Jones and Johnston, 1989, 1991; Haygarth *et al.*, 1993; Nicholson *et al.*, 1993). The various treatments to the plots on the Classical Experiments enable inputs associated with specific sources to be identified. For example, cadmium is present as an impurity in phosphate fertilizers, so studies comparing trends on the control (untreated) and P-treated plots have proved fruitful (Rothbaum *et al.*, 1986; Johnston and Jones, 1992). Similarly, the effects of varying amounts of organic matter and soil pH on the retention of cadmium in soil and cadmium concentrations in crops can be studied by judicious selection of appropriate plots for study.

More recently, our attention has switched to examine historical trends of trace organic contaminants and their fate and behaviour in soil–plant systems. This chapter will focus specifically on trace organic compounds in soils. The groups of compounds studied to date are the polynuclear aromatic hydro-carbons (PAHs), the polychlorinated dibenzo-p-dioxins and -furans (PCDD/Fs) and the polychlorinated biphenyls (PCBs). Research interest in the environ-mental fate and behaviour of these compounds began in the 1960s and has been sustained because of their persistent (recalcitrant) nature, the propensity for some of them to accumulate through the foodchain and concerns over their toxicity. All these groups of compounds are ubiquitous in the environ-ment. Consequently, our work has had a number of goals; one has been to investigate the nature of air–soil exchanges by analysis of soil samples col-lected from the control plots; another has been to relate retrospective infor-mation about production and use of these compounds to their environmental persistence. It is important to appreciate that little reliable environmental ana-lytical data exist for these compounds prior to the 1970s. Contemporary con-centrations of these compounds in normal agricultural soils range from pg kg^{-1} for PCDD/Fs, to mg kg^{-1} for PAHs. It is only recently that sufficiently sensitive and reliable analytical procedures have become available to determine them. As a result, retrospective analysis of archived samples from the Classical Exper-iments has provided a unique and valuable insight into the influence of human activity on the inputs, environmental cycling and time trends of these contaminants.

The Compounds of Interest

Polynuclear aromatic hydrocarbons (PAHs)

PAHs are primarily formed by the incomplete combustion of organic materials. Hundreds of homo- and heterocyclic PAHs can potentially be formed, but typically about 16–20 compounds are routinely analysed. These have been listed by the United States Environmental Protection Agency and the European Commission as priority pollutants. Natural combustion pro-cesses (e.g. forest fires) and man-made fires will have released PAHs into the atmosphere, but the more recent large-scale consumption of fossil fuels has enormously increased emissions over the last century or so. Here the contrast-ing behaviour of two PAH compounds – phenanthrene and benzo(a)pyrene (Fig. 9.1) – is discussed. Phenanthrene is a 3-ringed aromatic of sufficiently low molecular weight (mol. wt. = 178) that it has a moderate aqueous solubility (1.2 mg l^{-1}) but exists in the atmosphere primarily in the vapour phase. Indeed, phenanthrene dominates the total (Σ) PAH content of air (Halsall et al., 1993). Benzo(a)pyrene (B(a)P), a known carcinogen, is a 5-ringed molecule (mol. wt. = 252) of low aqueous solubility (1.6 µg l^{-1}), which binds strongly to particu-

PAHs

Phenanthrene

Benzo(a)pyrene

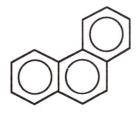

PCDD/Fs

2,3,7,8-tetrachlorinated
dibenzo-*p*-dioxin

Octachlorinated
dibenzo-*p*-dioxin

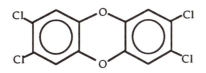

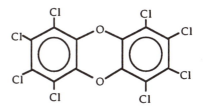

PCBs

2,4,4'-trichlorobiphenyl

2,2',3,4,4',5'-hexachlorobiphenyl

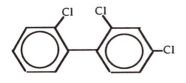

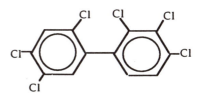

Fig. 9.1. The pairs of compounds representative of the three groups compared in this chapter.

late matter and rarely exists in the gaseous state under ambient conditions. Phenanthrene can be degraded by soil microflora, whereas B(a)P is very recalcitrant (Wild *et al.*, 1990). PAHs can be quantified by either high performance liquid chromatography with fluorescence or diode array detection, or capillary gas chromatography with flame ionization or mass selective detection.

Polychlorinated dibenzo-p-dioxins and -furans

PCDD/Fs are formed in two ways; by combustion and 'accidentally' during the production of organochlorine chemicals, such as chlorophenols and phenoxyacetic acid herbicides (Harrad and Jones, 1992). Over 200 different PCDD/F compounds can exist, containing one to eight chlorine atoms. One particular PCDD, 2,3,7,8-tetrachlorinated dibenzo-p-dioxin (2,3,7,8-TCDD) has been called 'the most toxic chemical known to man', but its concentrations in environmental samples are usually very low. The PCDD/Fs can be substantially accumulated through foodchains (particularly in aquatic environments), so toxic effects have been reported in freshwater and marine top predators (Jones and Bennett, 1989). Two PCDD/Fs will be compared here; 2,3,7,8-TCDD and the octachlorinated dioxin (OCDD) (Fig. 9.1). PCDD/Fs are determined in soils by high resolution gas chromatography–mass spectrometry after extensive clean-up of solvent extracts using liquid chromatography.

Polychlorinated biphenyls

PCBs were first synthesized in 1929. About 2×10^9 kg of PCBs have been manufactured worldwide, primarily for use as electrical insulating fluids in transformers and capacitors (Jones et al., 1991b). They are very persistent compounds and are mobile through the environment because of their ability to volatilize and exist in the vapour phase in air. They are lipophilic and concerns over high concentrations in top predators in the late-1960s led to restrictions and ultimately a ban on their use in the UK and elsewhere in the mid- to late-1970s. Nonetheless, they are still being linked to subtle ecotoxicological effects on reproductive potential and the immune system in marine mammals (Jones et al., 1991b). Two PCB congeners are considered here: the low molecular weight 2,4,4'-triCB (IUPAC number 28; PCB-28) and the 2,2',3,4,4',5'-hexaCB (IUPAC number 138; PCB-138) (see Fig. 9.1). Both are among the most abundant congeners in environmental samples (Jones, 1988). Following Soxhlet extraction and sample clean-up, PCBs are quantified by capillary gas chromatography with electron capture and/or mass selective detection.

Net Changes in Soil Composition due to Air–Soil Interactions

In soils, organic chemicals are subject to: (i) sorption to the soil – either reversibly or irreversibly; (ii) removal from the plough layer either to the subsurface zone by leaching or in the harvested crops following uptake; (iii) biotic or abiotic degradation; or (iv) volatilization. Hence, analysis of the archived samples yields information about *net changes* in the amounts extracted from soil *after* these various processes have occurred. Soils generally have a capacity to store contaminants by sorption. Hence, the soil can become a significant long-term sink for contaminants which come into contact with it (Jones, 1991). Indeed, budget calculations for the UK show that surface soils represent the most substantial 'reservoir' of PAHs, PCDD/Fs and PCBs in the environment (Harrad and Jones, 1992; Harrad *et al.,* 1994; Wild and Jones, 1994) so that large-scale net changes in soil composition can have a profound effect on the total amount of these chemicals in the environment as a whole.

Studies on PAH, PCDD/F and PCB trends in the Classical Experiments have focused on the control plots where changes in plough layer (0–23 cm) depth soil and vegetation will be influenced by air–soil exchanges (Jones *et al.,* 1989a, b, 1992b; Wild *et al.,* 1990; Kjeller *et al.,* 1991; Alcock *et al.,* 1993). Table 9.1 presents information on the six selected compounds in the soils taken at various times from control plots of the Broadbalk winter wheat experiment (started 1843) and the Market Garden experiment which started at Woburn in 1942. The latter is included because samples have been taken more frequently from this site, giving better temporal resolution of trends in recent decades. The data highlight the contrasting behaviour of the different compounds, both within and between homologue groups. Figure 9.2 presents the trends from Table 9.1 diagrammatically, by normalizing the average concentration for each compound from the two experiments to its peak concentration over time. Four of the chemicals (phenanthrene, B(a)P, 2,3,7,8-TCDD and OCDD) were detectable in the earliest samples indicating that these compounds occur naturally at low concentrations in soils. This observation is consistent with combustion as a source. All four compounds also increase in concentration through the time series – again supporting the assertion that anthropogenic activities have increased their atmospheric burden (and hence deposition) during this century. However, phenanthrene concentrations clearly peaked in the 1960s and have declined subsequently, whereas B(a)P, 2,3,7,8-TCDD and OCDD concentrations have either continued to increase up to the present or have stabilized. This observation can be explained from our knowledge of the potential volatility of phenanthrene (Wild and Jones, 1992) and/or its susceptibility to biological degradation (Table 9.2; Wild and Jones, 1993), but it implies that the inputs of PAHs have declined in recent years. Various lines of evidence indicate that this is indeed the case. Determination of PAHs in archived vegetation samples, for example, showed a decline in ΣPAH concentrations since the 1960s (Jones *et al.,* 1989b, 1992b), whereas data from dated lake sediment cores indicate maximum ΣPAH deposition fluxes in the 1950s (Sanders *et al.,* 1993). Coal consumption is thought to be the largest contemporary source of PAHs to the UK environment (Wild and Jones, 1994), and total UK consumption peaked in the 1950s. In addition, there has been a shift in coal usage away from the relatively inefficient domestic combustion of

Table 9.1. Concentrations of the six selected compounds in soils (0–23 cm depth) taken at various times from the control plots of Broadbalk and Woburn experiments.

Broadbalk

	1846	1856	1881	1893	1914	1944	1956	1966	1980	1986
Phenanthrene μg kg⁻¹	46		68	45	89	110	120	160	140	48
Benzo(a)pyrene μg kg⁻¹	18	33	6.7		12	23	73	28	120	72
2,3,7,8-TCDD pg kg⁻¹				29	40	40	49	60	79	110
OCDD ng kg⁻¹		7.6		11	11	10	13	32	20	25
PCB-28 μg kg⁻¹	—	—	—	—	—	1.7	3.4	106	1.9	
PCB-138 μg kg⁻¹	—	—	—	—	—	0.26	<0.15	3.4	0.21	

Woburn

	1942	1951	1960	1966/67	1972	1980	1984	1992
Phenanthrene μg kg⁻¹	17	110	340	250	130	160	150	
Benzo(a)pyrene μg kg⁻¹	17	18	30	38	45	39	35	
PCB-28 μg kg⁻¹	9.7	50	154	146	183	110	4.1	0.38
PCB-138 μg kg⁻¹	0.6	<1.2	4.5	3.9	5.5	2.7	2.3	1.0

— indicates not detected; space indicates not analysed.

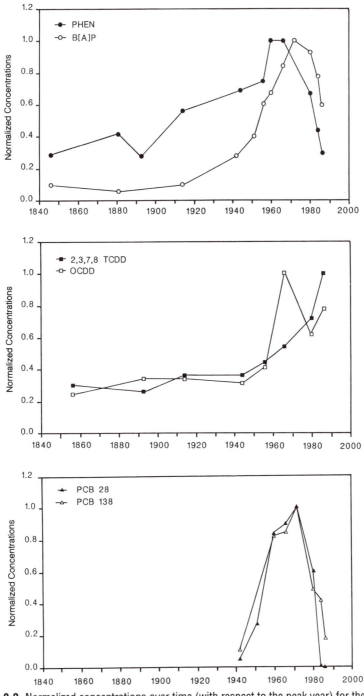

Fig. 9.2. Normalized concentrations over time (with respect to the peak year) for the six compounds studied. Data from both the Broadbalk and Woburn Market Garden experiments have been combined.

Table 9.2. Summary properties of the six selected compounds.

Compound	Volatilization potential[a]	Log octanol: water partition coefficient[b]	Susceptibility to biodegradation[c]
Phenanthrene	High	5.6	Yes
Benzo(a)pyrene	None	6.3	No
2,3,7,8-TCDD	Low	6.8	Very slow
OCDD	None	8.2	Very slow
PCB-28	High	5.6	Yes, but slow
PCB-138	Medium	6.7	Slow

[a] Criteria applied as defined in Wild and Jones (1992).
[b] Generally the higher the log Kow, the greater the strength of compound binding to soil. Data obtained from Mackay *et al.* (1991, 1992).
[c] Data summarized from Mackay *et al.* (1991, 1992).

coal at lower temperatures to high-temperature combustion in very large quantities at power stations (Jones *et al.,* 1989a). It is noteworthy, therefore, that, despite the reductions in input, the concentrations of B(a)P have declined only slightly in the soils at Broadbalk and Woburn (Table 9.1 and Fig. 9.2). As Table 9.2 suggests, however, once in the soil the potential loss pathways for B(a)P are very limited. In other words, B(a)P has a long residence time in soils (Wild *et al.,* 1990).

The continued increase in concentrations of 2,3,7,8-TCDD and OCDD throughout the 20th century again appear consistent with our knowledge of the likely inputs (Kjeller *et al.,* 1991) and their physicochemical properties (Table 9.1). PCDD/Fs can be formed and released from a variety of sources (Harrad and Jones, 1992) and it seems likely that contemporary inputs have remained high, notably from municipal waste incineration and coal combustion. The temporal trends of the PCBs in the two experiments most closely resemble that for phenanthrene (see Table 1). There is a clear peak in concentrations of both PCB-28 and PCB-138 in the 1960s and 1970s. Contemporary concentrations are much lower (Table 9.1). These trends are consistent with our knowledge of the production and use history of these compounds in the UK (Alcock *et al.,* 1993); the peak in worldwide usage was in the late 1960s (Harrad *et al.,* 1994).

One interesting aspect of the PCB time trend data is the rapid loss of these congeners from the soil since about 1970. Table 9.2 indicates that the likely predominant loss pathway for these compounds from soil is volatilization. In other words, transfer back to the atmosphere from where they came. In fact, the air–soil exchange can be regarded as a process of equilibrium partitioning (Mackay, 1991). PCBs are very 'dynamic' in the environment and may continue to re-cycle between air and soil (as well as water and air over long periods of time (Achman *et al.*, 1993) as they move towards equilbrium partitioning between different environmental compartments (Mackay, 1991). However, the kinetics of these exchange processes will presumably differ between congeners. PCB-138 is bound more firmly to soil organic matter than PCB-28 and will volatilize less readily because of its lower vapour pressure (see Table 9.2).

In summary, the PCDD/Fs and high molecular weight PAHs can be expected to reside in soils over many years/decades, firmly retained by soil constituents. In contrast, UK soils – such as those at Rothamsted and Woburn – now appear to be acting as a *source* emitting PCBs and low-molecular-weight PAHs back to the atmosphere. This is of significance for the global cycling of these compounds.

Implications of the Trends in Rothamsted Soil for the Global Cycling of PCBs

One fascinating feature of the time trend data is that ΣPCB concentrations (defined as the sum of several individual congeners measured by Alcock *et al.*, 1993) at Broadbalk have now apparently reverted to levels close to those in the mid-1940s, albeit with a somewhat different congener composition, with the heavier homologues constituting a greater proportion of the ΣPCB content. Clearly, manufactured Aroclor (commercially synthesized mixtures of PCBs) inputs now exert far less influence on ΣPCB levels at these sites than during the years of peak use. As an illustration of the scale of the reduction in levels of PCBs, if it is assumed that concentrations at Broadbalk and Woburn (and other sites studied by Alcock *et al.*, 1993) were/are representative of those in the UK generally (total surface area is 2.475×10^{11} m^2), that soil density averages 1300 kg m^{-3}, and that the soils were sampled to plough layer depth (23 cm), the ΣPCB burden of UK soils has declined from ca. 26,600 t to ca. 1500 t in the last two decades.

It is interesting to speculate on the ultimate fate of the PCBs lost from these soils. As already stated, the most likely loss process is volatilization. Biodegradation is likely to be a minor loss process because aerobic degradation is slow, especially for the more recalcitrant congeners (Abramowicz, 1990). Moza *et al.* (1979) showed that volatilization losses of ^{14}C-labelled tri, tetra-, and penta-chlorinated PCBs from soils was substantial and accounted for the majority of the compounds lost. Volatilization fluxes are temperature dependent (Hoff *et al.*, 1992) and may result in PCBs from temperate latitudes, such as the UK, migrating by 'cold condensation' processes to the sub-Arctic and Arctic regions (Fig. 9.3) where large concentrations have been observed far from local sources (Oehme, 1991).

If it is assumed that volatilization has been the only loss mechanism over the last 20 years, the above estimates of 26,600 t in UK soils in 1970 reducing to a burden of 1500 t in 1990 imply that the net annually averaged volatilization

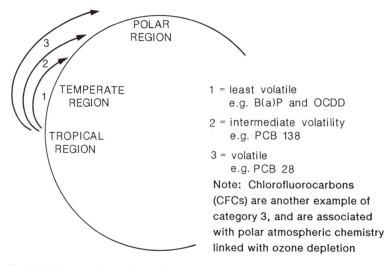

Fig. 9.3. Schematic illustration of the 'global fractionation' or 'cold condensation' hypothesis (after Wania and Mackay, 1993) In a process resembling distillation, organic compounds become latitudinally fractionated according to their volatility as they condense at different ambient temperatures.

flux of PCBs over this 20-year period has been $\sim$ 5 mg m^{-2} year^{-1} or about 14 μg m^{-2} day^{-1}. To put this in context, Achman *et al.* (1993) reported volatilization rates to the air above the highly contaminated Green Bay, Lake Michigan sediment water system of 0.013–1.3 μg m^{-2} d^{-1} in the height of summer. This is relevant, because it has been argued that the Great Lakes are a significant source of PCBs to the overlying air and hence to the Canadian Arctic (Swackhamer and Eisenreich, 1991). Clearly this implies that outgassing from soil is potentially a major contributor to the atmospheric burden of PCBs, which may account for the majority of PCBs entering UK air at the present time (Harrad *et al.*, 1994). Presumably PCBs entering UK air will be subject to long range atmospheric transport. It is therefore of considerable importance to establish the equilibrium partitioning between the soil and that in the overlying atmosphere. This will enable predictions of future trends to be made. The Rothamsted archive is proving useful in this respect too, as shown below.

The Possible Natural Production of PCBs

In recent years there has been some debate about whether PCBs existed in the environment prior to 1929 (i.e. prior to their industrial synthesis), perhaps produced in trace amounts by combustion processes, in the same manner as

PCDD/Fs (Harrad *et al.*, 1994). The archived soil collection has also provided an opportunity to investigate this issue.

Subsamples had been taken from the collection and transferred to Lancaster where they were milled for heavy metals analysis in 1985 (Jones *et al.*, 1987a). In the early 1990s, these subsamples were used for PCB analysis. Readily detectable levels of PCBs were present in the pre-1929 soils. In order to confirm these observations, a previously unopened (wax-sealed) glass sample jar containing soil from 1914 was subsampled in 1992 and immediately extracted and analysed by capillary gas chromatography with electron capture detection (GC-ECD). The chromatographic trace for this sample is shown in Fig. 9.4a, no PCB congeners were detected. A subsample of the 1914 sample was then spread out on an aluminium tray and left on the laboratory bench for 40 days. The chromatographic trace for this sample is shown in Fig. 9.4b. Exposure to the contemporary atmosphere led to detectable levels of PCBs in the soil after only a few hours. The air sampled in the laboratory contained 3.8 ng ΣPCB m^{-3}, somewhat more than has been measured in ambient inner city locations (Halsall *et al.*, 1993). Hence, the Rothamsted archive has been useful in showing that (i) there is little or no 'natural' production of PCBs and (ii) extreme caution needs to be exercised in preparing environmental samples containing low concentrations of PCBs for analysis if problems of sample contamination from the ambient atmosphere are to be avoided.

Preliminary Experiments on Air–Soil Equilibrium Partitioning of PCBs

By extending the experiment described in the previous section, it is possible to gain an understanding of the dynamics and mechanisms of air–soil exchange identified earlier as a controlling factor on the global cycling of PCBs. Equilibrium partitioning will vary between soils with different properties. In addition, desorption may be kinetically limited such that the characteristic biphasic nature of desorption may result in the PCB concentration of a soil with a previous history of exposure approaching a different 'equilibrium concentration' than a previously uncontaminated soil, when exposed simultaneously to the same air concentration. In other words, a proportion of the PCBs bound in soils may take a very long time to be desorbed (the 'recalcitrant fraction' in Fig. 9.5). This hypothesis is shown schematically in Fig. 9.5. Again, the archived soil collection provides a unique opportunity to investigate such processes and to move towards a predictive model for the relationship between air and soil concentrations. Such experiments are currently in progress at Lancaster University.

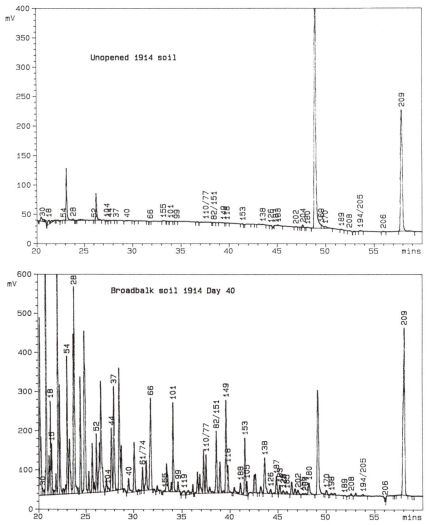

Fig. 9.4. Chromatographic traces of soil extracts analysed by GC-ECD (chromatographic conditions as described in Alcock *et al.* (1993)). Top: Extract of a 1914 soil taken from a sealed jar; bottom: Extract of the 1914 soil exposed to laboratory air for 40 days.

Implications of Long-term Changes for Transfer of Trace Organic Compounds Through the Foodchain

Obviously one of the main reasons for studying recalcitrant xenobiotics in the environment relates to concerns over foodchain contamination and potential human exposure. Human exposure to PAHs, PCBs and PCDD/Fs largely arises from dietary intake (as opposed to inhalation) with fatty foods such as dairy

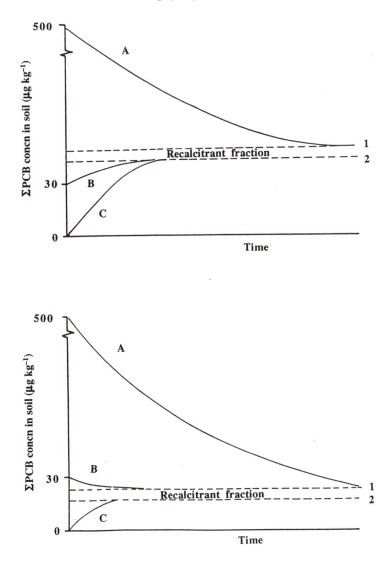

Fig. 9.5. Hypothetical response of PCB concentrations in a soil containing different starting concentrations of PCBs and then brought into contact with air of two different PCB concentrations. Top: 'Dirty' air (e.g. ΣPCB concentration of 5 ng m^{-3}); bottom: 'Clean' air (e.g. ΣPCB concentration of 0.1 ng m^{-3}). Broadbalk soil sampled in 1966 (A), 1993 (B) and 1914 (C) 1 = equilibrium concentration approached by 'net desorption'. 2 = equilibrium concentration approached by 'net adsorption'.

products and some plant-based foodstuffs making important contributions to total intake (Jones and Bennett, 1989; Duarte-Davidson and Jones, 1994). Livestock grazing on pasture grassland, with subsequent transfers of lipophilic compounds to animal fat and milk products is of significance (Fries, 1991). Although plant uptake of PAHs, PCBs and PCDD/Fs from soils is generally very inefficient (Wild and Jones, 1992), direct ingestion by grazing animals of surface soil which contains atmospherically deposited compounds can make a major contribution to total human exposure to PCBs and PCDD/Fs (Wild *et al.*, 1994). PAHs are generally metabolized by grazing animals (Wild and Jones, 1994). Clearly dietary intake of PCDD/Fs via this pathway can be expected to have increased through this century, and intake of PCBs will have changed markedly over time, given the changes in the soil burden (Fig. 9.2 above).

Concluding Remarks

In this chapter we have shown how archived samples from the Classical Experiments have helped us look back in time to gain information about chemicals/substances which could not be analysed until recently (indeed were unheard of a few decades ago). Some of these were present in the environment in the last century, others have only been manufactured this century and their release to and presence in the environment is comparatively recent. In addition, we have shown how the Rothamsted Classical Experiments can help the move towards making reliable predictions about the behaviour of chemicals on a regional scale in the future. In the process, we have demonstrated that the Rothamsted archive may contain samples which are unique in another respect. The old unopened bottles, unexposed to modern air, are uncontaminated with respect to chemicals which are ubiquitous in the contemporary environment. This is another example of how Lawes and Gilbert's foresight is still yielding insights into processes of fundamental importance which affect our modern environment.

Acknowledgements

Various research workers at Lancaster have been involved with the collaboration discussed here. We are grateful to Ruth Alcock, Vicky Burnett, Phil Haygarth, Fiona Nicholson, Gordon Sanders, Keith Waterhouse and Simon Wild for their input. Funding from the Natural Environment Research Council, the Agricultural and Food Research Council, the Ministry of Agriculture, Fisheries and Food and the Water Research Centre is also gratefully acknowledged.

References

Abramowicz, D.A. (1990) Aerobic and anaerobic biodegradation of PCBs: a review. *Critical Reviews in Biotechnology* 10, 241-251.

Achman, D.R., Hornbuckle, K.C. and Eisenreich, S.J. (1993) Volatilisation of polychlorinated biphenyls from Green Bay, Lake Michigan. *Environmental Science and Technology* 27, 75-87.

Alcock, R.E., Johnston, A.E., McGrath, S.P., Berrow, M.L. and Jones, K.C. (1993) Long-term changes in the polychlorinated biphenyl content of United Kingdom soils. *Environmental Science and Technology* 27, 1918-1923.

Duarte-Davidson, R. and Jones, K.C. (1994) PCBs in the UK population: estimated intake, exposure and body burdens for the general population. *Science of the Total Environment* (in press).

Fries, G.F. (1991) Organic contaminants in terrestrial foodchains. In: Jones, K.C. (ed.) *Organic Contaminants in the Environment: Environmental Pathways and Effects.* Elsevier Applied Science Publishers, Barking, Essex, pp. 207-236.

Halsall, C., Burnett, V., Davis, B., Jones, P., Pettit, K. and Jones, K.C. (1993) PCBs and PAHs in U.K. urban air. *Chemosphere* 26, 2185-2197.

Harrad, S.J. and Jones, K.C. (1992) A source inventory and budget for chlorinated dioxins and furans in the United Kingdom environment. *Science of the Total Environment* 126, 89-107.

Harrad, S.J., Sewart, A.S., Alcock, R.E., Boumphrey, R., Burnett, V., Duarte-Davidson, R., Halsall, C., Sanders, G., Waterhouse, K.S., Wild, S.R. and Jones, K.C. (1994) Polychlorinated biphenyls (PCBs) in the British environment: sinks, sources and temporal trends. *Environmental Pollution* 85, 131-147.

Haygarth, P.M., Cooke, A.I., Jones, K.C., Harrison, A.F. and Johnston, A.E. (1993) Long term change in the biogeochemical cycling of atmospheric selenium: deposition to plants and soils. *Journal of Geophysical Research* 98, D9, 16,769-16,776.

Hoff, R.M., Muir, D.C. and Grift, N. (1992) Annual cycle of polychlorinated biphenyls and organohalogen pesticides in air in southern Ontario. 2. Atmospheric transport and sources. *Environmental Science and Technology* 26, 276-283.

Johnston, A.E. and Jones, K.C. (1992) The cadmium issue: long-term changes in the cadmium content of soils and the crops grown on them. In: Schultz, J.J. (ed.) *Phosphate Fertilizers and the Environment.* International Fertilizer Development Center, Muscle Shoals, Alabama, USA, pp. 255-270.

Jones, K.C. (1988) Determination of polychlorinated biphenyls in human foodstuffs and tissues: suggestions for a selective congener analytical approach. *Science of the Total Environment* 68, 141-159.

Jones, K.C. (1991) Contaminant trends in soils and crops. *Environmental Pollution* 69, 311-325.

Jones, K.C. and Bennett, B.G. (1989) Human exposure to environmental polychlorinated dibenzo-*p*-dioxins and dibenzofurans: an exposure commitment assessment for 2,3,7,8-TCDD. *Science of the Total Environment* 78, 99-116.

Jones, K.C. and Johnston, A.E. (1989) Cadmium in cereal grain and herbage from long-term experimental plots at Rothamsted, UK. *Environmental Pollution* 57, 199-216.

Jones, K.C. and Johnston, A.E. (1991) Significance of atmospheric inputs of lead to grassland at one site in the United Kingdom since 1860. *Environmental Science and Technology* 25, 1174-1178.

Jones, K.C., Symon, C.J. and Johnston, A.E. (1987a) Retrospective analysis of an archived soil collection. I. Metals. *Science of the Total Environment* 61, 131-144.

Jones, K.C., Symon, C.J. and Johnston, A.E. (1987b) Retrospective analysis of an archived soil collection II. Cadmium. *Science of the Total Environment* 67, 75-89.

Jones, K.C., Stratford, J.A., Waterhouse, K.S., Furlong, E.T., Giger, W., Hites, R.A., Schaffner, C. and Johnston, A.E. (1989a) Increases in the polynuclear aromatic hydrocarbon content of an agricultural soil over the last century. *Environmental Science and Technology* 23, 95-101.

Jones, K.C., Grimmer, G., Jacob, J. and Johnston, A.E. (1989b) Changes in the polynuclear aromatic hydrocarbon content of wheat grain and pasture grassland over the last century from one site in the UK. *Science of the Total Environment* 78, 117-130.

Jones, K.C., Symon, C.J., Taylor, P.J.L., Walsh, J. and Johnston, A.E. (1991a) Evidence for a decline in rural herbage lead levels in the UK. *Atmospheric Environment* 25A, 361-369.

Jones, K.C., Burnett, V., Duarte-Davidson, R. and Waterhouse, K.S. (1991b) PCBs in the environment. *Chemistry in Britain* 27, 435-538.

Jones, K.C., Jackson, A. and Johnston, A.E. (1992a) Evidence for an increase in the cadmium content of herbage since the 1860s. *Environmental Science and Technology* 26, 834-836.

Jones, K.C., Sanders, G., Wild, S.R., Burnett, V. and Johnston, A.E. (1992b) Evidence for a decline of PCBs and PAHs in rural vegetation and air in the United Kingdom. *Nature* 356, 137-140.

Kjeller, L-O., Jones, K.C., Rappe, C. and Johnston, A.E. (1991) Increases in the polychlorinated dibenzo-*p*-dioxin and -furan content of soils and vegetation since the 1840s. *Environmental Science and Technology* 25, 1619-1627.

Mackay, D. (1991) *Multi-media Environmental Models: The Fugacity Approach.* Lewis Pub.Co., Michigan, USA.

Mackay, D., Shiu, W.Y. and Ma, K.C. (1991) *Illustrated Handbook of Physical-Chemical Properties and Environmental Fate for Organic Chemicals. Vol. 1. Monoaromatic Hydrocarbons, Chlorobenzenes, and PCBs.* Lewis, Michigan, USA.

Mackay, D., Shiu, W.Y. and Ma, K.C. (1992) *Illustrated Handbook of Physical-Chemical Properties and Environmental Fate for Organic Chemicals. Vol. 2.* Lewis, Michigan, USA.

Moza, P., Weisgerber, I. and Klein, W. (1979) Studies with 2,4′,5-trichlorobiphenyl-^{14}C and 2,2′,4,4′,6-pentachlorobiphenyl-^{14}C in carrots, sugar beets, and soil. *Journal of Agricultural Food Chemistry* 27, 1120-1124.

Nicholson, F.A., Jones, K.C. and Johnston, A.E. (1993) The effect of phosphate fertilizers on long-term changes in the cadmium content of crops and soils in the UK. In: *International Conference on Heavy Metals in the Environment,* Toronto, September 1993.

Oehme, M. (1991) Dispersion and transport paths of toxic persistent organochlorines to the Arctic - levels and consequences. *Science of the Total Environment* 106, 43-53

Rothbaum, H.P., Goguel, R.L., Johnston, A.E. and Mattingly, G.E.G. (1986) Cadmium accumulation in soils from long-continued applications of superphosphate. *Journal of Soil Science* 37, 99-107.

Sanders, G., Jones, K.C., Hamilton-Taylor, J. and Dörr, H. (1993) Concentrations and deposition fluxes of polynuclear aromatic hydrocarbons and heavy metals in the dated sediments of a rural English lake. *Environmental Toxicology and Chemistry* 12.

Swackhamer, D.L. and Eisenreich, S.J. (1991) Processing of organic contaminants in lakes. In: Jones, K.C. (ed.) *Organic Contaminants in the Environment*. Elsevier Applied Science Publishers, Amsterdam, pp. 33-86.

Wania, F. and Mackay, D. (1993) Global fractionation and cold condensation of low volatility organochlorine compounds in polar regions. *Ambio* 22, 10-18.

Wild, S.R. and Jones, K.C. (1992) Organic chemicals entering agricultural soils in sewage sludges: screening for their potential to transfer to crop plants and livestock. *Science of the Total Environment* 119, 85-119.

Wild, S.R. and Jones, K.C. (1993) Biological and abiotic losses of polynuclear aromatic hydrocarbons (PAHs) from soils freshly amended with sewage sludge. *Environmental Toxicology and Chemistry* 12, 5-12.

Wild, S.R. and Jones, K.C. (1994) Polynuclear aromatic hydrocarbons in the United Kingdom environment: a preliminary source inventory and budget. *Environmental Pollution* (in press).

Wild, S.R., Waterhouse, K.S., McGrath, S.P. and Jones, K.C. (1990) Organic contaminants in an agricultural soil with a known history of sewage sludge amendments: polynuclear aromatic hydrocarbons. *Environmental Science and Technology* 24, 1706-1711.

Wild, S.R., Harrad, S.J. and Jones, K.C. (1994) The influence of sewage sludge amendment to agricultural land on human exposure to polychlorinated dibenzo-*p*-dioxins and -furans. *Environmental Pollution* 83, 357-369.

10 Statistics and the Long-term Experiments: Past Achievements and Future Challenges

V. BARNETT

Institute of Arable Crops Research, Rothamsted Experimental Station, Harpenden, Hertfordshire AL5 2JQ, UK.

Introduction

The theme of this volume is the long-term experiments (the 'Classicals') at Rothamsted. They were started in the 1840s: a crucial time throughout the world in so many respects. England was riven over the repeal of the Corn Laws, whereas France was torn apart with the 'Miserables' revolution in Paris. North America was in turmoil in its efforts to implement the union of the states; the Republic of Texas was annexed and war raged with Mexico. Karl Marx was seeking to set the world to rights with his communist manifesto.

In contrast, science and technology were making great strides. Charles Babbage had, a few years earlier, proposed his 'analytical engine' as a first move towards automated computing. As the decade progressed, Joule advanced the principle of the conservation of energy, Kelvin suggested a new approach to taking temperatures, Foucault measured the speed of light, Boole explained how to be logical and Justus Liebig showed that chemical reactions were ubiquitous and held the answers to so many questions. Someone noticed Neptune in the sky, another put people out with ether and Charles Singer produced the first major consumer appliance.

In the midst of such a maelstrom, agricultural research was in its infancy when John Bennet Lawes began his pioneering work on the use of chemical fertilizers to improve crop yield and initiated in 1843 the first long-term experiments (with wheat on Broadbalk and turnips on Barnfield) as the vital empirical base for his studies.

©CAB INTERNATIONAL, 1994. From R.A. Leigh and A.E. Johnston (eds), *Long-term Experiments in Agricultural and Ecological Sciences.*

The science of Statistics was also hardly a gleam in anyone's eye, notwith-standing the establishment in 1834 of the Statistical Society of London by the Belgian, Adolphe Quetelet. The Society motto *aliis exterendum* (inscribed under a sheaf of wheat) expressed the prevailing view: that statistical data should be collected and assembled but should be 'threshed by others'. Statistical inference and interpretation were thus no part of the contemporary science. Statistics was little more than what Karl Pearson in the 1920s described as 'shop arithmetic' (Pearson, 1978) and in spite of mention of a primitive notion of correlation (Bravais, 1846) no serious attempts at conceptualization, inferential statistical reasoning or statistical methods were to see the light of day for 50 years or so.

Reference to 'agriculture' is not entirely absent from the statistical literature of the time as represented by the *Journal of the Statistical Society*, but such riveting themes as 'prejudice against the use of agricultural machinery' and 'the religious improvement of agricultural workers' were restricted at most to the presentation of relevant tabulated data.

Over the ensuing years, Rothamsted was to become a focus of agricultural research. It also became a stimulus for vital elements in the fields of statistical principles and methods starting with the uniquely important developments arising from the Rothamsted Statistics Department, which was established in 1919, before any University department of Statistics.

This chapter will explore this dual aspect of progress arising from the interaction of the 'Classicals' with the new-born science of Statistics from the turn of the century – and assess the present situation. Does statistics have more to contribute to our interpretation of the 'Classicals' – do the 'Classicals' in turn have challenges still for the sophisticated panoply of modern statistical methods?

The Early Years

Lawes was much influenced by the work of Liebig on the centrality of organic chemistry to an understanding of the nutrition of plants and animals, but it was from his tutor at Oxford, C.G. Daubeney that he received the most direct stimulus for his work on superphosphates as an agricultural fertilizer. In setting out the long-term experiments he saw these as the raw material on which to test the effects on crop development not only of chemical agents but of climatological factors. The results from these experiments were clearly intended to provide a statistical basis for investigating such influences – reflecting Lawes' dual prescience in the importance of the two infant sciences of agricultural chemistry and of statistics for advancing agriculture.

From the outset, he collected year-by-year the most detailed data on yields, crop diversity, soil properties, weather conditions augmented by physical samples of the soils, seeds and other substances: a process that continues to the present day. A recent biography (Dyke, 1993) carefully chronicles Lawes' work over the half-century of his guardianship of Rothamsted. Lawes clearly regarded climate as of major importance; evidenced for example in the pioneering work of Lawes and Gilbert (1871) on how crop yields on the 'Classicals' depended on rainfall and temperature.

The real extent and nature of climatological effects remains important but elusive even to the present day. Statistical investigations have always attracted attention, as in the work of Hooker (1907) on effects of residual temperature above 42°F and the extensive studies by Fisher, Wishart and others through the 1920s and 1930s using Rothamsted data (e.g. Fisher, 1924 on the influence of rainfall on wheat yield, Wishart and Mackenzie, 1930). Later, Cashen (1947) sought to relate yield and species abundance on Park Grass to rainfall levels – finding influence on yield but less effect on the biodiversity. Other crucial work was by Smith (1960) and Buck (1961).

Wider reference was introduced; to sunshine, windspeed, soil moisture deficit, etc. Although the statistical models and methods have become more and more sophisticated with time, we are still far from really understanding the full effect of climate, as we shall see later in terms of some very recent efforts.

An intriguing aspect of the earlier studies on crop yield *per se* is found in the early work of Fisher (1921). Having developed means of separating different variational components, he set out to apply this approach in detailed studies of the wheat yields over the initial period of about 70 years on Broadbalk. He exhibited a strange and still unexplained pattern of behaviour in the form of slow long-term cyclic variations across all fertilized plots even after he had made allowance for the 'annual effects' of weather (including rainfall) and for an apparent steady deterioration over the years. The patterns showed relative maxima in about 1860 and 1900 and relative minima around 1880 and 1915. To investigate this soundly, he introduced the idea of orthogonal polynomials (to remove the effects of correlation) and applied his newly developed notions of statistical significance and analysis of variance. The 'slow-changes' effects were highly significant even compared with the 'annual causes', and oddly were not apparently present on the other 'Classicals'. This remains a bizarre and unexplained phenomenon. We might ask if it is still evident in the yields for the second 70 years of results on Broadbalk.

Rothamsted and Statistics

Joan Box wrote a book (Box, 1978) about her father, entitled 'R.A. Fisher: The Life of a Scientist'. She remarked of him that 'In 1919 he was hired as statistician at Rothamsted Experimental Station, where the work he had already done was immediately applicable': a sublime understatement of the significance of Fisher and of Rothamsted in the development of statistical science.

We note in Lawes' election as a Fellow of the Royal Statistical Society in 1880, and subsequent service on its Council, that Rothamsted was present at the very dawning of statistics but it was Sir John Russell (Director 1911–43) who formalized the link after the First World War by seeking a bright young Oxbridge graduate 'who would be prepared to examine our data and elicit further information that we had missed'. In the event he identified the unemployed and allegedly flighty Ronald Aylmer Fisher whom he was told could be first class if he only 'stuck to the ropes'. Russell ventured £200 for Fisher to stay as long as 'that should suffice' and to advise on whether there was much that required 'proper statistical examination'. He soon discovered that Fisher not only had great ability but 'was in fact a genius who must be retained' (Russell, 1966).

This was indeed a happy conjunction. The Rothamsted Statistics Department was one of the earliest departments in the world, set up at a time when the subject was poised for rapid formal development. It had as its first Head of Department someone who was stimulated by the detailed empirical base of the long-term experiments and who is now seen as the person who in so many respects 'laid the foundations, coined the language and developed the methods' on which the rich complexity of modern statistical science is based. Of course he was not alone in this crusade and names of 'Student' (W.S. Gosset, the statistician of the Guinness company) and of Pearson and Neyman must be included in this phrenetic development phase.

Fisher's interests were wide; they encompassed vast areas of biological science and of genetics and eugenics, all of which added to the stimulus of the empirical resource in Rothamsted to promote his rapid expansion of research in statistical concepts and methods. His major contributions included the crucial notion of efficiency (as a means of assessing how good a statistical answer could be) and the method of maximum likelihood for constructing formal estimators of population parameters and determining their accuracy. His genetical and eugenic interests led him, *inter alia*, to unravel the mystery of the Rhesus factor in human blood and to investigate statistically the case for and against natural selection. He made major contributions to curve fitting of data, refining notions of regression and analysis of variance, urged on by the apparent long-term trends in the Rothamsted meteorological data and crop yields – a topic of great current concern in the global warming debate as will be considered later.

As has been remarked, much of Fisher's early activities at Rothamsted focused on the 'long-term' data, starting with the Broadbalk wheat yields and leading to many important publications on 'crop variation' and to the development over a 7-year period of major innovative discoveries on analysis of variance and the vitally important topic of experimental design. The Park Grass data were also a major formative stimulus, particularly in the intriguing changes in the species composition of the swards due to the treatment over the years.

Fisher's contributions to the design of experiments in the 1920s were most influential. At that time interest was spreading in the use of statistical methods in different fields of application. In particular, the need was being expressed for statistical procedures to investigate how industrial and chemical processes reacted to changes in controlling variables such as temperatures or pressures. The prevailing attitude was that readings should be taken at different temperatures to examine the effect of temperature, then at different pressures to see how pressure affected the process, and so on.

With hindsight, we see the immediate dangers of such a 'one factor at a time' approach. Fisher demonstrated this clearly, introducing the notion of multifactor, or factorial, experiments where all factors were allowed to vary within the same experiment and for which the concept of interaction measured the way effects of one factor were influenced by the other factors. This was only part of the story. The notion of statistical significance made it possible to answer the question of whether some observed effect was statistically meaningful, or could have arisen by chance, and promoted the importance of the principle of randomization. The analysis of variance provided a basis for separating out the effects of different factors, of their interactions, and of residual variation components.

The fact that the long-term experiments were of a comparative form led to the vital idea of blocking as a means of reducing extraneous (but systematic) variation. However, it is highly likely that the influence of the long-term experiments on statistical design and analysis of experiments was even more central. It could not have escaped Fisher's notice that the multi-crossed-factor feature was present from the outset – 70 years before it was formalized and developed within statistical methodology. Many of the 'Classicals' show this feature clearly. With Park Grass, for example, we have the factor nitrogen (at six levels) in combination with five other two-level factors (phosphorus, potassium, sodium, magnesium and farmyard manure) on the usual convention that one of the levels is 'absence' of that treatment. So it had the features of a 6×2^5 design and became even more sophisticated when lime was introduced, creating a 'split-plot' element to the design. Of course not all factor combinations were present (see Fig. 10.1), and there was no randomization and (almost) no replication, but many of the essential features of the designed experiment were clear to see (if one had the wit to recognize them).

Interaction and blocking were vital concepts not only for field work but for all designed experiments in any practical field, and their form needs to be well understood and appreciated. Let us start with interaction. For example, suppose we conduct a field experiment in which we have two wheat varieties,

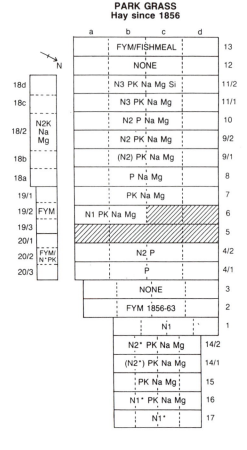

PARK GRASS
Hay since 1856

Fig. 10.1. The plan of Park Grass. Treatments (every year except as indicated).
Nitrogen (applied in spring): N1, N2, N3, sulphate of ammonia supplying 48, 96, 144 kg N ha⁻¹; N1*, N2*, nitrate of soda supplying 48, 96 kg N ha⁻¹ (N2), (N2*), last applied 1989. *Minerals* (applied in winter): P 35 kg; K 225 kg; Na 15 kg; Mg 10 kg ha⁻¹; Si Silicate of soda at 450 kg ha⁻¹ of water soluble powder; Plot 20, Rates of manuring in years when FYM not applied: 30 kg N*, 15 kg P, 45 kg K ha⁻¹. *Organics* (each applied every fourth year since 1905): FYM, 35 t ha⁻¹ farmyard manure (bullocks) (1989, 1993); fish meal (about 6.5% N) to supply 63 kg N ha⁻¹ (1991, 1995). *Lime*: a,b,c, Lime applied as needed to maintain pH 7,6 and 5 respectively; d, No lime applied (pH range 3.5 (plot 11–1) to 5.7 (plot 17).

A and B, and two levels N_1 and N_2 of nitrogenous fertilizer. The results of the experiment might look like this:

| | | \multicolumn{2}{c}{Level of N} | |
		N_1	N_2	mean
	A	64	82	73
wheat variety	B	80	62	71
	means	72	72	72

Looking at either factor separately, the mean responses 73 and 71 for wheat varieties A and B, and 72 and 72 for nitrogen levels N_1 and N_2, suggest no effects on single factor bases. But this is obviously not so; variety A shows a clear nitrogen effect (64, 82) as does variety B (80, 62). It is merely that these effects are different for the two varieties. This is precisely what is meant by interaction, and is the crucial component of a cross-factor design.

Blocking is readily illustrated by one of Fisher's early experiments on which varieties of trees were best for withstanding exposure to the elements. The experiment was conducted on a mountain in Wales. It was clear that systematic effects could arise which did not directly reflect the intrinsic robustness of the tree varieties. For example, trees planted higher on the mountain, or those at edges of the stand, would be more exposed. To cater for such systematic differences, a design was used in which each variety appeared equally often at each height and in each lateral position. Although this would 'even out' the effects of height and lateral position it would increase the apparent variability and possibly conceal the true effect of variety. Thus two extra factors (called blocking factors) for height and lateral position should be directly measured and allowed for, so as to most clearly enable any variety effect to be exhibited. With five varieties, a 5×5 square design (with five heights and five lateral positions) allows this to be done economically. The five varieties are allocated at random but subject to the constraint that each variety A, B, C, D and E appears once in each row (height) and in each column (lateral position).

The design – a randomized 5×5 Latin Square – could appear as follows:

C	D	E	A	B
E	A	B	C	D
B	C	D	E	A
A	B	C	D	E
D	E	A	B	C

Such a design is blocked for height and lateral position, i.e. allows the effects of these extraneous systematic influences to be measured and eliminated for better examination of the factor of main interest: tree variety.

In terms of statistical methodology, the 'Classicals' continue to provide a major challenge as a unique long-term basis for highly complex multivariate environmental statistical analysis (see Barnett, 1993), even when viewed in the context of present-day sophisticated statistical methodology. This is not only because of the lack of balance, or incompleteness, of the crossed-factor statistical design of the experiments. In fact, it is the apparent advantage of constancy of treatment over the long period that causes the major difficulty. We essentially have no randomization or replication (this being confounded with time). Rather than a designed experiment it is necessary to view the data

from the Classicals as high-dimensional time-series data with nearly 150 observations (say, for Park Grass) each of which is, in terms of yields, of dimension
about 80, with an internal quasi-designed experiment structure. The problems
of analysing such complex multivariate time series have not yet been solved,
although work is progressing in this field (e.g. Young and Minchin, 1991).

Global Warming and the Greenhouse Effect

Global warming and the greenhouse effect are matters of current scientific
interest and social concern. Two particular manifestations of such changes are
increases in greenhouse gases (such as atmospheric CO_2) and in temperatures;
admittedly on vastly different scales of about 50 ppmv (c. 20%) for CO_2, and of
about 0.5°C (c. 2.5%), per century.

The wide-ranging research on global warming falls into various categories.
Much of it takes for granted the climate changes predicted by certain models
(such as General Circulation Models, GCMs) and explores the possible effects
of such changes; crop development studies in controlled environment experiments are of this type. More direct approaches seek to answer the questions:

1. can we observe the affects of global warming directly in the physical
environment (e.g. in crop growth on the 'Classicals')?
2. can we confirm that the greenhouse effect is really operating, e.g. that global
warming is taking place?

Thus we have a trichotomy of interests:

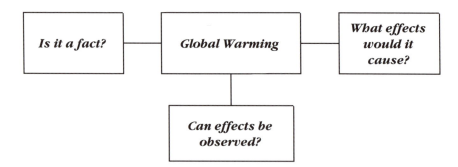

These questions are crucial, and give the 'Classicals' yet another boost of
importance in the contemporary situation. Many environmentally important
matters can now be addressed using their unique historical resource – how soil
and seeds, or yield and biodiversity, have changed over the last 150 years, and
how such changes might relate to climatological factors.

Rothamsted has collected its own comprehensive meteorological data some of it over long periods and all data relating to the 'Classicals' are now being assembled in a sophisticated computer database called ERA (Electronic Rothamsted Archive) which already contains the meteorological and the Park Grass data.

The Meteorological data in ERA have not yet been fully analysed in the spirit of question (2) above, to see if they provide confirmation of global warming, but some work has been done on this general issue using other data sets with some ambiguous results. Smith (1993) reports on a major study using three important data sets providing temperatures for Central England from 1659 to 1989 (the UK data), for 95 stations over the USA from 1890 to the present time (the US data) and IPCC (Intergovernmental Panel on Climate Change) data for both hemispheres from 1854 (the global data). Work on the GCMs predicts various values for the temperature rise that might occur from a doubling of atmospheric CO_2 levels under idealized conditions – the values range from c. 2°C to c. 5°C, although the IPCC has adopted 2.5°C as its best estimate.

A rise of about 20% in atmospheric CO_2 over a century would support a temperature rise of about 0.5°C, a commonly assumed value. But has this happened? There is some evidence of increases in the range 0.3-0.6°C but the patterns are far from consistent from one location to another, and the rise is far from steady (with apparent rapid increases in some periods, and even falls in others). Smith (1993) seeks to find powerful and efficient models which can resolve such uncertainties and ambiguities. These are needed because we are trying to detect statistical significance of small effects in the presence of large variation – a notoriously difficult situation to handle.

Models for temperature variations have often taken the form of stationary time series in the ARMA (autoregressive moving-average) family fitted to deseasonalized data. Thus if

$$\{Y_n; n = 1,2, ...,N\}$$

is a set of N consecutive (deseasonalized) temperature readings, we are interested in whether they can be regarded as coming from a stationary ARMA (Autoregressive Moving-Average) time series with constant mean value μ (i.e. no trend over time).

Such an ARMA time series

$$\{Y_n; n = 1,2, ...,N\}$$

depends on $(p + q)$ parameters (θ, Φ), taking the form

$$(Y_n-\mu) = \phi_1(Y_{n-1}-\mu) + ... + \phi_p(Y_{n-p}-\mu) + a_n - \theta_1 a_{n-1} ... - \theta_q a_{n-q} \qquad (10.1)$$

where the series $\{a_n\}$ consists of independent random components. A time series is characterized by its autocovariance function

$$\gamma_k = E\{(Y_n - \mu_n)(Y_{n+k} - \mu_{n+k})\}$$

which for an appropriate ARMA process typically decays exponentially. Equivalently, the process can also be represented by its spectral density

$$f(\omega) = \frac{1}{2\pi} \sum_{n=-\infty}^{\infty} \gamma_n e^{-n\omega} = \left\{\gamma_o + 2\sum_{n=1}^{\infty} \gamma_n \cos(n\omega)\right\}/2\pi$$

which usually provides a more convenient form for statistical processing.

Many attempts to demonstrate global warming using such models have been equivocal due to good fit but large variability. This, and the many hints of long-term dependence (see above) led Smith (1993) to employ a different form of time-series model with spectral density

$$f(\omega) \sim \omega^{1-2H} \text{ as } \omega \rightarrow 0 \ (0.5 < H < 1) \tag{10.2}$$

which brings in directly the low-frequency (long-term dependence) characteristics. He proceeds to examine the UK, US and global data sets for upward trend ('global warming') in the presence of long-term dependence, in the hope that this might provide a more efficient approach to estimating the trend. Some evidence of long-term dependence is found, with correspondingly more efficient estimation of the trend, but the outcomes are still ambiguous, with estimates of 100-year rises ranging from 0.21°C to 0.87°C which were none-the-less not statistically significant for the UK or the US data overall. For the global data, long-term dependence is also evident, but an estimated 100-year rise of 0.4°C is now in fact significant.

It would be most valuable to conduct such model-based analyses on the Rothamsted temperature (and other) data, and plans are in train to do so. Figure 10.2 (adapted from Chmielewski and Potts, 1994) shows annual average maximum and minimum air temperatures from 1850.

Returning to question (1) (p. 172), we might well consider the evidence from the 'Classicals': in particular, does yield reflect the acknowledged increase of about 20% in atmospheric CO_2 over the last 100 years or so? (Fig. 10.3; Boden *et al.*, 1990). This is precisely the question posed by Jenkinson *et al.*, (1994) in the more specific form: having allowed for any effects of climate, do the yields on the unmanured plots of Park Grass echo the increase in atmospheric CO_2 levels? (Two fertilized plots, and some measures of biodiversity were also considered.)

We considered above some of the early work on climate effects using data from the Classicals. Jenkinson *et al.* (1994) examine in detail the yields for two unfertilized plots (3d and 12d) on Park Grass. As shown in Perry *et al.* (1993) they show marked correlation of yields, dominance of one plot over the other (12d tending to produce the higher yields, and a change of structure from 1960 onwards (when the harvesting method was changed) (see Fig. 10.4; from

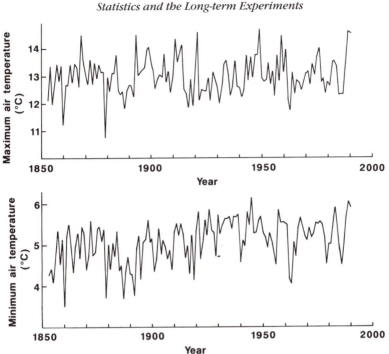

Fig. 10.2. Maximum (a) and minimum (b) air temperatures at Rothamsted from 1878 and augmented to 1856 by estimating the 1856–77 values using data from Oxford and the relationship between Rothamsted and Oxford values after 1878. (Adapted from Chmielewski and Potts, 1994.)

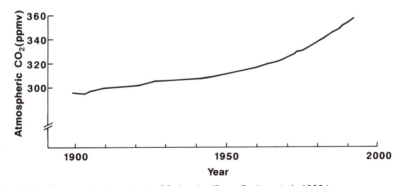

Fig. 10.3. Changes in atmospheric CO_2 levels. (From Boden *et al.*, 1990.)

Perry *et al.*, 1993). Before looking for evidence of increase of yield with increase in atmospheric CO_2 – mooted by Gifford (1992) and others, and predicted from short-term controlled environment experiments to be likely to lead to c. 6–7% yield increase for a 20% increase in CO_2 levels – it was necessary

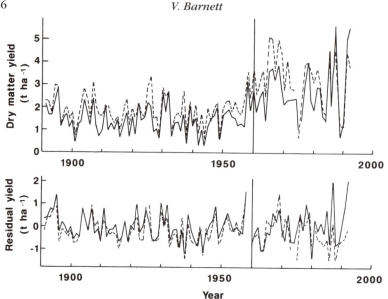

Fig. 10.4. Park Grass yields on those parts of the two unfertilized plots, 3 and 12, which have never received any $CaCO_3$ and the residual yield after fitting the regression model described in equation 10.3 (method of estimating dry matter yields changed in 1959, see text). ——— Plot 3, - - - - - Plot 12.

to seek to take out the effects of climate on yield. A change-point regression model was fitted to total yield in the form

$$y_t = \alpha_{i(t)} + \beta_{i(t)}s_t + \gamma_{i(t)}r_t + \eta c_t + \epsilon_t \qquad (10.3)$$

with $\quad i(t) = \begin{cases} 1, t \leqslant 1960 \\ 2, t > 1960, \end{cases}$

where y_t is total yield in the year t, s_t is total sunshine in April and May, r_t is total rainfull in June, July and August and c_t is the average atmospheric CO_2 concentration (from Boden *et al.*, 1990). We want to estimate the parameters α_1, β_1, γ_1, ; α_2, β_2, γ_2 and η representing the effect of CO_2 in the combined data set.

Some interesting information arose from fitting the model (10.3). The residuals, plotted for both plots in Fig. 10.4 still show the same high degree of cross correlation found in the yields themselves. This is remarkable and indicates that some strong common explanatory factor remains unaccounted for. Dates of harvest are an obvious candidate, but further analysis does not show this to be so and we are left with an important anomaly (see also Barnett, 1993).

What then of the effect of atmospheric CO_2 on yield? Again, no clear effect is evident. There was no statistically significant evidence of an increase in herbage yields with atmospheric CO_2.

It is important to place in perspective these various negative results of global warming and the greenhouse effect under the categories (**1**) and (**2**).

We have noted that the increase in greenhouse gases is not in dispute. Its possible effects have prompted a great deal of important 'what if' research. The questions of climate change and its observed effects are of different status and we have noted above the paucity of statistically significant observations of these matters. This is not to say that temperatures are not rising and consequential influences on crop development are not arising. What is happening is that the relatively large natural variation does not at present allow even the sophisticated statistical models and methods to confirm the expected patterns. At the same time until such confirmation can be achieved we must act as if 'the jury is still out'!

Sustainability

A topic which is currently of great concern to environmentalists, scientists and even politicians is that of *sustainability*, which is essentially the ability to maintain agricultural output in quantity and quality year after year without degradation; particularly in relation to environmental concerns. The importance of sustainability is not in doubt; what is less well developed is how precisely this concept should be defined and more specifically how it should be quantified (measured) in the practical situation.

Again the 'Classicals' provide vital potential information on this issue. After all, most of them have been successfully maintained over a very long time; other experiments, in contrast, after being continued for different periods were terminated for various reasons. How are these to be differentiated in terms of their defining characteristics of crop type, soil features, costs of operation, cultural practices, and so on. This was precisely the question posed by the Rockefeller Foundation in New York, who commissioned six long-term-experiment sites to conduct major research studies with the brief to produce

1. an operational definition of sustainability
2. a statistical/economic way of measuring it based on their experiences with their long-term experiments.

Inevitably, Rothamsted was a principal site with others in Canada, India, the Phillippines and the US. Each site worked for 6–9 months using a multidisciplinary team for detailed statistical and economic analysis. The reports from each team provided the basis for an International Working Party at Rothamsted in Spring 1993, and will form the core of a future publication (Barnett *et al.*, 1994b).

The Rothamsted study (see Barnett *et al.*, 1992, for details) used a dozen or so plots from three of the 'Classicals' chosen to span the range of what might be expected to be different levels of sustainability (or non-sustainability). A substantial database was developed with yield and culture variables augmented by extensive economic information (including prices of output products, of soil treatments, of labour, machinery and land-use) stretching back to the middle of the last century. This database is a vital resource in its own right for further research, and has been assimilated into ERA.

It is clear that any measure of sustainability will show statistical variation from year to year and needs to be expressed in quantitative (economic) terms. A detailed statistical/economic examination was given by Barnett *et al.* (1994a). Essentially we need to contrast output value with input cost. The principal output is the yield of the relevant agricultural product (e.g. wheat or turnips) for the year in question. The inputs are manifold: quantities and costs of seeds, machinery, labour, soil treatments (fertilizers, pesticides, etc.), cost of land use. Additionally there are many environmental costs ('externalities') relating to such matters as degradation, pollution, transfer, etc., which will be far less tangible, and difficult to cost. As examples, we have nitrate leaching or soil erosion.

Thus we need a means of measuring the relationship between output value and input costs, and the many ideas and methods of economic indexes can be considered for this purpose. Armed with an appropriate measure we adopted the stance that an agricultural system is sustainable if, in statistical terms and in the long run

1. output value exceeds input costs
2. the measure of relative value is non-decreasing over time.

Monetary value was an obvious common standard for measurement and the comparison of output value to input costs was carried out in terms of total (social) factor productivity *TFP* (see Lynam and Herdt, 1989) defined as the ratio of aggregated output to aggregated input in the form

$$TFP(t) = \frac{\sum\limits_{j=1}^{m} P_{jt} Q_{jt}}{\sum\limits_{i=1}^{m} W_{it} X_{it}} \quad (t = 1, 2, ..., T) \qquad (10.4)$$

where X_{it} is the quantity and W_{it} the price of input factor i used at time t and Q_{jt} is the quantity and P_{jt} the price of output factor j used at time t.

The quantity *TFP(t)* is an appropriate basis for comparing output value and input cost in criterion **1** above.

To contrast the situation at two timepoints r and s, an arithmetic index *TFP(r;s)* is considered: formed as the ratio of *TFP(r)* and *TFP(s)* using constant prices. Thus using prices at time s, we have

$$TFP(r;s) = \frac{\sum P_{js} Q_{jr} / \sum P_{ks} Q_{ks}}{\sum W_{js} X_{jr} / \sum W_{ks} X_{ks}} = \frac{output\ index}{input\ index} \qquad (10.5)$$

Both the numerator and the denominator indexes can be re-expressed as linear combinations of factor shares F_{js} (based on prices at time s) and ratios of quantities.

For example, the output index is

$$\text{output index} = \frac{\sum\limits_{j=1}^{m} P_{js} Q_{jr}}{\sum\limits_{k=1}^{m} P_{ks} Q_{ks}} = \sum_{j=1}^{m} \frac{P_{js} Q_{js}}{\sum\limits_{k} P_{ks} Q_{ks}} \cdot \frac{Q_{jr}}{Q_{js}} = \sum_{j=1}^{m} F_{js} \cdot \frac{Q_{jr}}{Q_{js}} \qquad (10.6)$$

where F_{js} is the proportion of value due to each of the output factors at time s. The input index is defined similarly. Thus the factor shares provide the appropriate basis for comparing values of different components.

For criterion **2** above we need a relative index measure of general form (10.5), but spanning the whole time period of study (not just restricted to two time points). There are various choices. We can use constant prices for the whole period, e.g. those prevailing at the beginning or at the end. However, the period of study is long (more than 100 years perhaps) and price comparisons may well have changed noticeably over the period. To take account of this we employed the principle of chainlinking often used in calculating economic indexes from long time series. The particular form which seemed most appropriate was to divide our time series into a sequence of subseries of similar periods of time (9 years in this case, using average prices within each subseries and holding quantities constant at the end time point) and then chainlinking subseries to construct the relevant index, both for input and output components.

This yielded our relative output/input index *I* for considering the criterion (**2**) above, although full-period indexes using initial or final prices provided little difference in their implications.

The values of *TFP* and *I* were determined for each of the chosen experimental plots over the periods of their use (from about 1870 to the present time in most cases, but one experiment had been terminated after about 50 years). The aim was to see if what was believed to be true about the various plots was borne out by these more formal quantitative measures thus giving them added credence, and also to obtain more refined comparative assessments.

To illustrate the results, Fig. 10.5 shows the factor shares, the *TFP* (output/input ratio) and the index *I* for Broadbalk plot 22 from 1871 to 1989. Plot 22 has had only farmyard manure applied and it was assumed that it would show sustainability. This is borne out by the *TFP* being typically in excess of 1 in value and the index *I* showing an increasing trend.

In contrast, Fig. 10.6 shows the *TFP* and index *I* for plot 8a of the Woburn Continuous Wheat experiment. This plot had no lime and the soil became increasingly acidic (due to use of sulphate of ammonia) with drastic falls in yield before the experiment was terminated in 1926. This plot was clearly

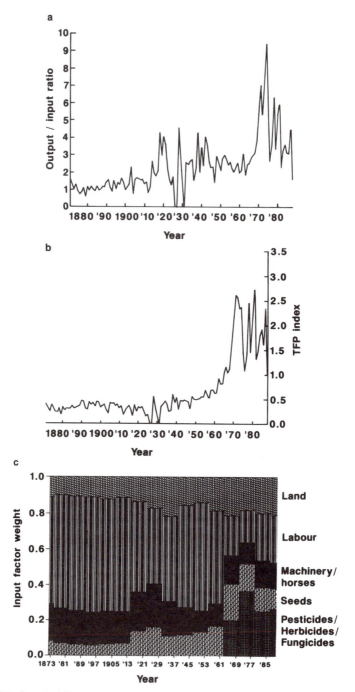

Fig. 10.5. Sustainability measures for the FYM treated plot on Broadbalk 1871–1988. (a) Output/input ratio (*TFP*); (b) *TFP* index; (c) input factor weights, *TFP* index.

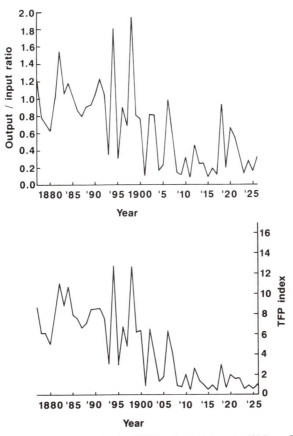

Fig. 10.6. Sustainability measures for the NPK fertilized plot on the Woburn Continuous Wheat Experiment 1877–1926. Top: output/input ratio (*TFP*); bottom: *TFP* Index.

unsustainable; Fig. 10.6 shows the *TFP* moving typically below 1 in value and the index *I* on a markedly downward trend.

The statistical variational properties of the *TFP* and *I* measures and appropriate tests of trends are discussed in Barnett *et al.* (1994).

The reported results have not at this stage fully incorporated the crucial 'externalities': environmental costs and implications. There are major difficulties in taking proper account of the externalities for a variety of reasons. Pollution effects due to fertilizer and pesticides are in principle the ones most readily quantifiable, but even in these cases the prevailing local circumstances can produce quite different costs for similar amounts of applied chemicals. Soil erosion is not a major problem at Rothamsted. Questions of effects on wildlife habitats, aesthetic features of landscape, biodiversity, and so on, are bound to be most difficult to assess. These matters are discussed in more detail in Barnett *et al.* (1992). The only quantitative evaluation in the present study took the form of a sensitivity analysis, increasing the fertilizer and pesticides in varying proportions (up to the unrealistically high levels of 120% and 200%, respectively). The clearly sustainable plots remained remarkably robust even up to the high levels of enhancements of fertilizer and pesticide cost.

There is clearly much more important work to be done on sustainability using the data from the Rothamsted long-term experiments, and the economic/quantity data assembled in ERA from this study will prove a most valuable resource.

Indeed, many substantial statistical challenges of various kinds remain to be explored in the long-term-experiment data for the mutual benefit of statistics and of agriculture.

References

Barnett, V. (1993) Multivariate environmental statistics in agriculture. In: Patil, G.P. and Rao, C.R. (eds) *Multivariate Environmental Statistics*. Elsevier, Amsterdam, pp. 1–32.

Barnett, V., Landau, S., Payne, R.W. and Welham, S.J. (1992) *Sustainability – The Rothamsted Experience*. Internal Report, Rothamsted Experimental Station.

Barnett, V., Landau, S. and Welham, S.J. (1994a) Measuring sustainability. *Journal of Ecological and Environmental Statistics* 1, 21–36.

Barnett, V., Payne, R.W. and Steiner, R. (eds) (1994b) *Agricultural Sustainability: Economic, Environmental and Statistical Considerations*. J. Wiley and Sons, Chichester.

Boden, T.A., Kanciruk, P. and Farrell, M.P. (1990) *Trends '90 : A Compendium of Data on Global Change*. Oakridge National Laboratory, Tennessee.

Bravais, A. (1846) Analyse mathématique sur les probabilités des erreurs de situation d'un point. *Mémoires de l'Institut de France, Séances mathématiques et physiques*, **IX**.

Box, J.F. (1978) *R.A. Fisher: The Life of a Scientist*. Wiley, New York

Buck, S.F. (1961) The use of rainfall, temperature, and actual transpiration in some crop-weather investigations. *Journal of Agricultural Science* 57, 355–365.

Cashen, R.O. (1947) The influence of rainfall on the yield and botanical composition of permanent grass at Rothamsted. *Journal of Agricultural Science* 37, 1–9.

Chmielewski, F-M and Potts, J.M. (1994) The relationship between crop yields from an experiment in Southern England and long-term climatic variations. *Agricultural and Forest Meteorology* (in press).

Dyke, G.V. (1993) *John Lawes of Rothamsted*. Hoos Press, Harpenden.

Fisher, R.A. (1921) Studies in crop variation. I. An examination of the yield of dressed grain from Broadbalk. *Journal of Agricultural Science* 11, 107–135.

Fisher, R.A. (1924) The influence of rainfall on the yield of wheat at Rothamsted. *Philosophical Transactions B* 213, 89–142.

Gifford, R.M. (1992) Interaction of carbon dioxide with growth-limiting environmental factors in vegetation productivity: implications for the global carbon cycle. *Advances in Bioclimatology* 1, 24–58.

Hooker, R.H. (1907) Correlation of the weather and crops. *Journal of the Royal Statistical Society* 70, 1–42.

Jenkinson, D.S., Potts, J.M., Perry, J.N., Barnett, V., Coleman, K. and Johnston, A.E. (1994) Trends in herbage yields over the last century on the Rothamsted Long-term Continuous Hay Experiment. *Journal of Agricultural Science, Cambridge,* 122, 365-374.

Lawes, J.B. and Gilbert, J.H. (1871) Effects of the drought of 1870 on some experimental crops at Rothamsted. *Journal of the Royal Agricultural Society,* 2nd Series, 7, 91-132.

Lynam, J.K. and Herdt, R.W. (1989) Sense and sustainability: sustainability as an objective in international agricultural research. *Agricultural Economics* 3, 381-398.

Pearson, E.S. (1978) *The History of Statistics in the 17th and 18th Century.* (Lectures by Karl Pearson), Griffen, London.

Perry, J.N., Potts, J.M. and Welham, S.J. (1993) The classical experiments and the environment. *ARIA Newsletter, 1993,* IACR, 11-16.

Russell, E.J. (1966) *A History of Agricultural Science in Great Britain.* George Allen and Unwin, London, 493pp.

Smith, L.P. (1960) The relationship between weather and meadow-hay yields in England. *Journal of the British Grassland Society* 15, 203-208.

Smith, R.L. (1993) Long-range dependence and global warming. In: Barnett, V. and Feridun Turkman, K. (eds) *Statistics for the Environment.* Wiley, Chichester, pp. 141-161.

Wishart, J. and Mackenzie, W.A. (1930) Studies in crop variation. VII. The influence of rainfall on the yield of barley at Rothamsted. *Journal of Agricultural Science* 20, 417-439.

Young, P.C. and Minchin, P.E.H. (1991) Environmetric time-series analysis: modelling natural systems from experimental time-series data. *International Journal of Macromolecules* 13, 190-201.

III

CURRENT NEEDS FOR LONG-TERM EXPERIMENTS IN THE DEVELOPMENT OF AGRICULTURE

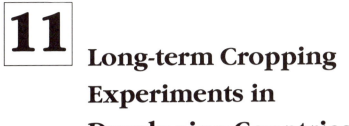

Long-term Cropping Experiments in Developing Countries: The Need, the History and the Future

D.J. GREENLAND
Department of Soil Science, University of Reading, London Road, Reading, Berks RG1 5AQ, UK.

The Need

Long-term experiments are, of necessity, often large and therefore expensive to maintain and manage. They require careful management on a continuing basis, supplementary studies based on critical analyses of the changes taking place, and a recognition that the results they produce are more relevant to long-term rather than short-term concerns. They require a stable social and political environment if they are to produce results commensurate with their costs. Most of these requirements are missing in developing countries, and the question may well be asked, do we need such experiments in developing countries? The answer is of course yes. Not only are they needed, but the importance of them is often greater than in other parts of the world.

It has been recognized since Malthus wrote his essay (1798) that increasing population pressure posed a threat to world supplies of food, fuel and fibre. Fertilizers and the other 'green revolution' ingredients plus the levelling off in population growth have removed the threat from the developed world and also created surpluses of much agricultural produce. In most countries in the developing world the threat to food security remains, as rapid population expansion continues.

To support the continuing increase in demand for food, fuel and fibre it is essential that agricultural production increases at a rate comparable to, or in developing countries, ahead of population growth, and does so not only on a sustainable basis but in such a way that further increases in productivity are possible.

The basic ingredients for greater productivity are well known – they are the same 'green revolution' ingredients that have produced the surpluses in developed countries, namely the use of fertilizers and other soil management practices to maintain soil fertility, the use of improved crop varieties, pest control measures, and better water management. Where uncertainty remains is in how to combine these ingredients in an economically sustainable agricultural system.

To test the chemical, physical and biological sustainability of a cropping or land management system long-term experiments are essential. To realize the significance of the experiments they must be linked to other studies of the factors involved in maintaining and increasing yields, and other trials where the validity of the conclusions reached is tested in a range of different agroecological environments, which include different economic and social conditions.

Whereas the prime need for long-term experiments in developing countries relates to sustainability and increased productivity of soils, environmental issues are also important. For instance, long-term experiments are needed to determine how changes in land use affect water supplies and water quality. Similar studies are needed to assess how soil erosion affects the environment at sites away from the point where the erosion originates. This implies that some experiments at least need to be conducted on a catchment basis, and include an opportunity for socioeconomic factors to be considered.

Historical

Traditional systems

Throughout the world the traditional stable system for cultivating the uplands has been a natural fallow or shifting cultivation system (Richardson, 1946; Nye and Greenland, 1960; Allan, 1965; Greenland, 1974; Watts Padwick, 1983). The sustainability of such systems fails if the restoration of fertility in the fallow phase is inadequate to restore the losses of fertility in the cropping phase. Mostly this has occurred as population pressure has caused the fallow period to be shortened, but it also occurs if the cropping period is prolonged and when cultivators are driven to open unsuitable land.

There are of course other traditional systems of agriculture, notably rice production systems in the wetlands. These have been sustained for many years although at a generally low level of productivity in which rice yields have remained close to 1 t ha^{-1} for many years. Green revolution methods have now raised the national rice yields in China and Indonesia to well over 4 t ha^{-1} and questions have been raised of the sustainability of such intensive production, and whether the methods used are giving rise to adverse environmental effects. In north-west India there is at present serious concern about declining rice yields, possibly due to increasing soil salinity. Long-term experiments are necessary to confirm this and to answer questions about the sustainability of this and other less intensive systems.

Long-term experiments initiated pre-1930

In spite of the example from Rothamsted there were few long-term experiments established in developing countries prior to 1930. India was one exception. The Report for the Cawnpore Agricultural Station for 1906–7 states, regarding the experiment on continuous maize and wheat, grown singly and in rotation

> This experiment has now been in progress for some 24 years. Started on the plan of the famous Rothamstead (sic) experiments, it has passed through many changes of supervision, and has never had the advantage of the continuous interest of two well known experts as was the case at Rothamstead, and can hardly claim in comparison, and as a mere copy, the same scientific interest which the original held . . . I think the experiment might now well be closed.

Although this experiment may not have survived, the 'Imperial Agricultural Chemist', J.W. Leather, in his Report to the Government of India for 1904–5 (p.59) states

> I submitted in 1899 those (results) on the continuous growth of maize and wheat at Cawnpore to comparison with similar ones at Rothamsted and Woburn in England (vide Northwestern Provinces and Oudh Bulletin No.9 of 1900). A study of the figures is of considerable interest and illustrates well the value of manures, especially nitrogenous manures, in India.

Leather appears to have initiated drainage as well as cropping experiments on the lines of those at Rothamsted (Leather, 1911). Further long-term cropping experiments were established at Pusa in 1908, and at Coimbatore in 1909. Sree Ramulu (1990) has reported some results from the 'old' experiment at Coimbatore in southern India which was started to study the effects of N, P, K fertilizers and cattle manure on cereals, pulses, grasses and cotton, and of a new experiment started in 1925. Both experiments at Coimbatore continue. Sadly Nambiar *et al.* (1989) record that the experiments at Kanpur (the former Cawnpore) and Pusa 'were lost to posterity', together with nine others started between 1935 and 1950. However, some information from these and other long-term experiments has been retrieved by Krishnamoorthy (1982).

Mukerjee (1965) notes that the Royal Commission on Agriculture in India reported in 1929 that in spite of many years of fertilizer experiments on experiment stations, advice to farmers on fertilizer use was still not forthcoming – a report which led to a concentration on simple multiple trials on farmer's fields and the thousands of FAO fertilizer demonstration trials (Richardson, 1962).

Another early long-term trial, which also continues, was established in Egypt in 1912. It examines the effects of fertilizers and manures on cotton, grown continuously or in rotation with food crops, and, again was inspired by the Rothamsted experiments. Recent results have been reported by El-Sweely *et al.* (1983).

In the 1920s and 1930s there was considerable interest in the possibility that legumes grown as green manures might maintain the productivity of many soils in the tropics. Nicol (1935) suggested that legumes grown in mixed cropping systems in India could take the place of animal manures. In Nigeria an experiment was initiated in 1922 to assess the significance of a green manure in a rotation of several crops. Early reports (Faulkner, 1934; Doyne, 1937) were only moderately encouraging but the work was maintained until 1950 (Vine, 1953). Rotation experiments with legumes, cereals and cotton were started in the Sudan in 1931 (Crowther and Cochran, 1942; Dutta Roy and Kordofani, 1961).

Perennial crops such as cocoa, coffee, tea, rubber and oil palm are of vital importance to the economies of many developing countries. It is now recognized that the long-term productivity of these crops requires appropriate fertilization practices. Given the life span of most perennials it might well have been expected that some long-term experiments would have been initiated at an early stage following the establishment of the crop as a commercial species. In fact this does not seem to have happened. The first long-term manurial experiment on a tropical perennial appears to be that on rubber in Sumatra, Indonesia, started in 1919, followed by a similar experiment in Malaysia in 1928 (Haines and Guest, 1936).

Long-term experiments initiated 1930–50

The comments in the 1929 report of the Royal Commission on Agriculture in India fortunately had no immediate influence outside of India and the period following 1930 saw the initiation of several long-term experiments in Africa. The need for a system to replace shifting cultivation was already recognized, and various experiments were started on possible methods to raise and maintain soil fertility. In East Africa the advantages of grass fallows in fertility restoration were being urged and a large experiment was initiated at Serere in Uganda in 1945. The main objective of the experiment was to compare the relative advantages of rotations in which one or two years of 'rest' under a grass or a legume fallow were included in a 5-year cycle with those where no rest crop was included. The experiment was made more elaborate by the inclusion of cutting and grazing regimes for the rest crops, and the inclusion of three rates of animal manure, although there were no treatments with inorganic fertilizers. The results have been reported by Jameson and Kerkham (1960) and McWalter and Wimble (1976). They show that there was almost always a response to the animal manure, and a tendency for yields to increase after the rest crops, whether grass or legume, for the first 20 years (four cycles) but a decline after that. Only meagre data on soil analyses are reported, and no data are given on pest incidence or yield potential and other characteristics of the crop varieties grown. Thus little more than a pragmatic interpretation of the results was possible, although some studies of the changes in soil structure under grass were made (Pereira *et al.* 1954). More detailed studies of the effects of the grass fallows were reported by Stephens (1967).

Although the experimental results available were still very limited, E.M. Crowther, then Head of the Chemistry Department at Rothamsted, was able to publish a remarkably perceptive paper on Soil Fertility Problems in Tropical Agriculture (Crowther, 1949a), in which he defined most of the research agenda which is concerning soil scientists working in the tropics today. He included the importance of agroforestry to crop production systems, and the need for soil and site characterization linked to work on soil survey and classification, to enable the results of experimental work to be extrapolated on a sound basis. His final paragraph states 'In general fertilizer experiments should be planned in series which will bring out the average responses of individual crops over a range of soils and of individual soils over a variety of crops. Long-term experiments are needed to analyse the nutrient cycles in various systems of cropping'.

Long-term experiments in the period 1950–66

Following Crowther's paper there was a rash of activity in which many fertilizer experiments were established on experiment stations in Africa and elsewhere in the tropics. Russell (1968) has described the results of several of these. He stressed the importance of soil analyses to the interpretation of the results.

Experiments in Ghana were established at several sites in each of the four major agro-ecological zones in the country (Nye, 1951). All of these were maintained for the first nine years (Djokoto and Stephens, 1961a,b) although subsequent reports (Ofori, 1973; Kwakye *et al.*, 1994) only provide data for one site, at the Central Research Station (now the Soil Research Institute), Kwadaso, near Kumasi. This experiment provides the longest record of continuous cultivation in West Africa. Site mean yields are shown in Fig.11.1, together with the calculated 'control' yield and the response to grass mulch. Where mulch has been used it is clear that yields have been maintained or, with maize, increased. Without the mulch the small amounts of fertilizer (28 kg N, 10 kg P, 32 kg K ha^{-1}) have failed to maintain yields, although they are much better than the near zero yields obtained without fertilizer or mulch. As expected the continued use of ammonium sulphate led to a fall in pH from 6.3 to 4.4, and lime

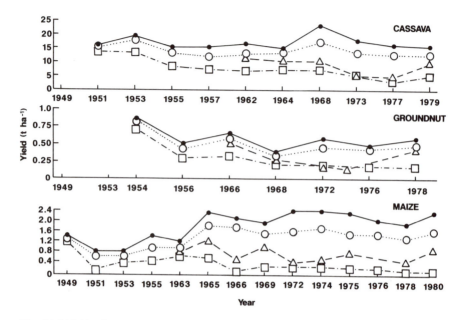

Fig. 11.1 Yields of cassava, groundnuts and maize in the 2^5 factorial experiment (NPK Lime Mulch) at Kwadaso, Ghana, 1949–80. (Compiled from data in Djokoto and Stephens, 1961a; Ofori, 1973; Kwakye, *et al.*, 1994). Control, □; Site mean, ○; without mulch, △; with mulch ●.

(5 t ha^{-1} as quicklime applied every fourth year) maintained pH between 5.5 and 6.5. The results, especially in the early years, were influenced by pest incidence. For instance rust on the maize was severe in 1951 and 1953. Nevertheless the yield trends are clear.

The experiments in northern Nigeria provide an excellent example of the importance of maintaining such trials on an indefinite basis. When first analysed (Obi, 1959) it was concluded that inorganic fertilizers could not maintain yields at the levels given by animal manure. However, modifications in the treatments later showed that potassium could raise yields to those being obtained with manure (Heathcote, 1973). Subsequently, Heathcote and Smithson (1973) were able to show that maize needed zinc, groundnuts required molybdenum and cotton needed boron. Singh and Balasubramanian (1979) also established that a more general need existed for magnesium. Given that sulphur deficiency also occurred, it seems likely that yields would be better maintained by the use of manures, as the earlier work at Samaru had indicated, and as the work in the francophone countries in West Africa, reviewed by Jones and Wild (1975) and Pieri (1992) also indicated.

The experiment at Ukiriguru in Tanzania (Le Mare, 1972) was another where later additions to the experimental treatments were made as phosphate, nitrogen and compost failed to maintain yields after the first nine years. Soil analyses showed the need for lime, which was then confirmed experimentally.

Another example of the importance of subsidiary studies is provided by Le Mare's work on the response of cotton to phosphate in Uganda. Field experiments (Le Mare, 1968) showed an initial depression of yield to applied P, only changing to a positive response at high levels of addition. Much further study was required to establish that the problem arose in soils of relatively high manganese level, and low cation exchange capacity, where the extra calcium added with the larger rates of fertilizer was needed to correct the manganese toxicity which was expressed only when P was added (Le Mare, 1977).

A long-term experiment in Zambia was initiated in 1962 (Lungu, 1987). Rates of ammonium sulphate up to 1793 kg ha^{-1} were applied each year to maize grown continuously. At the largest rate of application soil pH fell to 3.9 and yields to 26 kg ha^{-1} by 1967. Over the next four years when lime was applied on split plots, yield increased to 1713 kg ha^{-1} where the largest N rate had been used, but was larger, 4324 kg ha^{-1}, where only 112 kg ha^{-1} of ammonium sulphate had been applied.

From 1960 onwards a series of long-term experiments were initiated in francophone West Africa. There were 25 in all lasting from 6 to 28 years. The results have been well summarized by Pieri (1992). As in the experiments in anglophone countries in Africa they have been almost entirely concerned with crop responses to fertilizers, manures and resting periods when the soil is left under a grass or other non-harvested crop. It is not possible in a short chapter to do justice to the considerable effort which has gone into these experiments. An example of the results obtained is given in Fig. 11.2, which shows the yields of monocropped sorghum obtained in an experiment at Saria, Burkina Faso, between 1960 and 1983. Without manure or fertilizers yields fell close to zero after the second year of the experiment. With fertilizers they were maintained at a low level, and with manure, especially manure and fertilizers, they were maintained at moderate to good levels for 16 years. When the pH fell below 5, yields fell to below 1 t ha^{-1}. Liming then restored yields to satisfactory levels. The amount of organic matter in this soil, as in many of the other soils in the Sahelian region, was very low (0.24–0.31% C) unless manure was added, and even with 40 t ha^{-1} applied annually it only rose to 0.66% C in 1978. In other long-term trials various crop rotations were studied, different cultivation procedures, different straw management techniques and various restorative fallows. The most important conclusion was that a critical organic matter level exists below which it becomes extremely difficult to maintain yields with inorganic fertilizers alone. Continued long-term studies are necessary to establish what the critical levels of nutrients and organic matter are for different soils and environments.

As well as soil nutrient levels, acidity and organic matter, crop yields are much affected by pests (including in that term weeds, disease organisms and

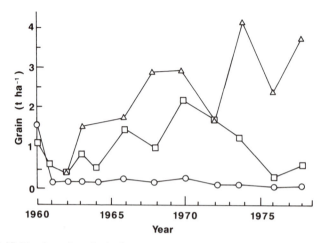

Fig. 11.2. Yields of sorghum in the long-term trial with NPK and cattle manure at Saria, Burkina Faso, 1960–78 (Pichot *et al.*, 1981). Control, ○; NPK, □; NPK + cattle manure, △.

insect pests), and by water availability. Many of the reports of the long-term experiments refer to the problems in interpretation caused by pest attacks. Strangely there are no discussions in any of the papers cited above of the experiments being used for systematic studies of the changes in pest incidence. Major occurrences are mentioned to explain occasional very low yields, and control measures receive an occasional mention, although they are mostly ignored.

The experiments in the Sahel have been used to develop and test water use models (Fig. 11.3). A more ambitious study to evaluate the effects of land use changes on the water balance was established in East Africa in 1955. This experiment was continued until 1978 (Pereira *et al.*, 1962; Blackie *et al.*, 1979; Edwards and Blackie, 1981). It provided valuable data on the calculation and modelling of water run-off and deficits.

Soil and fertilizer management methods for rice and plantation crops differ from those for annuals grown on upland soils; rice because of the special conditions in which it is grown, and perennial crops because of the changing nutrient demand over the life span of the crop. Long-term experiments on flooded rice in the tropics were relatively few in the period between 1950 and 1966. The largest amount of information came from India and it has been summarized by Mahapatra *et al.* (1972).

During this period several long-term experiments were also being conducted on perennial crops, for instance on rubber in Malaysia and Sri Lanka (Bolton, 1964; Yogaratnam and Weerasuriya, 1984), oil palm in Nigeria (May, 1956; Haines and Benzian, 1956), coffee in Kenya (Jones and Wallis, 1963), tea (Eden, 1944) and coconuts in Sri Lanka (Eden and Gower, 1963) and cocoa in Ghana (Acquaye and Smith, 1965). These experiments provided the basis for

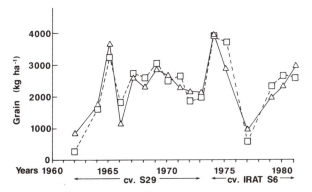

Fig. 11.3. Real and simulated sorghum yields calculated from a water balance model at Saria, Burkina Faso (Forest and Lidon, 1984). Computed yield, △; measured yield, □; Error of model 8.5%

many ancillary studies which have been important to understanding the management of the soils for economic crop production. One example is the study (Tinker, 1964) of the importance of maintaining appropriate cation balances for oil palms grown in the 'acid sands' of eastern Nigeria. Kemmler and Malicornet (1976) reviewed many of these experiments with plantation crops, and the need to conduct such studies on a long-term basis.

Long-term experiments established since 1966

In 1966 the International Rice Research Institute (IRRI) released IR8, the semi-dwarf variety with much enhanced yield potential. For the following 20 years IR8 has been used for comparison with the yields obtained from other elite varieties. The comparison has been made in long-term trials at IRRI, and from 1968 at three national research stations in the Philippines. The yields for IR8 and the highest yielding variety among others included in the trial over 20 years and 40 crops are discussed by Flinn and De Datta (1984). The plots were continuously flooded until shortly before harvest. At IRRI they received P and K sufficient to ensure no deficiencies of those nutrients, and regular sprays with a wide-spectrum insecticide. At all four stations two crops were grown each year, and five rates of nitrogen application were used in the dry season and four rates in the wet season.

At IRRI there was a declining yield trend for IR8 in the wet seasons when disease pressure was severe. The highest yielding variety was then that with sufficient resistance to offset this. In the dry seasons yields of IR8 are still high, although from about 1973 to 1982 they declined quite severely. The cause was traced to an accumulation of boron from the deep well water used for irrigation (Flinn and De Datta, 1984). By using better quality water in subsequent years the decline appears to have been reversed (IRRI Annual Reports, 1981–

85). The varieties giving high yields throughout presumably had resistance to adverse soil conditions.

At the other stations phosphorus and potassium were not applied, and deficiencies, and a strong interaction between them, subsequently became apparent (De Datta and Gomez, 1975; De Datta *et al.*, 1988).

Another long-term experiment was established at IRRI in 1963. In this experiment three crops were grown each year, and the experiment continues. The dry season results in Fig. 11.4 (Cassman and Pingali, 1994) show that although there is clearly a declining yield trend, after 22 years and 66 crops the highest yielding variety was still producing more than 5 t ha⁻¹. In spite of the declining yield trend, these trials are important compared with earlier long-term trials because they show that for flooded rice very high annual production can be sustained over an extended period. They also provide the opportunity to continue intensive studies of factors responsible for the declining trend in yield. The factors under study include pest problems and their interaction with nitrogen supply, the role of soil organic matter, and water quality (Cassman and Pingali, 1994). The experiments at IRRI were established to monitor the yield potential of new rice varieties. They have succeeded in doing so, and also provided important information regarding other factors determining the sustainability of high yields. They also show that whereas considerable progress has been made in developing varieties with tolerance to adverse conditions, there has been little change in the yield potential since the original semi-dwarf, plant type was developed. Hence there is a renewed emphasis at IRRI on the development of a new higher yielding plant type (IRRI, 1990.)

A long-term experiment was also initiated at the International Institute of Tropical Agriculture (IITA) in Nigeria shortly after its establishment. Kang and

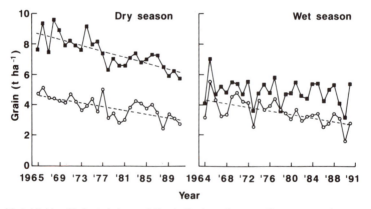

Fig. 11.4. Yields of irrigated rice varieties in the long-term continuous cropping experiment at the International Rice Research Institute in the Philippines, 1964–90 (Cassman and Pingali, 1994). NPK, ■; Control, ○.

Balasubramanian (1990) have summarized the results of the soil fertility experiments, conducted on conventional small plots where erosion and physical
damage to the plots was avoided. Yield trends on large blocks of land cleared
and cultivated by different methods, and with small catchments instrumented
so that erosion and run-off could be measured as well as changes in other soil
properties, have also been followed at IITA (Lal, 1992). Yields of maize grown
in the first season in each year in rotation with cowpeas or cassava, are shown
in Fig. 11.5a and b. Both experiments show a decline in yield, with the decline
more significant in the larger experiment. Many data on soil degradation were
obtained from the catchment experiment. Alley cropping and mucuna fallows
were among the treatments used in the catchment experiments, but these did
not arrest the serious yield decline. The size and scale of the experiment was
such that it was not possible for it to be maintained beyond 10 years. It leaves
many unanswered questions, as does the simpler fertilizer experiment.

A most important series of long-term trials were initiated in India in 1971
(Nambiar and Ghosh, 1984; Nambiar and Abrol, 1989; Nambiar *et al.*, 1989).
These were established at 11 centres on a range of different soils and in widely
differing environments. The objectives were: (i) to study the effect of continuous application of plant nutrients, singly and in combination, in inorganic and
organic forms including secondary and micronutrients, on crop yield, nutrient
composition and uptake in multiple cropping systems; (ii) to monitor the
changes in physical, chemical and microbiological properties of soils as a

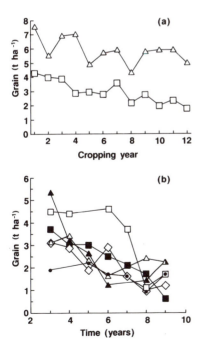

Fig. 11.5. Yields of maize at IITA, Ibadan,
Nigeria, (a) with and without 'fertilizer' on small
plots, 1971–82 (Kang and Balasubramanian,
1990) and (b) with fertilizer following different
methods of land clearing, on small catchments,
1979–88 (Lal, 1992). Treepusher, no till, ▲;
treepusher, ploughed, ●; shearblade, ■; manual
clearing, no till, △; manual clearing, ploughed,
◇; subwatershed, □.

result of continuous intensive use of chemical fertilizers and manures along with biocidal chemicals. Again it is impossible to do justice to these experiments in a short review. One example of the results must suffice, the yield trends for maize, wheat and cowpea fodder at Ludhiana in the Punjab (Fig. 11.6). Reviews of the results of these and other long-term experiments in India were published in *Fertiliser News* (Nambiar and Abrol, 1989).

An important long-term experiment was established at Yurimaguas, Peru, on the highly acid soils of the llanos in the upper Amazon basin. The experiment was the responsibility of INIPA, the Peruvian national agricultural research service, supported by scientists from North Carolina State University. Without fertilizer, yields of upland rice, maize, and soyabean grown in rotation fell almost to zero after the first cycle (Fig. 11.7). By the end of the first ten years of the experiment it had been established that maintaining moderate yields required N, P, K, Mg, Cu, Zn, B and for legumes Mo, as well as lime equivalent to 3 t ha^{-1} of CaCO$_3$ every third year (Sanchez *et al.*, 1982). Given the necessary nutrients at adequate levels yields have tended to increase slowly.

In other experiments on similar soils in Brazil and Indonesia, also conducted by the national agricultural research services in association with North Carolina State and other American universities, similar results have been obtained. These experiments have benefited from many supplementary soil studies. The extrapolation of the results to other areas has also been aided by appropriate site characterization (Tropsoils, 1987).

The Symposium on Long-term Experiments held at the International Soil Science Congress in Japan in 1990 included brief reports on recently established long-term experiments in China, Bangladesh, Thailand, Taiwan and Korea in addition to those conducted in several of the countries already mentioned. There are many experiments managed by national and international research organizations which have yielded and are yielding results important to the understanding of the basis of sustainability in agricultural production although a full analysis of the broader significance of most of these has yet to be made.

Lessons from the history of long-term experiments in developing countries

The management of soils for continuous arable cultivation in the tropics is more difficult than in temperate conditions. Changes in many soil properties tend to occur more rapidly than in temperate climates. The majority of soils in developing countries are dominated by low activity clays, and the rate at which organic matter is lost during cultivation is greater than in soils of the temperate zone. Thus as organic matter and pH fall, the cation exchange capacity also falls, and falls to very low levels, so that deficiencies arise from both the lack of exchangeable cations and their imbalance. There are also many soils in developing countries where the inherent levels of plant nutrients

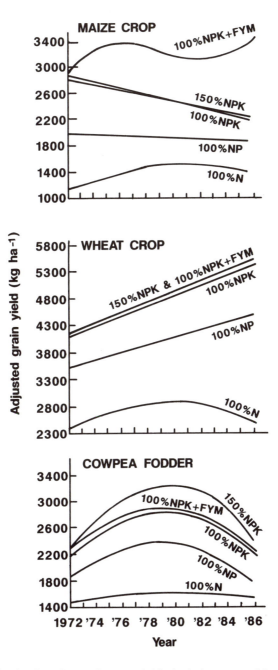

Fig. 11.6. Yields of maize, wheat and cowpea fodder in the long-term trials with NPK and cattle manure at Ludhiana, India, 1972–86 (Nambiar and Ghosh, 1984).

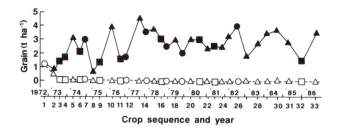

Fig. 11.7. Yields of dryland rice (o●), maize (△▲) and soyabeans (□■) grown in rotation in the long-term trial with and without complete fertilizers at Yurimaguas, Peru, in 1972–86 (Tropsoils, 1987). Solid symbols, complete fertilizers; open symbols, no fertilizer.

are low, so that deficiencies of phosphorus, sulphur and micronutrients are common (Greenland, 1981; IRRI, 1981). Almost all long-term experiments in developing countries show the expected deficiencies of N, P and K. Many show other deficiencies arising with continued cropping, complicated by increasing soil acidity. The Tropsoils trials indicate that very careful and scientifically supported management can enable yields to be maintained with lime and necessary additions of missing nutrients in inorganic form. However in most trials the long-term sustainability of yields has only been achieved where organic material in some form has been added (cf. Figs 11.1, 11.2 and 11.6). As Pieri (1992) noted the essential requirement is to establish the level of organic matter which must be maintained. Long-term experiments can enable this critical level to be established, especially if they are used to determine quantitative rates of change of organic matter. What they cannot do is to provide data on the rates and economics of fertilizer use, or the economics of collecting and using organic materials. Both of these require experiments on economically representative farms, and across a range of representative soils and environments. Many of the earlier trials were in fact complemented by simpler trials conducted on farmers' fields (see for example Nye, 1953; Le Mare, 1959; Stephens, 1960; Foster, 1978, 1980; Pieri, 1992). Although these indicated the extent to which the results of the long-term experiments might be extrapolated they still failed to provide any indication of the economic, social and other constraints which determine the advantage that the farmer may gain from the use of fertilizers. This requires more complete studies of farming systems, and appropriate soils information. If the farming system studies are to be informed by the results from long-term experiments it is essential that those experiments are conducted on sites which have been carefully characterized, and that the farming systems studies include site descriptions (Zandstra *et al.*, 1981).

The Future

The experiments which have been conducted have established that the productivity of many soils in the tropics can be maintained by appropriate use of fertilizers and organic manures. The bibliographical study by Robison and McKean (1992) highlights the need for studies of soil management to be associated with socioeconomic assessments and further shows that what is now needed is that the constraints to adoption of better management practices become a principal focus for future research. A few large, well-managed and well-instrumented experiments need to be conducted at carefully selected sites to evaluate proposed innovations or modifications to management practices. Such long-term experiments provide the only direct method by which the chemical, physical and biological sustainability of an agricultural system can be measured, and the rates of change in soil properties associated with changing agricultural practices quantified (Johnston and Powlson, 1994; Powlson and Johnston, 1994). The rapid loss of productivity of many tropical soils which arises when technology developed for soils of the temperate zone is transferred unthinkingly to the tropics makes the need for long-term experiments on soil management even more important than it is in developed countries. The importance of sustainable land management is now recognized by most countries. The richer and larger countries can afford to maintain long-term experiments as part of their continuing research programmes, recognizing their importance not only to agricultural production but also to many environmental aspects, such as the contributions of soils to the greenhouse effect (Bouwman, 1990; Lal and Kimble, 1994). The governments of many of the smaller developing countries recognize more immediate demands for their limited financial and research capacities. Thus work in this area is likely to require support from various international agencies for some time to come.

To quote from Crowther (1949b); 'There is an urgent need for international cooperation in developing the field experimental basis for planning the more efficient use of natural resources in land and raw materials for fertilizers'. Many years later this call for international cooperation was echoed by Swindale (1980), a call which resulted in the establishment of the International Board for Soil Research and Management (IBSRAM). The mandate and mode of operation of IBSRAM require that it collaborate with national agricultural research systems and with institutions in developed countries with relevant expertise in long-term trials of soil and land management (Latham and Syers, 1994). It has given special emphasis to proper site characterization to enable the results of the experiments to be extrapolated to other environments on a sound basis (IBSRAM, 1988), and reviewed much work on soil management based on long-term experiments (IBSRAM, 1986, 1987, 1989, 1990).

Two parts of the International Geosphere–Biosphere Programme (IGBP) are also concerned with monitoring soil performance (IGBP, 1990). The Global Change and Terrestrial Ecosystems (GCTE) Project involves a range of studies on land use which could benefit by making use of the sites of existing long-term experiments (IGBP, 1992). Under the Global Change System for Analysis, Research and Training (START) project (IGBP, 1991) it is proposed to establish a number of Regional Research and Training Centres (RRCs). It is to be hoped that these will work in close collaboration with IBSRAM and the Ecoregional Centres funded by the Consultative Group on International Agricultural Research (CGIAR). As already mentioned it will also be essential that any internationally supported long-term experiments are conducted in close collaboration with national programmes, as are those in which IBSRAM is involved. As most of the long-term experiments so far conducted in the tropics have shown, the management of soil organic matter, through proper use of plant and animal residues, is critical to the sustained productivity of tropical farming systems. It is therefore appropriate that special attention is given to soil organic matter and soil organisms, and the Tropical Soil Biology and Fertility (TBSF) programme has an important role to play in assisting the various organizations concerned with the conduct and maintenance of long-term experiments. Further cooperation between TSBF and IBSRAM could usefully be encouraged.

Nye (1992a,b) has shown that computer models can now be used to predict the behaviour of crops in different soil management systems. Much of the work which has been conducted in developing countries, was at its initiation based on educated guesses at what would be a sustainable soil-crop management system. As Nye (1992b) argues, models can now provide 'theoretical insight against which the feasibility of many possible cropping and management practices may be assessed'. If this insight is to guide future research it is essential that the conduct and management of long-term experiments is supported by the necessary skills in system modelling, and in other related research areas. The value of the experiments at Rothamsted compared with many others arises not only from their longevity but also from the scientific support that has been provided to them.

For long-term experiments to be maintained in the tropics they must be provided with secure long-term funding. Where this cannot be provided by the national agricultural research programme international support will be needed. Several international organizations will need to be involved, but much more value is likely to be obtained from the results of long-term experiments if one organization has responsibility and accountability for their conduct. At the present time the mandate of IBSRAM, which requires that it work as a partner with national programmes, and coordinate support from leading soil science research centres to the work of the national programmes on sustainable soil and land management, is that which seems best suited to the adoption of such a role. It will only be able to fulfil the role successfully by close and active collaboration with other organizations.

Summary and Conclusions

Long-term field experiments are of greatest value in relation to the productivity and sustainability of different soil management systems. Hence they are particularly important where soil management practices are changing. In most developing countries increasing population pressure is forcing land use intensification and necessitating the development of more productive soil management systems. Long-term experiments are needed to determine the sustainability of the new systems of management which are being introduced, and to determine the efficacy of measures used to promote greater productivity and sustainability, to ensure long-term food security and avoid environmental damage.

Factors that must be studied in such long-term experiments include yield trends, the dynamics of changes in such soil properties as organic matter and nutrient content, the development of acidity or in drier areas salinity and sodicity, and the incidence of pests, including weeds, pathogens and insects. The relationship of each of these to weather and climate is also important.

Long-term experiments are of necessity rather large and expensive to maintain and manage. Hence they are likely to be rather few in number. Results obtained from them provide an essential reference for many other less elaborate experiments of shorter duration. If the long-term experiments are to be of general utility it is essential that they are documented as fully as possible, and the sites where they are conducted adequately characterized. Most value is likely to arise when they can be related to trials in which social and economic constraints to adoption of the practices under study are evaluated.

Long-term manure and fertilizer trials were started around the turn of the century in India. Rotation experiments with grass fallows and green manures were conducted at several locations in Africa from about 1930, and long-term experiments with inorganic fertilizers on a range of crops were initiated in Africa and other tropical countries from 1940 onwards. Some of these were maintained for over 25 years.

A few long-term catchment studies were conducted in which the effects of land management on run-off and erosion were measured. There were also some studies of tree crops and of various practices involving trees in relation to the farming system.

The value of many of these experiments as far as current practice is concerned is limited due to the fact that soil and other site factors were often inadequately recorded, problems due to pests were neither identified nor remedied, variations in management practices were not always recorded, and crop varieties of relatively low yield potential were usually used.

Experiments initiated more recently remedy many of these deficiencies. Rather few developing countries other than the very large (China, India) have the financial resources and the research capacity to maintain long-term soil management experiments. Thus support from aid agencies has been and still is needed. The policies of many aid agencies restrict the period of support for any one project to ten years or less. Such a time scale is inadequate for proper evaluation of soil management practices.

The International Agricultural Research Centres and the International Board for Soil Research and Management (IBSRAM) have an important role to play in support of long-term studies. It is important that the International Centres and their supporting agencies appreciate the need for such work, and that adequate funds are provided on a sufficiently secure long-term basis.

References

Acquaye, D.K. and Smith, R.W. (1965) Effect of ground covers and fertilisers on establishment and yield of cocoa on clear felled land in Ghana. *Experimental Agriculture* 1, 131–139.

Allan, W. (1965) *The African Husbandman*. Oliver and Boyd, Edinburgh.

Blackie, J.R., Edwards, K.A. and Clarke, R.T (1979) Hydrological Research in East Africa. *East African Agricultural and Forestry Journal* 43, 1–313.

Bolton, J. (1964) The manuring and cultivation of *Hevea brasiliensis. Journal of the Science of Food and Agriculture* 15, 1–8.

Bouwman, A.F. (ed.) (1990) *Soils and the Greenhouse Effect*. Wiley, Chichester. pp.1–575.

Cassman, K.G. and Pingali, P.L. (1994) Extrapolating trends from long-term experiments to farmers' fields: the case of irrigated rice systems in Asia. In: Barnet, V., Payne, R. and Steiner, R. (eds) *Agricultural Sustainability: Economic, Environmental and Statistical Considerations*. Wiley, Chichester.

Crowther, E.M. (1949a) Soil fertility problems in tropical agriculture. *Technical Communication, No.46*. Commonwealth Bureau of Soil Science, Harpenden, pp. 134–141.

Crowther, E.M. (1949b) Field experiments as the basis for planning fertilizer practice. *Proceedings UN Conference on the Conservation and Utilisation of Resources*, Vol.6, pp. 221–224.

Crowther, F. and Cochran, W.G. (1942) Rotation experiments with cotton in the Sudan Gezira. *Journal of Agricultural Science* 32, 390–405.

De Datta, S.K. and Gomez, K.A. (1975) Changes in soil fertility under intensive rice cropping with improved varieties. *Soil Science* 120, 361–366.

De Datta, S.K., Gomez, K.A. and Descalsota, J. (1988) Changes in yield response to major nutrients and in soil fertility under intensive rice cropping. *Soil Science* 146, 350–358.

Djokoto, R.K. and Stephens, D. (1961a) Thirty long-term fertilizer experiments under continuous cropping in Ghana. 1 Crop yields and responses to fertilizers and manures. *Empire Journal of Experimental Agriculture* 29, 181–195.

Djokoto, R.K. and Stephens, D. (1961b) Thirty long-term fertilizer experiments under continuous cropping in Ghana. 2 Soil studies in relation to the effects of fertilizers and manures on crop yields. *Empire Journal of Experimental Agriculture* 29, 245–258.

Doyne, H.C. (1937) Green manuring in southern Nigeria. *Empire Journal of Experimental Agriculture* 5, 248-253.

Dutta Roy, D.K. and Kordofani, A.Y. (1961) Study of long-term rotation effects in the Sudan Gezira. *Journal of Agricultural Science* 57, 387-392.

Eden, T. (1944) Studies on the yield of tea. Part 5. Further experiments on manurial response with special reference to nitrogen. *Empire Journal of Experimental Agriculture* 12, 177-190.

Eden, T. and Gower, J.C. (1963) A factorial fertilizer experiment on coconuts. *Empire Journal of Experimental Agriculture* 31, 283-295.

Edwards, K.A. and Blackie, J.R. (1981) Results of the East African catchment experiments. In: Lal, R. and Russell, E.W. (eds) *Tropical Agricultural Hydrology*. Wiley, Chichester, pp. 163-188.

El-Sweely, A.M., Abou-el-Ezz, N. and Abd-El-Nour, A.S. (1983) Results of the permanent manuring experiment at Bahtim of crop yields from 1970-1981. I Cotton. *Agricultural Research Review*, April 1983.

Faulkner, O.T. (1934) Some experiments with leguminous crops at Ibadan, Southern Nigeria, 1925-33. *Empire Journal of Experimental Agriculture* 2, 93-102.

Flinn, J.C. and De Datta, S.K. (1984) Trends in irrigated rice yields under continuous cropping at Philippine research stations. *Field Crops Research* 9, 1-15.

Forest, F. and Lidon, B. (1984) Simulation du bilan hydrique pour l'explication du rendement et l'appui aux producteurs. In: *La secheresse en zone intertropicale*. CIRAD-CILF, Paris, pp. 55-65.

Foster, H.L. (1978) The influence of soil fertility on crop performance in Uganda. 1 Cotton. *Tropical Agriculture* 55, 255-268.

Foster, H.L. (1980) The influence of soil fertility on crop performance in Uganda. 3. Finger millet and maize. *Tropical Agriculture* 57, 123-132.

Greenland, D.J. (1974) Evolution and development of different types of shifting cultivation. In: *Shifting Cultivation and Soil Conservation in Africa. Soils Bulletin* 24, FAO, Rome, pp. 5-13.

Greenland, D.J. (ed.) (1981) *Characterisation of Soils in Relation to their Classification and Management for Crop Production: Examples from Some Areas of the Humid Tropics*. Oxford University Press, London, pp. 1-446.

Haines, W.B. and Benzian, B. (1956) Some manuring experiments on oil palm in Africa. *Empire Journal of Experimental Agriculture* 24, 136-160.

Haines, W.B. and Guest, E. (1936) Recent experiments on manuring Hevea and their bearing on estate practice. *Empire Journal of Experimental Agriculture* 4, 300-324.

Heathcote, R.G. (1973) The use of fertilizers in the maintenance of soil fertility under intensive cropping in northern Nigeria. *Proceedings, 10th Colloquium*, International Potash Institute, pp. 335-340.

Heathcote, R.G. and Smithson, J.B. (1973) Boron deficiency in cotton in northern Nigeria 1. Factors influencing occurrence and methods of correction. *Experimental Agriculture* 10, 199-208.

IBSRAM (1986) Land development - Management of Acid Soils. *IBSRAM Proceedings, No.4.* IBSRAM, Bangkok.

IBSRAM (1987) Land development and Management of Acid Soils in Africa II. *IBSRAM Proceedings No.7.* IBSRAM, Bangkok.

IBSRAM (1988) Site selection and characterisation. *IBSRAM Technical Notes No. 1*. IBS-
RAM, Bangkok.

IBSRAM (1989) Soil management and smallholder development in the Pacific Islands.
IBSRAM Proceedings, No.8. IBSRAM, Bangkok.

IBSRAM (1990) The establishment of soil management experiments on sloping lands.
IBSRAM Technical Notes No.4. IBSRAM, Bangkok.

IGBP (1990) The International Geosphere–Biosphere Programme: a Study of Global
Change. The initial core projects. *Report No.12*. IGBP, Stockholm.

IGBP (1991) Global change system for analysis, research and training. *Report No.15*.
IGBP, Stockholm.

IGBP (1992) Global change and terrestrial ecosystems: the operational plan. *Report
No.21*. IGBP, Stockholm.

IRRI (1981) *Priorities for Alleviating Soil Related Constraints to Food Production in
the Tropics*. IRRI, Los Banos.

IRRI (1981–1985) *Annual Reports*. IRRI, Los Banos.

IRRI (1990) *Program Report*. IRRI, Los Banos.

Jameson, J.D. and Kerkham, R.K. (1960) The maintenance of soil fertility in Uganda. I.
Soil fertility experiment at Serere. *Empire Journal of Experimental Agriculture*
28, 179–192.

Johnston, A.E. and Powlson, D.S. (1994) The setting-up, conduct and applicability of
long-term continuing field experiments in agricultural research. In: Greenland,
D.J. and Szabolcs, I. (eds) *Soil Resilience and Sustainable Land Use*. CAB International, Wallingford, pp. 395–422.

Jones, M.J. and Wild, A. (1975) Soils of the West African Savanna. *Technical Communication No. 55*, Commonwealth Bureau of Soils, Harpenden, pp. 1–246.

Jones, P.A. and Wallis, J.A.N. (1963) A tillage study in Kenyan coffee. III. The long-term
effects of tillage practices upon yield and growth of coffee. *Empire Journal of
Experimental Agriculture* 31, 243–254.

Kang, B.T. and Balasubramanian, V. (1990) Long-term fertilizer trials on Alfisols in West
Africa. *Transactions 14th International Congress of Soil Science, Kyoto, Japan* 4,
20–25.

Kemmler, G. and Malicornet, H. (1976) Fertilizer experiments – the need for long-term
trials. *IPI Research Topics, No.1*. International Potash Institute, Berne, pp. 1–36.

Krishnamoorthy, K.K. (1982) Long-term fertiliser experiments. In: *Review of Soil
Research in India*, Part 1, pp. 453–464.

Kwakye, P.K., Dennis, E.A. and Asmah, A.E. (1994) Management of a continuously
cropped forest soil through fertiliser use. *Experimental Agriculture* (in press).

Lal, R. (1992) Tropical Agricultural Hydrology and Sustainability of Agricultural Systems – A ten year watershed management project in southwestern Nigeria. IITA
and Ohio State University, Columbus, Ohio, pp. 1–303.

Lal, R. and Kimble, J. (1994) *Greenhouse Gas Emissions and Carbon Sequestration.
Advances in Soil Science*, special issue (in press).

Latham, M. and Syers, J.K. (1994) Using collaborative research networks to promote
sustainable land use. In: Greenland, D.J. and Szabolcs, I. (eds) *Soil Resilience and
Sustainable Land Use*, CAB International, Wallingford, pp. 513–520.

Leather, J.W. (1911) Records of drainage in India. *Mem. Department of Agriculture in
India (Chem.)* 2, 63–140.

Le Mare, P.H. (1959) Soil fertility studies in three areas of Tanganyika. *Empire Journal of Experimental Agriculture* 27, 197-222.

Le Mare, P.H. (1968) Experiments on the effects of phosphate applied to a Buganda soil III. A chemical study of the soil phosphate, the fate of fertilizer phosphate and the relationship with iron and aluminium. *Journal of Agricultural Science* 70, 281-285.

Le Mare, P.H. (1972) A long-term experiment on soil fertility and cotton yield in Tanzania. *Experimental Agriculture* 8, 299-310.

Le Mare, P.H. (1977) Experiments on the effects of phosphorus on the manganese nutrition of plants. *Plant and Soil* 47, 593-630.

Lungu, O.I.M. (1987) Monitoring soil fertility. In: *IBSRAM Proceedings, No.7*, pp. 161-173.

Mahapatra, K., Raheja, S.K. and Bapat, S.R. (1972) Response of rice varieties to NPK fertilizer and micronutrients. *Revista Il Riso* 21, 349-362.

Malthus, T.R. (1798) An essay on the principle of population as it effects the future improvement of society. *Reprinted by Royal Economic Society*, London.

May, E.B. (1956) The manuring of oil palms. *Journal of the West African Institute for Oil Palm Research* 2, 6-73.

McWalter A.R. and Wimble, R.H. (1976) A long-term rotational and manurial trial in Uganda. *Experimental Agriculture* 12, 305-317.

Mukerjee, H.N. (1965) Fertilizer tests in cultivators' fields. In: *Mineral Nutrition of the Rice Plant*. IRRI and Johns Hopkins University Press, Baltimore, pp. 329-354.

Nambiar, K.K.M. and Ghosh, A.B. (1984) Highlights of research of a long-term fertilizer experiment in India (1971-1982). *LTFE Research Bulletin No.1*, Indian Agricultural Research Institute, New Delhi, pp. 100.

Nambiar, K.K.M. and Abrol, I.P. (1989) Long-term fertiliser experiments in India. An Overview. *Fertiliser News* 34, 11-20.

Nambiar, K.K.M., Soni, P.M., Vats, M.R., Sehgal, D.K. and Mehta, D.K. (1989) *All India Coordinated Research Project on Long-Term Fertilizer Experiments, Annual Report, 1985-86/1986-87*. Indian Agricultural Research Institute, New Delhi.

Nicol, H. (1935) Mixed cropping in primitive agriculture. *Empire Journal of Experimental Agriculture* 3, 189-195.

Nye, P.H. (1951) Studies on the fertility of Gold Coast soils. Part I. General account of the experiments. *Empire Journal of Experimental Agriculture* 19, 217-223.

Nye, P.H. (1953) A survey of the value of fertilizers to the food farming areas of the Gold Coast. *Empire Journal of Experimental Agriculture* 21, 176-183.

Nye, P.H. (1992a) Towards the quantitative control of crop production and quality. I. The role of computer models in soil and plant research. *Journal of Plant Nutrition* 15, 1131-1150.

Nye, P.H. (1992b) Towards the quantitative control of crop production and quality. II. The scientific basis for guiding fertilizer and management practice, particularly in poorer countries. *Journal of Plant Nutrition* 15, 1151-1173.

Nye, P.H. and Greenland, D.J. (1960) The soil under shifting cultivation. *Technical Communication No.51*, Commonwealth Bureau of Soils, Harpenden, pp. 1-156.

Obi, J.K. (1959) *Samaru Technical Report No.8*. Ministry of Agriculture, Northern Nigeria, Kano.

Ofori, C.S. (1973) Decline in fertility status of a tropical forest ochrosol under continuous cultivation. *Experimental Agriculture* 9, 15-22.

Pereira, H.C., Chenery, E.M. and Mills, W.R. (1954) The transient effects of grasses on the structure of tropical soils. *Empire Journal of Experimental Agriculture* 22, 148-160.

Pereira, H.C., McCulloch, J.S.G., Dagg, M., Kerfoot, O., Hosegood, P.H. and Pratt, M.A.C. (1962) Hydrological effects of changes in land use in some East African catchment areas. *East African Agricultural and Forestry Journal* (special issue) 27, 1-131.

Pichot, J., Sedogo, M.P., Poulain, J.F. and Arrivets, J. (1981) Evolution de la fertilité d'un sol ferrugineux tropical sous l'influence de fumures minérales et inorganiques. *Agronomie Tropicale* 28, 751-766.

Pieri, C.J.M.G. (1992) *Fertility of Soils: A Future for Farming in the West African Savannah.* Springer-Verlag, Berlin, pp. 1-348.

Powlson, D.S. and Johnston, A.E. (1994) Long-term field experiments: their importance in understanding sustainable land use. In: Greenland, D.J. and Szabolcs, I. (eds) *Soil Resilience and Sustainable Land Use.* CAB International, Wallingford, pp. 367-394.

Richardson, H.L. (1946) Soil fertility maintenance under different systems of agriculture. *Empire Journal of Experimental Agriculture* 14, 1-17.

Richardson, H.L. (1962) Developments in the FAO Freedom from Hunger campaign. *Proceedings, No.73.* The Fertiliser Society, London, pp. 1-46.

Robison, D.M. and McKean, S.J. (1992) *Shifting Cultivation and Alternatives: An Annotated Bibliography.* CAB International, Wallingford.

Russell, E.W. (1968) The place of fertilisers in food crop economy of tropical Africa. *Proceedings, No.101.* The Fertiliser Society, London, pp. 1-48.

Sanchez, P.A., Bandy, D.E., Villchica, J.H. and Nicholaides, J.J. (1982) Amazon basin soils:management for continuous crop production. *Science* 216, 821-827.

Singh, L. and Balasubramanian, V. (1979) Effects of continuous fertilizer use on a ferruginous soil (Haplustalf) in Nigeria. *Experimental Agriculture* 15, 257-265.

Sree Ramulu, U.S. (1990) Long term effects on soil chemical properties due to continuous application of inorganic fertilizers and cattle manure to 122 crops. *Transactions 14th International Congress of Soil Science, Kyoto, Japan* 4, 386-387.

Stephens, D. (1960) Fertilizer trials on peasant farms in Ghana. *Empire Journal of Experimental Agriculture* 28, 1-15.

Stephens, D. (1967) Effects of grass fallow treatments in restoring fertility of Buganda clay loam in South Uganda. *Journal of Agricultural Science* 68, 391-403.

Swindale, L.E. (1980) Toward an internationally coordinated program for research on soil factors constraining food production in the tropics. In: *Priorities for Alleviating Soil Related Constraints to Food Production in the Tropics.* IRRI, Los Banos, pp. 5-22.

Tinker, P.B.H. (1964) Studies on soil potassium I. Cation activity ratios in Nigerian soils II. Equilibrium cation activity ratios and responses to potassium fertilizer of Nigerian oil palms. *Journal of Soil Science* 15, 24-41.

Tropsoils (1987) *Technical Report for 1986-87.* Department of Soil Science, North Carolina State University, Raleigh, pp. 1-360.

Vine, H. (1953) Experiments on the maintenance of soil fertility at Ibadan, Nigeria, 1922-1951. *Empire Journal of Experimental Agriculture* 21, 65-85.

Watts Padwick, G. (1983) Fifty years of Experimental Agriculture II. The maintenance of soil fertility in tropical Africa: a review. *Experimental Agriculture* 19, 293-310.

Yogaratnam, N. and Weerasuriya, S.M. (1984) Fertilizer responses in mature Hevea under Sri Lankan conditions. *Journal of the Rubber Research Institute of Sri Lanka* 62, 19–39.

Zandstra, H.G., Price, E.C., Litsinger, J.A. and Morris, R.A. (1981) A Methodology for On-farm Cropping Systems Research. IRRI, Los Banos, pp. 1–146.

12

Long-term Agricultural Experiments in Eastern Europe

12.I: Long-term Continuous Experiments in Poland, Bulgaria, Czech Republic and Slovakia

S. MERCIK
Department of Agricultural Chemistry, Warsaw Agricultural University, Rakowiecka 2630, 02-528 Warsaw, Poland.

Introduction

Two long-term fertilizer experiments in Skierniewice, Poland, have been continued without interruption since 1923. One with field crops is in the Experimental Field of Warsaw Agricultural University, the other with vegetable crops is in the Experimental Field of the Research Institute of Vegetable Crops.

Skierniewice is situated in the Great Valley with climatic conditions typical of the central region of Poland. Precipitation is relatively low (520 mm year^{-1}) and its distribution is somewhat unfavourable for cereals because water deficiency often occurs in May or June. The mean annual temperature is 7.8°C.

Soils of both Experimental Fields are podsols of the very good ryeland complex. The content of silt and clay is 15–17% in the A1 layer (0–25 cm) and 25% in the layer below 40 cm. The humus content in the cultivated layer ranges from 1.2 to 1.4%.

Field Crops

The experiment with field crops tests different crop rotations, with and without farmyard manure (FYM) and with and without legumes. The land occupied by each rotation is divided into plots for different fertilizer and CaCO$_3$ treatments (Table 12.1); plot treatments have been constant since 1923, but the amounts of fertilizer used have been increased in recent years.

Soil properties

Both fertilization and crop rotation have affected considerably the availability of macro- and microelements and some physical properties of the soil. Consequently there are now large differences in yields and chemical composition of the plants grown on the different plots. Using FYM increased exchange capacity, sum of exchangeable bases, humus content, capillary water capacity, permeability coefficient and number of crumbs >0.25 mm in diameter. Liming (1.6 t ha^{-1} CaO every 4 years) maintained pH$_{KCl}$ between 5.6–6.0. On unlimed plots without N (O, PK) soil pH is now about 4.5, whereas with N as ammonium nitrate and ammonium sulphate it is about 4 and 3.2–3.5 respectively.

The humus content of the soil is now only slightly less on plots without nitrogen than it is on plots with nitrogen fertilization (Table 12.2). The concentration of readily soluble phosphorus and potassium are both much higher where these nutrients were applied than where they were not given (Table

Table 12.1. The crop rotations and fertilization tested in the long-term experiments with arable field crops, Skierniewice, Poland.

Rotation			Fertilization[a]	
A		Arable rotation without farmyard manure (FYM) or legume	0 PK NP	
	A1–4	N as NH$_4$NO$_3$	NK	
	AF1-3	N as (NH$_4$)$_2$SO$_4$	NPK	CaNPK
A	A5–8	Arable rotation without FYM or legume		Ca CaPK
	A9–11	Arable rotation without FYM but with legume		CaNP CaNK
E		5-course rotation: potatoes (30 t FYM), spring barley, legume, winter wheat, rye	NPK	CaNPK
D	D5	Potatoes in monoculture		
	D6	Rye in monoculture		

[a] Rates per ha since 1967: N, 90 kg; P, 26 kg; K, 91 kg; Ca, 1.6 t CaO.
NPK annually; CaO every 4th year.

Table 12.2. Effect of treatment on some soil properties in 1992 in the long-term field experiment with arable crops, Skierniewice, Poland.

Soil property	Crop rotation[a]	Fertilization since 1923[a]					
		CaNPK	NPK	CaPK	CaNP	CaNK	Ca + FYM
pH_{KCl}	A5-11	5.9	4.0	6.1	6.0	5.8	
	E	5.2	4.2	5.4	5.2	4.8	
	D6	6.1	4.4	6.1	5.5	5.6	6.2
Humus, %	A5-11	1.06	1.01	0.92	1.01	1.07	
	E	1.56	1.59	1.60	1.56	1.50	
	D6	1.17	1.07	0.97	1.13	1.08	1.54
Readily soluble P mg kg^{-1}	A5-11	84	78	119	100	17	
	E	63	50	60	53	29	
	D6	67	48	76	50	14	50
Readily soluble K mg kg^{-1}	A5-11	116	98	152	40	115	
	E	109	102	135	54	125	
	D6	82	82	113	36	104	202

[a] See Table 1

12.2). Between 1957 and 1992 readily soluble P and K did not change in soil to which these elements were not applied. Thus phosphorus and potassium taken up by crops during this 35-year period came from reserves of P and K.

Yields of the crops

Only yields from rotations A 5–11, E and D are discussed here. Largest potato yields were obtained in rotation E (with FYM and legumes), they were much less in rotation A (without FYM or legumes) and were lowest where potatoes were grown in monoculture (rotation D) (Table 12.3). Potatoes grown in monoculture yielded more when given the same amount of N, P and K in mineral fertilizers than in manure. In all rotations, mineral fertilizers always had the largest effect on potatoes when they were grown without FYM.

Yields of rye with CaNPK and NPK were no larger in rotation E (with FYM and legumes) than in rotation A (without FYM) (Table 12.3). Without NPK the yields of rye were much lower in rotation A than in rotation E (Mercik, 1989). Rye in monoculture (rotation D) gave yields only about 10–15% lower than in rotation A. Differences in yields between the fully fertilized crop and that without NPK increased with time, especially where nitrogen was withheld. In the last 20 years, when the same amount of NPK was applied in FYM and mineral fertilizer, yields were less (by 25–35%) with FYM than with mineral fertilizers.

Table 12.3. Yields (1984–91) of potato tubers and rye grain (t ha[-1]) in the long-term experiment with field crops, Skierniewice, Poland.

Fertilization[a]	Potatoes				Rye			
	Crop rotation[a]				Crop rotation[a]			
	A	E	D	Mean	A	E	D	Mean
CaNPK	28.9	36.3	15.8	27.0	4.76	4.70	3.95	4.47
NPK	23.4	34.1	13.9	23.8	4.93	4.63	4.07	4.54
CaPK	12.5	28.8	10.0	17.1	1.93	2.96	2.27	2.38
CaNP	11.4	29.9	12.1	17.8	3.92	4.23	3.34	3.84
CaNK	11.8	31.0	13.5	18.8	2.91	3.93	2.81	3.22
Ca	10.9	23.8	7.9	14.2	1.59	2.77	1.83	2.07
LSD		1.38				0.16		
Mean	16.5	30.6	12.2		3.34	3.88	3.03	
LSD		0.46		0.8		0.06		0.10

[a] See Table 12.1.

Spring barley gave much larger yields in rotation E (FYM and a legume) than in rotation A. The differences were especially large on unlimed plots and those without NPK (Table 12.4). Like barley, winter wheat was adversely

Table 12.4. Yields (1984-91) of winter wheat and spring barley (grain t ha[-1]), in the long-term experiment with field crops, Skierniewice, Poland.

Fertilization[a]	Winter wheat			Spring barley		
	Crop rotation[a]			Crop rotation[a]		
	A	E	Mean	A	E	Mean
CaNPK	3.52	4.39	3.95	4.31	4.93	2.62
NPK	2.31	3.75	3.03	2.24	4.26	3.25
CaPK	2.00	3.48	2.74	2.11	3.56	2.83
CaPN	2.74	4.02	3.38	3.15	4.31	3.73
CaKN	2.22	3.34	2.78	2.93	3.65	3.29
Ca	1.73	2.66	2.19	1.57	2.94	2.07
LSD		0.19			0.16	
Mean	2.42	3.61		2.72	3.94	
LSD		0.05	0.14		0.05	0.01

[a] See Table 12.1.

affected by soil acidification, especially in rotation A. In rotation A, winter wheat responded most to nitrogen and least to potassium (Table 12.4). The strong response to nitrogen was observed early in the experiment but that to acidification and P and K fertilization occurred only after 30 years (Mercik, 1989).

Vegetable Crops

Long-term fertilization and rotation experiments with vegetable crops have been continued since 1923 by the Research Institute of Vegetable Crops in Skierniewice. Three amounts of farmyard manure and of mineral fertilizers were tested as shown in Table 12.5. Soil analysis data in Table 12.5 were obtained on samples taken in 1989 (Rumpel *et al.*, 1993). Nitrogen was determined by the Kjeldahl method. Other elements were determined after a two-step digestion: first a 3 h combustion in an electric furnace at 500°C followed by a wet digestion with a nitric–perchloric acid mixture (5:2 by vol.). After evaporation the residue was dissolved in 1:1 hydrochloric acid.

In 1980, all treatments were limed to bring soil pH within the range 6.7–7.1, but individual plots had values as low as 4.8 in 1989, 9 years later. The increase in acidity was mainly due to the mineral fertilizers and increased with increasing amounts of NPK. The acidification due to fertilizers was to some extent buffered by the application of FYM (Table 12.5).

The humus content in soil was strongly affected by the application of both organic manures and mineral fertilizers. In the manured soil, organic C content was appreciably larger (0.97–1.48%) than on the mineral fertilized soil (0.40–0.54%). Organic carbon content in soil increased with the amount of manure applied and decreased with the amount of mineral fertilizer used. The total nitrogen content in soil increased with increasing amounts of both fertilizers and manures. Plots with FYM contained, on average, over 1000 mg N kg^{-1}, whereas those with mineral fertilization had about 700 mg N kg^{-1} of soil. Soils fertilized with FYM plus NPK fertilizers had an N content similar to those fertilized with FYM only.

Phosphorus and potassium were both larger in soil given both FYM and fertilizers. When applied separately the two larger amounts of FYM left larger residues of P and K than did the mineral fertilizers. In the FYM treatments, each increment of FYM increased the Ca and Mg content of soil, whereas with mineral fertilization the calcium content of soil generally decreased.

There was little difference in the microelement content of soil according to past manuring with FYM or fertilizers. Data in 1989 show, however, that each manure increment increased the content of copper, manganese and zinc. Increasing mineral fertilizers decreased copper content and had no effect on the manganese and zinc contents of the soil.

Table 12.5. Effect of treatment on some soil properties in 1989 in the long-term field experiment with vegetable crops, Skierniewice, Poland.

Fertilization since 1923[a]	pH	% organic C	Content (mg kg^{-1} soil)							
			N	P	K	Ca	Mg	Cu	Mn	Zn
FYM										
20	6.3	0.97	934	441	695	909	469	8.5	95	26
40	6.4	1.13	1054	531	743	1282	511	9.6	111	31
60	6.0	1.48	1229	572	909	1677	601	11.0	128	37
FYM										
20 + 2NPK	5.5	0.87	872	559	787	940	458	10.1	125	28
40 + 2NPK	5.9	1.13	999	614	873	1203	501	9.1	130	34
60 + 2NPK	5.9	1.55	1221	720	831	1442	502	10.0	130	37
1NPK	5.8	0.54	668	438	623	813	428	9.9	99	29
2NPK	5.3	0.49	697	465	683	500	447	9.5	117	25
3NPK	5.0	0.40	799	458	727	495	399	8.0	96	35

[a] Treatment per hectare each year
FYM: farmyard manure, 20 40 60 t.
1NPK: 75 kg N, 50 kg P_2O_5, 100 kg K_2O.
2NPK: 150 kg N, 100 kg P_2O_5, 200 kg K_2O.
3NPK: 225 kg N, 150 kg P_2O_5, 300 kg K_2O.

Other Long-term Experiments in Poland

Five other fertilizer experiments have been continued for more than 25 years. They are

1. On a grey–brown podsolic soil, mineral and organic fertilization has been compared on a 5–course rotation: sugar beet, spring barley, red clover, winter rape, winter wheat since 1948 (Urbanowski, 1993).

2. Two permanent fertilizing experiments established at Chylice in 1955 on a black earth type soil; compare a four-crop rotation with legume and a three-crop rotation without legume. Fertilizer treatments: D, NPK, ½ NPK + ½ FYM, FYM (Lenart *et al.*, 1993).

3. On a light-textured loamy sand, FYM has been applied alone, with NPK, with lime or with clay in a 4-course rotation: potato, oat, lupin and rye since 1957 (Grzebisz *et al.*, 1993).

4. On a loamy sand, a comparison of mineral fertilizers without or with FYM for a 4-course rotation: potato, spring barley, winter rape, rye since 1960 (Kuszelewski and Labetowicz, 1992).

5. On permanent grass grown on a brown earth soil comparison of N, P or Ca since 1968 (Mazur *et al.*, 1993).

The Value of the Long-term Experiments in Poland

The long-term experiments with both field- and vegetable-crops at Skierniewice, and others in Poland are an invaluable asset for a number of investigations. These include

1. effect of differences in the nutrient content of soil on the chemical composition of plants and their quality;
2. modification of critical levels for nutrients in soils and plants;
3. response of particular varieties of cereals, potatoes and berry shrubs to different amounts of nutrients in soil;
4. investigations on the recovery of soil fertility after many years of inappropriate fertilization;
5. determination of both organic carbon and nutrient balances over long periods;
6. movement of nutrient elements in the soil profile;
7. the possibility of increasing potato yields in continuous cultivation by using varieties resistant and susceptible to nematodes;
8. effect of mineral nitrogen upon leguminous crops grown on soil infected with various nodule bacteria.

Experiments in Bulgaria, Czech Republic and Slovakia

The experiments in these three countries can be related to the long-term experiments in Poland.

In Bulgaria, at Plovdiv, a long-term experiment on saline soil has continued since 1968 (Gorbanov *et al.*, 1993). Crops were cultivated in rotation: maize, wheat, vetch–oat mixture, wheat, maize, spring barley. Phosphorus is applied every year, or once every 6 years to maize (rotational manuring), or once every six years in combination with manure, straw and green manure. The total amount of phosphorus taken up by plants during three 6–year rotations (18 years) and the coefficient of P utilization by crops (in %) were highest where P was applied every year and lowest where P was applied once in 6 years with green manure (Table 12.6).

Long-term (since 1955) experiments with field crops on clay-loam in Czechia are located in Praha-Ruzyne, Lukavec and Caslav. Plants grow in three-crop rotations: (i) 45% cereals + 33% root crops + 22% fodder; (ii) 50% cereals + 50% root crops; (iii) 67% cereals + 11% root crops and 22% fodder. Mineral NPK are applied in several combinations and different rates of N.

Research on grassland fertilization in the former Czechoslovakia became more intensive in the early 1960s. Three long-term experiments in Slovakia (Princina, Velka Luka, Chyzerovce) and three in Czechia (Cernikovice, Lety, Ollow) have been continued since then. Increasing amounts of N (0–200 kg

Table 12.6. The phosphorus taken up by plants and coefficient of P utilization (in %).

Treatments per hectare	P (kg P$_2$O$_5$ha^{-1}) taken up by plants in rotation number				P utilization (%)
	1	2	3	total	
P, 0	145	205	173	523	—
P, 67 kg P$_2$O$_5$ every year	309	359	345	1013	41
P, 400 kg P$_2$O$_5$ once in 6 years	309	367	317	993	39
P, 400 kg P$_2$O$_5$ once in 6 years + manure	303	345	332	980	38
P, 400 kg P$_2$O$_5$ once in 6 years + straw	289	346	326	961	36
P, 400 kg P$_2$O$_5$ once in 6 years + green manure	271	331	315	917	33

N ha^{-1}) in two ratios of N:P:K (1:0.3:0.83 and 1:0.15:0.41) are compared when applied to three valley meadows in Slovakia (Lihan and Jezikova, 1990). There was no decrease in yield with time, on the contrary, yields tended to increase. During the first 4 years differences between plots with and without nutrients increased, but they remained relatively stable as the population of various grass species stabilized.

References

Gorbanov, S., Matev, J., Tomov, T., Rachovski, G. and Kostadinova, S. (1993) Phosphorus balance under conditions of field crop rotation. *Proceedings of the International Symposium 'Long-term static fertilizer experiments I', June 15–18 1993, Warsaw, Cracow*, pp. 175–186.

Grzebisz, W., Kocialkowski, W. and Diatta, J. (1993) Effect of changes in the properties of light soil on copper sorption in the result of long-term fertilization with farmyard manure fermented with lime and clay. *Proceedings of the International Symposium 'Long-term static fertilizer experiments I', June 15–18 1993, Warsaw, Cracow*, pp. 139–150.

Kuszelewski, L. and Labetowicz, J. (1992) Long-term static fertilizer experiments in Lyczyn. P. 1, II, *Polish Agricultural Annual* 109(3), 81–108.

Lenart, S., Suware, J. and Gawroniska-Kulesza, A. (1993) The effect of long-term fertilization and crop rotation on formation of soil fertility and yielding of plants. Part II. Yielding of plants. *Proceedings of the international symposium 'Long-term Static Fertilizer Experiments II,' June 15–18, Warsaw, Cracow*, pp. 139–154.

Lihan, E. and Jezikova, O. (1990) Parameter changes of valley grasslands under long-term fertilization. Zaverecna sprava (Final report of GRI), *Banska Bistrica* p. 103.

Mazur, K., Mazur, B., Mazgaj, M. and Szczurowska, B. (1993) Twenty-five years of permanent fertilization in mountain meadow (Czarny Potok). Effect of different fertilization and liming on the crop and some soil properties. *Proceedings of the*

International Symposium 'Long-term static fertilizer experiments II', June 15-18, Warsaw, Cracow, pp. 5-22.

Mercik, S. (1989) Yielding of rye, wheat and potatoes depending on the many year fertilization and crop rotation. P. I rye, p. II wheat. *Soil Science Annual* Xl, 1s, 191-212.

Rumpel, J., Nowosielski, O. and Paul, M. (1993) The long-term fertilization and rotation experiment with vegetable crops in Skierniewice. 1. Effect on organic and mineral fertilization on nutrients and heavy metals content and pH of soil. *Proceedings of the International Symposium 'Long-term static fertilizer experiments II', June 15-18, Warsaw, Cracow*, pp. 183-200.

Urbanowski, S. (1993) The effect of long-term various fertilization treatments on crop yields and soil fertility. *Proceedings of the International Symposium 'Long-term static fertilizer experiments I', June 15-18 1993, Warsaw, Cracow*, pp. 151-173.

Long-term Agricultural Experiments in Eastern Europe

12.II: Some Recent Results from the Long-term Experiment at Fundulea, Romania

V. Mihaila[1] and C. Hera[2]

[1]*Research Institute for Cereals and Industrial Crops, 8264 Fundulea, Jud.Călărasi, Romania*; [2]*Department of Research and Isotopes, International Atomic Energy Agency, UNO-City, A2271, Wagramer Str.5, PO Box 100, 1400 Vienna, Austria.*

Introduction

Much research has established the importance of fertilizers in increasing the fertility of soil and in influencing its productivity (Bramao and Riquier, 1968; Hera and Eliade, 1978). It has been observed that applying fertilizers causes many changes in soil, including chemical changes that can negatively or positively influence its productiveness (Hera and Mihaila, 1981). Hera and Borlan (1975) showed that the long-term application of nitrogen fertilizers to the cambic chernozem at Fundulea lowered soil pH by 2.85×10^{-4} units pH kg^{-1} N, and that phosphorus fertilizers have influenced the quantity of the slightly soluble phosphates.

Paterson and Richter (1966) showed that applying potassium salts for more than 77 years increased the exchangeable K in soil down to 60 cm. At Rothamsted, UK, after 100 years the total nitrogen content of the soil has oscillated between 0.106 and 0.103% in the unfertilized plots, and between 0.121 and 0.115% where nitrogen, phosphorus and potassium fertilizers have been applied each year (Jenkinson and Rayner, 1977). Soil organic matter content has also been greater in soils fertilized regularly with larger rather than smaller amounts of nitrogen fertilizers (Blewins *et al.*, 1977). At Grignon, France, the nitrogen content of unfertilized soil was 0.118% after 73 years, and 0.142% in soil fertilized with nitrogen, phosphorus and potassium (Morel, 1976).

This chapter briefly describes the effects of fertilizers on yields of wheat and maize grown in a 2-year rotation since 1967 on a cambic chernozem at ICCPT (Research Institute for Cereals and Industrial Crops), Fundulea, Romania. The particle size distribution of the soil is approximately: fine sand, 35%; silt, 30%; clay, 35% with 250 me kg^{-1} total exchangeable bases. Initially, soil pH was about 6.5 and total N 0.155%.

Yields

Figure 12.1a shows yields of wheat and rainfall (November–June) and Fig. 12.1b those of maize and rainfall (November to August) between 1967 and 1992. On the control plot, yields of wheat grain have remained reasonably constant at about 1.5 t ha^{-1} whereas those of maize grain after remaining at about 5.5 t ha^{-1} until 1985, have decreased in recent years, perhaps due to a run of seasons with below average November–August rainfall.

On soils given 80 kg P$_2$O$_5$ ha^{-1} each year and 120 kg N for wheat and 150 kg N for maize, yields have also tended to be less in recent years compared to the first 20 years. Applying fertilizer gave a larger proportional increase in wheat yields than in maize yields.

Changes in Some Soil Properties

Soil samples from the tilled layer were taken in 1992, after 26 years, and analysed for total nitrogen (by the Kjeldhal method), pH in water (soil:solution ratio 1:2.5) and humus content (by the modified Schölemberger method). Available phosphorus and exchangeable potassium were determined after extraction with calcium lactate and ammonium acetate respectively. Changes in soil agrochemical indices, after 26 years of applying different amounts of nitrogen, phosphorus and potassium fertilizers are significant and there is a linear relationship between the agrochemical index and amount of nutrient applied.

Soil pH

Soil pH (Fig. 12.2) has declined, depending on the amount of N applied. It now ranges from 6.2 on the plot without N to 5.8 where 160 kg N ha^{-1} has been applied annually for 26 years. The decrease can be described by the following equation:

$$pH = 6.22 - 0.0025 \, N \qquad (12.1)$$

where N = amount of N (kg ha^{-1}) applied annually.

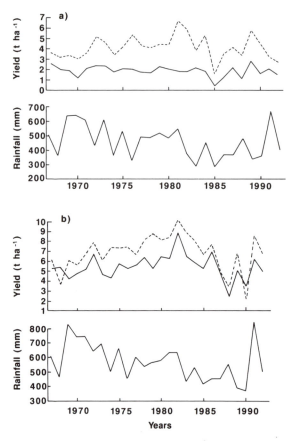

Fig. 12.1. (a) Wheat grain yields on the unfertilized control plot and that given 120 kg N and 80 kg P_2O_5 ha⁻¹ each year and total rainfall during November–June. (b) Maize grain yield on the unfertilized control plot and that given 150 kg N and 80 kg P_2O_5 ha⁻¹ each year and total rainfall during November–August.

If the total amount of N is taken into account, instead of the yearly application, then equation (12.1) becomes:

$$pH = 6.22 - 0.000096 \ QN \tag{12.2}$$

where QN is the total amount of nitrogen (kg ha⁻¹) used.

Equation (12.2) can be used to predict the change in pH of the cambic chernozem at Fundulea when a certain quantity of nitrogen is applied as ammonium nitrate to arable crops. For example, by applying 120 kg N ha⁻¹ for 9 years, the pH of this soil will decrease by 0.1 pH unit. Phosphorus and potassium fertilizers had less influence on soil pH.

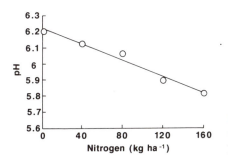

Fig. 12.2. Relationship between surface soil pH and annual N application after 26 years.

Soil organic matter

Nitrogen fertilizers have affected both the humus and total nitrogen content of the soil. The relationship between the quantity of nitrogen applied each year and the humus content of the soil is shown in Fig. 12.3. Soil humus now ranges from 3.42% without nitrogen to 3.63% with 160 kg N ha^{-1} each year for 26 years. The relationship between humus content and applied fertilizer N is described by equation (12.3):

$$H(\%) = 3.39 + 0.0013 \, N \tag{12.3}$$

where H = percentage humus, N = nitrogen (kg ha^{-1}) applied annually.

If the total amount of nitrogen applied is used, then equation (12.4) is obtained:

$$H(\%) = 3.39 + 0.00005 \, QN \tag{12.4}$$

where QN = the total quantity of applied nitrogen (kg ha^{-1}).

From equation (12.4), it can be shown that applying 2000 kg N ha^{-1} to crops over a number of years increases soil humus content by 0.1%. This could be achieved by applying 125 kg N ha^{-1} year^{-1}, for a period of 16 years.

Soil nitrogen

Previous research has shown that the total nitrogen content of the soil has been significantly increased where nitrogen fertilizers have been applied. Soil nitrogen now ranges from 0.148% without fertilizer nitrogen to 0.167% with

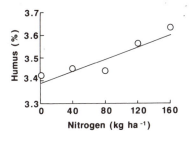

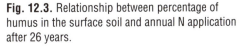

Fig. 12.3. Relationship between percentage of humus in the surface soil and annual N application after 26 years.

160 kg N ha^{-1} annually to maize and wheat grown in rotation for a period of 26 years. Equation (12.5) describes the change:

$$N\% = 0.145 + 0.00011\ N \tag{12.5}$$

where N = the quantity of nitrogen (kg ha^{-1}) applied annually for 26 years.

The average annual application of N needed to maintain the initial content of total soil nitrogen (Ni = 0.158%) was 118 kg N ha^{-1}. If less than this quantity was applied then soil N declined and where no nitrogen fertilizer was applied, the total nitrogen content of the soil at Fundulea decreased by 0.0005% yearly.

Readily soluble phosphorus

The readily soluble phosphorus content of soil is strongly influenced by the use of phosphorus fertilizers. After 26 years at Fundulea there was 126 mg kg^{-1} readily soluble P in the fertilized plot given 160 kg P_2O_5 ha^{-1} year^{-1} compared to 10 mg P kg^{-1} in the unfertilized plot. There was a strong linear correlation between the applied P fertilizer and the available phosphorus in soil (Fig. 12.4):

$$P = 3.6 + 0.7075\ P_2O_5 \tag{12.6}$$

where P is readily soluble P (mg kg^{-1}) and P_2O_5 (kg P_2O_5 ha^{-1} year^{-1}) is the phosphorus applied for 26 years to maize and wheat grown in rotation. Equation (12.6) shows that approximately 44 kg P_2O_5 ha^{-1} year^{-1} must be applied to maintain the initial readily soluble phosphorus content of the soil at 31 mg P kg^{-1}.

By introducing into Equation (12.6) the proportion of the added P that remains readily soluble, the initial content of available phosphorus (Pi, mg kg^{-1}), the total phosphorus applied (QP_2O_5, kg ha^{-1}) in n years and the annual maintenance quantity of phosphorus (44 kg P_2O_5 ha^{-1}), the change in the readily soluble phosphorus (P, mg kg^{-1}) after n years would be:

$$P\ (mg\ kg^{-1}) = Pi\ (mg\ kg^{-1}) + 0.0272\ (QP_2O_5 - n44) \tag{12.7}$$

From equation (12.7) it can be shown that to increase the readily soluble P

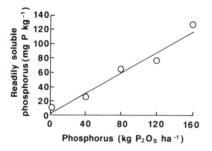

Fig. 12.4. Relationship between readily soluble phosphorus in soil and the amount of phosphorus applied each year.

content of the soil at Fundulea by 1 mg kg^{-1}, requires 81 kg P$_2$O$_5$ ha^{-1} year^{-1}, i.e. 37 kg P$_2$O$_5$ ha^{-1} must be applied annually in addition to the phosphorus required for maintenance (44 kg P$_2$O$_5$ ha^{-1} year^{-1}).

Readily soluble potassium

The initial content of exchangeable K (K$_i$) was 234 mg kg^{-1} and after 26 years exchangeable K had declined to 218 mg kg^{-1} in the unfertilized plot and increased to 327 mg kg^{-1} where 120 kg K$_2$O ha^{-1} had been applied annually (Fig. 12.5). The linear regression gives:

$$K = 209 + 0.8775 \, K_2O \tag{12.8}$$

where K is exchangeable K (mg kg^{-1}) and K$_2$O (kg K$_2$O ha^{-1}) is the amount of K applied annually for 26 years. From equation (12.8) it can be calculated that to maintain the initial soil K content (234 mg kg^{-1}) would have required 28.5 kg K$_2$O ha^{-1} year^{-1}. Taking into account the initial content (K$_i$) and the amount of K (28 kg) to maintain the initial level, the proportion of added K that remains exchangeable and the total quantity of potassium used (QK$_2$O) in a period of *n* years, the change in exchangeable soil K at Fundulea is:

$$K \, (mg \, kg^{-1}) = K_i \, (mg \, kg^{-1}) + 0.03375 \, (QK_2O - n28) \tag{12.9}$$

From equation (12.9) to increase readily soluble K by 1 mg kg^{-1} requires an annual addition of 31 kg K ha^{-1} in addition to the 28 kg K needed to maintain the initial level of soil K. This quantity is small and may be explained by the richness in potassium of the cambic chernozem at Fundulea and by the capacity of the soil to maintain the pool of exchangeable K from the non-exchangeable pool and mineral reserves.

Conclusions

The application of chemical fertilizers has had both direct and indirect effects on some agrochemical indices of the soil. Whereas nitrogen fertilizers have changed significantly the soil pH, the total nitrogen and humus content of the

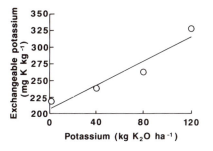

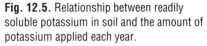

Fig. 12.5. Relationship between readily soluble potassium in soil and the amount of potassium applied each year.

soil, the phosphorus and potassium fertilizers have influenced significantly only the readily soluble P and K content of the soil, respectively.

The indirect effect of the application of a nutrient through an increase in the consumption of the other elements, has been much weaker than the direct effect. Thus, the application of a larger quantity of nitrogen increased the amount of phosphorus taken from the soil compared to where a smaller amount was given and this decreased the amount of this element in the soil. This decrease, however, was much less than when phosphorus fertilizers were not applied.

A decrease of 0.1 pH unit in the cambic chernozem at Fundulea occurred after 9 years when 120 kg N ha^{-1} year^{-1} was applied as ammonium nitrate.

An annual application of ammonium nitrate, supplying 118 kg N ha^{-1}, together with P maintained the total nitrogen content of the soil. Where nitrogen fertilizers were not applied total soil nitrogen decreased by 0.0005% N annually. An increase of 0.1% N was obtained after a period of 16 years by applying 125 kg N ha^{-1} year^{-1}.

To increase the readily soluble P content of the cambic chernozem at Fundulea by 1 mg kg^{-1} P required 37 kg P$_2$O$_5$ ha^{-1} in addition to the quantity (44 kg P$_2$O$_5$ ha^{-1} year^{-1}) required to maintain the initial level of readily soluble P.

To maintain the initial exchangeable K content of the cambic chernozem at Fundulea required only 28 kg K$_2$O ha^{-1} year^{-1} because of the richness of this soil in potassium and the ability of the non-exchangeable and mineral K to support the pool of exchangeable K.

Summary

The pH, the humus and total nitrogen contents and the readily soluble P and K in the cambic chernozem at Fundulea have been determined after 26 years of applying increasing amounts of nitrogen, phosphorus and potassium fertilizers to arable crops grown in a long-term experiment. Changes in each factor have depended on the quantity of fertilizer applied and with the aid of linear regression equations, some very significant correlations have been found. The quantities of applied nutrients needed to maintain the initial content of the soil have been determined from formulae for estimating change in soil pH, the humus and total nitrogen content and readily soluble P and K depending on the quantity of fertilizers applied.

References

Blewins, R.L., Thomas, G.W. and Cornelius, P.L. (1977) Influence of no-tillage and nitrogen fertilization on certain soil properties after 5 years of continuous corn. *Agronomy Journal* 69, 383–386.

Bramao, D.L. and Riquier, J. (1968) Matière organique et fertilité du sol. *Pontifica Academia Scientarium* 45-53.

Hera, C. and Borlan, Z. (1975) Ghid Pentru Alcatuirea Planurilor de Fertilizare. *Editura Ceres*, pp. 63-67.

Hera, C. and Eliade, G. (1978) Evolutia unor indici ai fertilitati solului in experiente de lunga durata lu imgrasaminte. *Analele ICCPT* 43, 175-179.

Hera, C. and Mihaila, V. (1981) The changing of some agrochemical indices of the soil by the application of the fertilizers. *Analele ICCPT* 47, 319-327.

Jenkinson, D.S. and Rayner, J.H. (1977) The turnover of soil organic matter in some of the Rothamsted Classical experiments. *Soil Science* 123, 5, 298-305.

Morel, R. (1976) Utilisation des résultats obtenus sur des essais de longue dureé dans l'étude des transferts de l'azote dans le sol. *Annales Agronomiques* 27, 5-5, 567-582.

Paterson, J.W. and Richter, A.C. (1966) Effect of long-term fertilizer application on exchangeable and acid soluble potassium. *Agronomy Journal* 58, 589-591.

13

Long-term Experiments in Africa: Developing a Database for Sustainable Land Use under Global Change

M.J. Swift[1], P.D. Seward[1], P.G.H. Frost[2], J.N. Qureshi[3] and F.N. Muchena[3]

[1]Tropical Soil Biology and Fertility Programme (TSBF), c/o UNESCO ROSTA, Nairobi, Kenya; [2]Department of Biological Sciences, University of Zimbabwe, Harare, Zimbabwe; [3]National Agricultural Research Laboratories (NARL), Kenya Agricultural Research Institute (KARI), Nairobi, Kenya.

Summary

There have been a substantial number of long-term experiments in Africa addressing a wide range of purposes. There is, however, no comprehensive inventory of these experiments. Initial analysis suggests that whereas a number are still extant and actively researched, some have been judged to have reached the end of their useful existence, and yet others have been discontinued or diminished in intensity because of lack of resources.

The increasing importance accorded to the development of sustainable management practices for tropical land-use systems and the apprehension of the potential impact of global climatic and environmental change has raised new interest in the datasets from these experiments as well as the possibilities for new initiatives in long-term monitoring and experimentation.

This chapter suggests developing a secure network of long-term experiment sites in Africa. Such a programme would need to draw on the experience of the past, and a sample of past and current experiments is reviewed in terms of the usefulness of the information generated, both in terms of the original objectives and in relation to current concerns. A proposal is made for action to promote:

1. publication of an inventory of past experiments and their results;
2. establishment of a database for long-term experiments in Africa;
3. development of a network of on-going experiments rigorously selected on the basis of representativeness in terms of environment, land use, and relevance to sustainable agricultural development. Experimental designs and measurement protocols in such experiments should also be capable of 'generic' interpretation.

Introduction

The importance of long-term experimental data sets to the development of agriculture and for the management of natural ecosystems is extensively reviewed elsewhere in this volume (e.g. Southwood, Chapter 1; Johnston, Chapter 2; Powlson, Chapter 6). The value of current data sets in Africa has, however, tended to be overlooked, as there is currently no published inventory of these long-term experiments. We have now, however, reached a point in the history of the biological and environmental sciences where the value of long-term experiments and observations is more apparent than it has ever been. In order to advance productively and economically in the area of long-term experimentation it is essential to learn from the lessons of the past and to determine a clear set of priorities for the future.

The purpose of this chapter is to propose the initiation of an international effort to establish a firm basis for long-term experimentation in Africa and to ensure the most effective utilization of results from past, current and future research of this type. In making this case we first outline what we perceive as some of the priorities for future research, we then outline a possible framework for a network of long-term experiments in Africa, and finally we review a small sample of past and current experiments to illustrate some of the main issues raised in the earlier sections.

The Research Agenda for African Agriculture

The present plight of African agriculture needs no detailed analysis here, but provides the context for which any research programme should be designed. The most fundamental fact is that per capita food production in SubSaharan Africa (SSA) has continued to decline over the last two decades, at a time when an increasing trend has been observed in the rest of the world, including the other tropical regions. Any research programme has to address this issue which is rooted in the circumstances and characteristics of African agricultural systems. Although the continent is highly heterogeneous in environment and the distribution of natural resources, as well as in patterns of social, cultural and political organization, a number of generalizations can be advanced.

With some important exceptions, the great majority of the farming population has farms of small size (0.5-5 ha) which are poorly, if at all, integrated with a market economy. In most countries of the continent the farmers operate within a weak infrastructure which gives little assistance or guidance to their farming activities. Conversely, however, they enjoy a strong communality at the village level which moulds the farming practice through joint activities and obligations (Izac and Swift, 1993). The farmers are thus primarily concerned with meeting subsistence goals, together with familial and social obligations. The farm product is usually diverse, based on intercropping or rotation of a variety of food crops, and plants of medicinal, ritual or cash value. Perennials are a common component of many systems. Livestock are integrated with arable cropping in many areas, although once again the production goals are diverse and rarely confined to meat or dairy produce. Biological production is influenced by significant constraints in many areas, ranging from low radiation and chronic pest problems in the humid zone to unreliable water supplies in arid areas. Low inherent soil fertility is widespread and exacerbated by intensification of cultivation without access to increased inputs of plant nutrients to replace those removed from the soil.

The primary focus of agricultural research remains that of providing the means to increase per capita production. In the last two decades, however, additional dimensions have been added to the research agenda by the increasing emphasis placed on the sustainability of farming practice. Although the sustainability agenda has a very broad context, embracing in some definitions, environmental and resource conservation, economic viability and the quality of life and human equity, its fundamental core rests on the ability to maintain or improve production trends over time (Lynam and Herdt, 1989; Spencer and Swift, 1992; Izac and Swift, 1993). Production can be diverse, economic viability can be defined in liberal terms and the quality of life may embrace simple or complex criteria but at the farming system scale the *sine qua non* is the capacity to meet the needs of the farmer both today and into the future. This implies *inter alia* the capacity to renew resources essential to production (a high efficiency of resource use) and the employment of 'low-risk' systems with the property of buffering the stresses and disturbances imposed by climate, pests and changing economic, political and social circumstances. Society's expectations of sustainable development may be more complex than those of individual farmers but the agricultural researcher may feel it sufficient to address those of the farmer as a first approximation (Izac and Swift, 1993).

The present and future impact of global change adds yet another dimension to this agenda. 'Global change' is shorthand for a diverse collection of factors which are perceived to embody the potential for significant modification to the patterns of distribution of different organisms to biological productivity and the loss of biotic diversity. The major driving forces are those of 'carbon fertilization' imposed by the increase in atmospheric CO_2, modifications of climate induced by the altered greenhouse gas composition, and the pollution of the environment by waste materials generated by industry (Jones *et al.*, Chapter 9; Tinker, Chapter 22). Interwoven with these physical forces of change are those imposed more directly by society itself, in particular, that of change in land use and land cover, which is often perceived as a response to the rate of human population increase.

The impact of these factors has been perceived in terms of changes in global patterns of agricultural production as well as in other areas of human activity (Nix, 1985; Parry, 1992). For regions dominated by small-scale subsistence farming the changes may mean the difference between success and failure in terms of emergence from the current poverty trap described above, although this is not an issue which has gained much attention either in the debate on global change or in that on sustainable agriculture.

Sustainable Agriculture, Global Change and Long-term Experiments

Even on the basis of the above brief discussion we suggest that it is evident that long-term studies are one essential key to any comprehensive research approach to the present and future problems of African agriculture. Information is required over time on the performance of current agricultural systems, and of the proposed sustainable alternatives. Furthermore, the performance must be assessed against the background of environmental change. Change includes the trends and fluctuations in 'natural' phenomena such as climate and chronic pest infestation, and in imposed pressures such as withdrawal of agricultural inputs and human-induced epidemics (Izac and Swift, 1993). Assessment of performance should be sufficiently penetrative to permit interpretation of the mechanisms of response of the system to the changing environment and the links between the driving and the state variables. It is indeed clear that any assessment of future sustainability or response to change beyond the highly site specific should not just be based on empirical evidence but be derived from evidence which has a genuinely predictive power.

Later in this chapter we make an appraisal of a small sample of current or past long-term experiments in Africa, that shows that there is a valuable body of information already available, a conclusion strengthened by the broader analysis given by Greenland (Chapter 11). We are also aware that there are some excellent experiments still extant on the continent which can serve many of the purposes of the current research agenda. We are also aware, sadly, that there are many experiments that have been discontinued for one reason or another, that would also have made a valuable on-going contribution to the present need for long-term data.

From this limited review we have also drawn a number of conclusions concerning actions that should be taken to ensure that there is an adequate long-term database for the needs of agricultural research in Africa through the next two decades.

Publication

First and foremost we recommend that action should be taken to ensure publication of the results of these studies in the open literature. There is a great deal of invaluable information from excellent experiments which is still in inaccessible data files or only published in the 'grey' literature of institutional reports or conference proceedings. Assistance will be needed from the international community for this, both in terms of funding support – to create the time for already overworked scientists to carry out this activity – and to provide the necessary infrastructure to facilitate this process.

An Africa-wide information system for long-term experimental data

Secondly there is an urgent need for an international database of long-term experiments. We now recognize that the 'normal' environment of agriculture fluctuates both seasonally and over longer cycles. There is, however, little clear evidence of how agricultural systems respond to such influences over the long-term. It is likely that there is already a good deal of valuable information available, but the accessibility, for comparative purposes, is still low. The availability of an interactive electronic database at both international and national levels would be of great value to those recommending or determining agricultural policy in Africa.

Such a database should be structured in such a way as to enable cross-reference between experiments in different countries and environments. Structural features might include:

- Georeferencing of the experimental site
- Reference of the site characteristics to an agreed agroecological classification (e.g. by the use of climatic, soil and vegetational data)
- Reference of the land-use practice to an agreed system of classification (e.g. of cropping, livestock, plantation system, etc.)
- Categorization of the types of treatment, environmental and response variables measured, and their patterns in time.
- Details of experimental design and experiment history

A long-term experimental network for Africa

We further advocate the need for a network of long-term experiments in Africa to ensure that records will be available to future generations to assist in making decisions in relation to sustainable land use. Such a network can be based, in part, on existing experiments but new ones may need to be established. In view of the resources involved in establishing and guaranteeing even a small number of sites a very rigorous set of criteria would need to be employed to select sites for the network.

We propose three categories of criteria: environmental representation; land use representation; experimental design and measurement.

Environmental representation

For the purposes of global change assessment, a geographical spread of sites, representative of major agroecological zones, will be needed. This issue is being addressed within the new ecoregional structuring in the CGIAR system and in the choice of Regional Research Centres in the International Geosphere Biosphere Programme. With respect to agricultural systems it is important that social, political and cultural criteria be used in the definition of zones in addition to the conventional biophysical parameters, like climate, soil, geology, vegetation. Not all defined zones need necessarily be represented. There may be good reasons for selecting one zone rather than another (e.g. on the basis of perceived likelihood of significant climatic change, or critical socioeconomic indicators) but this must be done within a comprehensive comparative framework, rather than on the basis of arbitrarily determined criteria no matter how important they seem in the short term.

Land-use representation

One way in which socioeconomic considerations can be incorporated into the selection criteria is by means of a land-use classification. Different agroecological zones embrace a variety of land uses, which differ in their relevance to sustainable development and, in particular, their probable pattern of response to the forces of change. A particularly challenging task in selecting sites for an international network would be the identification of a set of 'generic' land-use types that represent differing types of biological response to environmental fluctuation and change and management intervention. In addition to these biological aspects, however, selection criteria should also include socioeconomic judgments of the long-term probability of the relevance of land-use systems to the farmer and society. In this respect critical decisions have to be made as to whether the set should be predominantly represented by current systems, with or without minor incremental modifications, or by prototypes of alternative systems which are predicted to form part of the agriculture of the future on the basis of various models, like, for example, attempts to predict how to achieve an intensification of production.

Experimental designs and measurements

The requirement for 'generic' types of experiment, representative of a wide range of current or predicted land use practices requires a blend of both realism and compromise in the design. The realism should be embedded in the fundamental features of scale in space and time and in the nature of the resource base. In determining the size and location of plots the scale chosen should lie as close as possible to that actually managed by the farmer; the relationship to natural variation in resources (e.g. changes in soil character-istics) should also be taken into consideration. The time period for the exper-iment must be long enough to encompass the periodicity of any influencing variables. For instance many climates are now known to have cyclical periods of seven to eleven years. Any experiment that does not go through at least two or three cycles will still lack full predictive power concerning the relationship between climate and performance. The impact of climate change is expected to be observable at scales of decades to centuries. The available resource base should also reflect that which may be available to the 'average' land user. The intensive use of labour, tools, machinery and purchased inputs which is characteristic of many experiments on research stations may increase the efficiency of management but equally reduces the realism and opportunity for extrapolation of many experiments. On the other hand compromises have to be made in the details of the components such as the choice of plant vari-eties, and of specific management practices, because these vary enormously between one farmer and the next.

The choice of measurements is probably the most important factor in determining the usefulness of an experiment for purposes beyond those restricted to the specific site and treatment combination. An essential feature is that, irrespective of the range, frequency and intensity of subsequent measurements, the initial site characterization should be comprehensive and intense. It is extraordinary how often the potential to interpret the results of long-term experiments is limited by the lack of adequate initial character-ization.

Most experiments are established with a specific target in mind, a particu-lar hypothesis to test. For generic purposes, however, measurements should go beyond the minimum required to test the specific hypothesis. For instance an hypothesis linking the effect of grazing intensity to changes in sward quality and productivity may apparently require only measurements of plant compo-sition and biomass. The capacity to interpret the experiment, let alone extrapolate the results to other sites, is however greatly enhanced when measurements of climate, soil, fire and pest variables are made. Thus as a gen-eral rule the range of measurements should include not only the controlled treatment variables and the main (intended) response variables but also the non-controlled driving variables, and a range of additional state variables. The key to the choice of what might be seen as an infinitely long list lies in predicting the key interactions that may determine the outcome of the main effects. This necessitates a system level concept, in the form of a 'model' – whether verbal or mathematical – at the outset of the experiment. Measurements that reveal the interactions between components, particularly between driving and state variables, are likely to be of the greatest generic value.

The frequency and intensity of measurement should be determined by the natural, or imposed, variation in distribution over space and time of the variable in question. Thus not all factors need be measured with the same frequency or intensity.

Long-term Experiments in Africa: a Preliminary Appraisal

In the previous section we have advocated the need for an on-going network of long-term experiments in Africa, and indicated some of the criteria that could be used to select sites and determine assessment protocols. Is this network already catered for by the extant experiments? Could it be established with only minor modifications to existing site practices? The immediate answer is that it is very difficult to judge because of the first issue which we raised in the previous section; the lack of published information. We can, however, use some of the criteria we have advanced above to analyse a small sample for which there is readily available evidence.

Long-term experimentation has been a feature of both agricultural and ecological research in Africa throughout much of this century (Greenland, Chapter 11). A number of reviews have been published of agricultural experiments (Edwards and Blackie, 1979), rangeland experiments (O'Connor, 1985; O'Reagain and Turner, 1992) and experiments in natural ecosystems (McNaughton and Campbell, 1991). These reviews only cover a fraction of the experiments, however. A comprehensive analysis and database is still critically lacking. Furthermore results from long-term experiments have not been used to a satisfactory extent to influence policy or management recommendations. For instance O'Reagain and Turner (1992), in a review of over 50 experiments designed to provide an empirical basis for grazing management recommendations in southern African rangelands, concluded that some of the recommendations are not supported by experimental evidence. Despite this, these recommendations are often still made. For example, the type of grazing system appears to be of minor significance compared to stocking rate as the key determinant of both range condition and animal production (the same conclusion was reached by O'Connor, 1985). The differences in range condition and animal production between continuous and rotational grazing systems are minimal (O'Reagain and Turner, 1992). Nevertheless, rotational grazing is widely promoted. Likewise, there is no evidence of increased vegetation productivity and animal performance under multipaddock grazing systems (eight or more paddocks per herd) compared with grazing systems based on fewer numbers of paddocks. Given the higher costs of establishing and maintaining a multipaddock system, there does not seem to be any economic justification for this system either (Mentis, 1991). Nevertheless, some rangeland advisors continue to promote multipaddock schemes or variants thereof (e.g. Savory, 1988).

The empirical basis for other grazing management recommendations can also be queried (O'Reagain and Turner, 1992). Neither controlled selective grazing nor non-selective grazing, both of which have their advocates, have been adequately tested. The widely practised principle of separating different land units and their associated soil and vegetation types, to prevent selective over-utilization of preferred types, also remains untested.

Table 13.1 and Fig. 13.1 summarize information on two types of long-term experiments in Africa. Those in Fig. 13.1 are for rangeland management experiments whereas Table 13.1 summarises information on 21 arable cropping experiments published between 1937 and 1990, but is almost certainly incomplete because for many experiments results have not been published in the open literature.

Arable cropping experiments

These experiments may seem the most relevant to the research agenda described above. In our literature search we defined long-term experiments as those extending over a period of 5 years and more, even though this falls

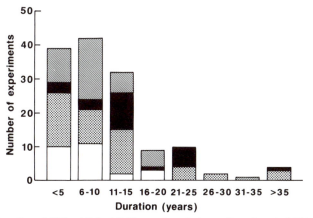

Fig. 13.1. Duration of 139 published field experiments in south and central African rangelands. Different patterns indicate different management systems: open, bush clearing; stipple, grazing; solid, fire; cross-hatched, other. Data from Barnes (1965), Geldenhuys (1977), O'Connor (1985), Calvert (1986), Chidumayo (1988), and O'Reagain and Turner (1992).

Table 13.1. Catalogue of long-term arable cropping experiments identified by a literature search.

Site	Duration	Agroecological zone[a]	Experimental aspects[a]	Measurements[a]	Reference[b]
Ife, Nigeria	1971–86	MSH	Fert	S,C	1
Embu, Kenya (2)	1952–56	DSH	Rot; Fert; OM	Y,S,C	2
	1952–66	DSH	Rot; Fert; OM	Y,S,C	3
Matuga (Coast), Kenya	1953–60	DSH	Fert; OM	Y,S	4
Mwanhala, W.Tanzania (2)	1956–61	DSH	Fert; OM	Y,S,C	5
	1957–66				6
Ukiriguru, Tanzania	1952–68	MSH	Fert; OM	Y,S,C	7
Saria, Burkina Faso	1960 –	DSH	Fert; OM	Y,S,C	8
Kabete, Kenya	1976–	DSH	Fert; OM	Y,S,C	9
Mingano, Tanzania	1981–	DSH	Fert	Y,S,C	10
Ghana (30 exp; 3–9 years)	1948/9–56/7	DSH	Fert; OM	Y	11
IFDC/ICRISAT, Niger	1983–	DSH	OM	Y	12
Serere, Uganda	1937–64	MSH	Rot; OM	Y,S,C	13 14 15
Gezira, Sudan	1941–66	MSH	Rot; OM	S,C	16
Samaru, Nigeria	1950–64		Fert; OM	S,C	17
	1967–76		Fert	Y,S	18
Sambwa, Tanzania	1962–67	DSH	Rot	Y,S	19
Kawanda, Uganda	1957/9–1963/4	DSH	Rot	Y,S	20
Mubuku, Uganda	1963–67	DSH	I; Fert	Y,S,C	21
Uganda (2 exps: 8 sites)	1959–63	MSH	Fert; OM	Y	22
Onne, Nigeria	1983–88	H	Tillage; OM	Y,S	23
Ibadan, Nigeria	Various	MSH	Rot	Y,S	24
Yambio, Sudan	1949–55	MSH	Rot	Y,S	25
N. Nigeria	1958–62	DSH	Rot	Y,S	26

[a] DSH = Dry subhumid (600–1200 mm rainfall); MSH = Moist subhumid (1200–1500 mm rainfall); H = Humid (>1500 mm rainfall); Fert = Fertilizer; OM = Organic Matter application; I = Irrigation; Y = Yield; S = Soil; C = Climate; Rot = Rotation.

[b] (1) *Expl. Agric.* 1989, 25: 207–215; (2) *E. Afr. Agric. For. J.* 1970, 246–253; (3) *E. Afr. Agric. For. J.* 1976, 42: 201–218; (4) *E. Afr. Agric. For. J.* 1962, 75–81; (5) *E. Afr. Agric. For. J.* 1963, 231–239; (6) *E. Afr. Agric. For. J.* 1971, 8–14; (7) *Expl. Agric.* 1972, 8: 299–310; (8) *Agronomie Tropicale*, 36: 122–133; (9) KARI, Nairobi; (10) Mingano Agric. Res. Inst.: Tanga, Tanzania; (11) *Emp. J. Exp. Agric.* 29: 181–195; (12) *Fertilizer Res.* 26: 327–339; (13) *Emp. J. of Exp. Agric.*, 1960, 28: 179–192; (14) *Expl. Agric.* 1976, 12: 305–317; (15) *J. Agric. Sci. Camb.* 1967, 68: 391–403; (16) *Expl. Agric.* 1970, 6: 279–286; (17) *Expl. Agric.* 1969, 5: 241–247; (18) *Expl. Agric.* 15: 257–265; (19) *Expl. Agric.* 1970, 1–12; (20) *J. Agric. Sci. Camb.* 1967, 68: 391–403; (21) *Expl. Agric.*1969; 5: 17–24; (22) *Expl. Agric.* 1990, 26: 235–240; (23) *Expl. Agric.* 1969, 5: 263–269; (24) *Emp. J. of Expl. Agric.* 1953, 21: 65–84; (25) *Emp. J. Exp. Agric.* 1956, 24: 75–88; (26) *Expl. Agric.* 1966, 2: 33–43.

below the desirable timescale of a decade as discussed above and elsewhere in this volume. We have also restricted our analysis to those experiments relating to annual crops; forestry and plantation crops have not been included for reasons of brevity.

An analysis of data from 21 sets of experiments (Table 13.1) shows that only three of the experiments have a life span of over 20 years. Of these, only one is still on-going, i.e. the Saria trial in Burkina Faso which was started in 1970. Others of shorter duration are still extant, however, for instance the Kabete trial in Kenya which commenced in 1976. Most of the other trials were either discontinued due to lack of financial/human resources, or because they were considered to have served their purpose.

Most of the experiments were designed to determine the effects of inorganic fertilizers and/or organic inputs on crop yields and/or soil fertility, although the effects of rotations have also been investigated particularly in the earlier trials (e.g. Martin and Biggs, 1937). Although yields were measured in all of the experiments, climatic variables were documented in only nine, and soil parameters in only 15. Of these, the highest frequency of soil analysis was biannual, with most reporting soil analysis at only the beginning and end of the trial. There is no evidence in the literature that other measurements of factors outside the treatment variable which would influence yields, such as pest and disease incidence were measured, and economic parameters were recorded only once. The fact that the non-controlled driving variables (precipitation, temperature, etc.), and state variables, such as soil parameters, are frequently not documented mean that the results of these experiments can not be extrapolated beyond the site, and beyond the time frame of experimentation. All the experiments were researcher-managed and conducted on research stations. Collectively they reflect very poorly the circumstances facing the African farmer as outlined earlier.

One of the most complete sets of measurements is that in the the on-going Kabete trial, in Kenya, which was established on a Nitisol in 1976. The soil, known locally as a Kikuyu Red Loam, is considered to be fertile, with a deep, well-drained profile and moderate amounts of available Ca and Mg, and adequate K, but low in total P and available N. The soil type occurs over a significant area of the East African Highlands (Fig. 13.2), which has traditionally supported the high density of populations in this region. The objective was to determine appropriate methods for maintaining and improving the fertility of the soil through the use of N and P fertilizers, farmyard manure (FYM) and crop residues following small-scale farmers' husbandry practices under a continuous cropping regime. Maize and beans are grown in the long and short rainy seasons, respectively. Yields and soil chemical parameters are measured annually and biannually respectively, and climatic variables are measured daily.

The field trial has, to date, yielded some important information, and the data (see selection of treatments, Fig. 13.3) clearly show the importance of

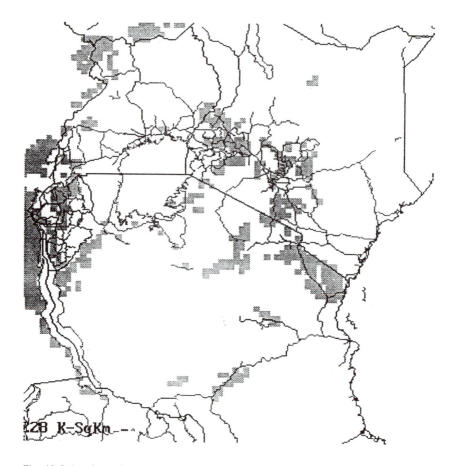

Fig. 13.2. Land area (shaded) occupied by Nitisols in East Africa. These soils cover 228,000 km².

long-term experimentation at this site. Data are presented as three-year sliding averages to diminish the effects of between-seasonal variability. The trough in 1982–86 was due to a drought in 1984 which resulted in total crop failure. If the experiment had been terminated after ten years, it could have been reasonably concluded that all the systems were sustainable and that there was no difference in the effect of using inorganic fertilizers only compared with that of FYM. After the tenth year, however, yields tended to decline more rapidly where only chemical fertilizers were applied compared to the FYM treatment. If recommendations based on these results had been given to farmers, then the consequences for long-term productivity could have been devastating, requiring large inputs of organic matter and/or lime (if available) to restore soil fertility. Other results worthy of note are the significant decline in soil

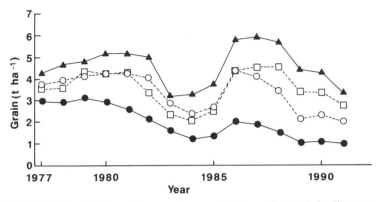

Fig. 13.3. Effect of applications of farmyard manure (FYM) and inorganic fertilizers on grain yields of maize in a long-term experiment at Kabete, Kenya. Values are calculated as sliding averages over three years. ●, no inputs; □, FYM = 5 t dry matter ha⁻¹; ○, NP = 60 kg N + 60 kg P_2O_5 ha⁻¹; ▲ FYM + NP.

organic matter in all treatments (Fig. 13.4), even in the treatments with large applications of organic material. It is interesting to compare the high rate of change in this tropical system with the slower rates observed in many temperate soils (Powlson, Chapter 6; Jenkinson *et al.*, Chapter 7). Is this simply an environmental effect or is it also related to the nature of the soils and the quality of the organic materials within them?

On the negative side the trial suffers from a number of deficiencies, the most significant being that it poorly represents farmers' activities. For example,

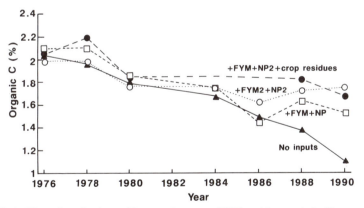

Fig. 13.4. Effect of applications of farmyard manure (FYM) and inorganic fertilizers on soil organic matter content in a long-term experiment at Kabete, Kenya. □, FYM + NP = 5 t dry matter + 60 kg N + 60 kg P_2O_5 ha⁻¹; ○, FYM2 + NP2 = 10 t dry matter + 120 kg N + 120 kg P_2O_5 ha⁻¹; ●, FYM2 + NP2 + crop residues (incorporation of stover of the preceding crop); ▲, no inputs.

most smallholder farmers do not grow simple seasonal rotations of maize and beans, intercropping being their preferred practice. Secondly, the importance of livestock in the agricultural systems means that *either* animal manure *or* maize stover is available for application, but rarely both. More importantly, it is not possible to infer the economically optimum rate of organic/inorganic application to sustain profitability on a long-term basis.

Secondly, the small plot size (6 m²) means that subdivision of plots is not possible to introduce other treatments which are, or may become, of interest. Also, as is the case with many experiments, it would have been better with hindsight to measure additional soil variables. For instance, it is not yet possible to determine the causes for the decline of yield due to continuous inorganic fertilization. One hypothesis is that it is due to the acidifying effects of the fertilizers, and/or the loss of soil structure due to a decline in the different fractions of soil organic matter. Measurements of exchangeable aluminium, soil carbon fractions, and soil physical parameters are required to test these hypotheses. Although archived soil samples exist, the amounts are so small that deciding which of these additional measurements to make will not be easy.

In spite of these deficiencies, it may be argued that the true value of this experimental data set is yet to be realized. The availability of uncontrolled driving variable data (precipitation and temperature data), and available records of all inputs, and outputs in terms of mass and nutrient content will enable the initialization of plant–soil models, such as Century (Parton *et al.*, 1987) or the current Rothamsted model (ROTHC 26.2, Jenkinson *et al.*, 1992) and a test of their ability to predict the effects of the different treatments on crop productivity, and soil parameters, under tropical conditions. Archived soil material could be used for analyses of important state variables to test the models. Such models, once validated, may be used to predict the effect of different management practices on productivity of different crops on different soils under widely differing climatic conditions, thereby diminishing the need for long-term experimentation in the distant future.

Rangeland management experiments in southern Africa

Rangelands comprise those uncultivated lands that can be browsed or grazed by wild or domestic animals. Most rangelands occur in arid and semiarid areas (< 700 mm precipitation p.a.) but large areas of uncultivated woodland on infertile soils occur throughout the higher rainfall areas of the tropics. Rangelands have traditionally supported both commercial and subsistence livestock husbandry, although production in the moist tropical woodlands has been constrained by the extremely poor nutritional quality of the grasses, especially during the dry season, and by disease, principally trypanosomiasis. In Africa at least, increasing recognition is being given to the multiple resource character of rangelands and moves to develop integrated resource use and management programmes, including ones based on free-living indigenous herbivores.

Long-term research, often involving some degree of experimentation, has been carried out on rangelands in southern and central Africa since the early part of this century. O'Connor (1985) identified 126 separate field experiments, ranging from two to 47 years in duration, aimed at investigating aspects of herbaceous layer dynamics in southern African savannas. O'Reagain and Turner (1992) identified 50 grazing experiments, ranging from two to 56 years duration, which were set up to provide an empirical basis for various grazing management recommendations. Specific issues include whether rotational or continuous grazing is economically more profitable and ecologically more sustainable; the optimum number of camps through which to move the herd in a rotational system; whether controlled selective grazing or non-selective grazing should be practised (Tainton, 1981); and the effects of burning and/or tree clearing on sward quality. Associated issues concern the utility of burning, as a means of reducing moribund material and altering species composition, particularly the reduction in density of woody plants that might compete with grasses, and the optimum vegetation composition, usually obtained by either partial or complete bush clearing. Particular attention has been directed towards determining the consequences of a given management regime in terms of changes in vegetation composition, soil loss and changes in productivity.

Detailed analysis of these experiments, however, reveals a number of shortcomings in design, measurement or interpretation, which severely limits their utility both for the stated objectives as well as for any more general purposes such as those addressed in this chapter (O'Connor, 1985; O'Reagain and Turner, 1992; Frost, unpublished). We select for comment here a number of the issues to which we gave prominence in the preceding section.

Scale in space and time

The issue of the spatial scale of an experiment in relation to the natural scale of the process being studied has seldom been properly considered, even though the scale at which a problem is studied determines both the patterns and the processes observed. For logistical reasons, many field experiments are carried out on small plots even though the ecological processes that they are intended to represent occur naturally over much larger areas.

For instance the geometric mean size of plots in 16 of the best known long-term fire experiments in Africa is 0.46 ha (range: 0.06–8.0 ha) and the distribution is strongly negatively skewed (Frost, unpublished). In contrast, 56 wild fires in the Hwange National Park, Zimbabwe, averaged 18 300 ha (range: 3–330,000 ha; Zimbabwe Department of National Parks and Wild Life Management, unpublished records). Fires may not reach peak rates of spread or intensities on small plots, potentially influencing their impact on the biota.

At the same time, species which are adversely affected by fire may be able to recruit from individuals in unburnt areas adjacent to small plots, increasing the rate of recovery or extending the period of persistence.

Scale effects are also apparent in many of the grazing experiments. Most of these experiments are carried out in paddocks that are at most only a few hectares in area, less than one-hundredth of the size of farm paddocks and thousands of times smaller than the area over which pastoralists range. At this experimental scale, livestock are unable to exploit the natural mosaic of the landscape by moving between patches in response to localized pulses in production. The few spatially extensive studies of both livestock and wildlife suggest that this pattern of use is essential in enabling the animals to overcome nutritional and energy constraints on production (McNaughton, 1985; Coppock *et al.*, 1986; Scoones, 1989).

The time scale of experiments is also often insufficient. Only 42% of 139 so-called long-term field experiments lasted longer than ten years; whereas 28% lasted for five years or less (Fig. 13.1 above). Experiments involving the use of fire have been maintained in Africa for many years and constitute some of the longest running of the experiments on the continent. The average duration, in 1993, of ten of the longest running fire experiments in central and southern Africa is 41 years (range 21–60 years: Frost, unpublished). Most of these experiments are still being run but a number are threatened with closure because of the cost of upkeep and apparent diminishing returns. Few results have been published in recent years.

Many of these studies have shown that annual or biennial late dry season fires reduce woody plant densities, either by killing saplings outright or by preventing their recruitment to the canopy, thereby reducing seed output. But in some cases the published conclusions reflect the duration of the experiment at the time of the survey. For example, Barnes (1965) analysed the results of the first nine years of a fire experiment designed to test whether burning and mattocking, separately and combined, could control the coppice regrowth of *Brachystegia spiciformis* and *Julbernardia globiflora*, the dominant species in many central African woodlands. Little difference between the plots burnt at any frequency was apparent at the end of the first decade; all were covered with dense woody regrowth. Barnes concluded that burning alone was ineffective as a control measure.

If the experiment was evaluated today, the conclusion would be different. Four decades of annual, late dry season burning have created an open grassland with the formerly dense woody coppice reduced to a few scattered, fire-suppressed, individuals (Frost, unpublished). In contrast, most of the four-yearly burnt plots and some of the three-yearly burnt plots have returned to woodland. This illustrates again the problem of scale. Given the long life spans of many savanna trees and their capacity to resprout from large, well established rootstocks, conclusions about the impacts of fire can only safely be made after an experiment has been running for some decades.

The choice of measurements

The choice of variables and the intensity of sampling also often left a lot to be desired. In 44 long-term studies reviewed in detail by O'Connor (1985) the intensity of sampling of the key response variables (herbaceous composition, plant density, total basal cover, herbaceous yield, and soil properties) differed greatly (Table 13.2). Most studies monitored herbaceous composition (86%), although not necessarily to species level, and total basal cover (61%), but both at relatively long intervals (3–4 years). Yield, in contrast, was measured annually, but only in about one-sixth of the studies. Plant densities and soil properties were seldom measured and then only at long intervals. This latter factor is one that becomes more significant when the question of long-term sustainability of management systems is an aim of the project. Although the main focus of most of the experiments was on grass sward development, it is difficult to envisage that some interaction with soil properties is not an essential part of the process of change in plant condition. Yet as Table 13.2 shows this has very rarely been taken into account.

Interactions of variables

There has also been a general failure to separate the effects of external driving forces, such as intraseasonal and interannual variations in rainfall, from treatment effects on both species composition and yield (O'Connor, 1985; Walker *et al.*, 1986; Dye and Walker, 1987). Most of the studies have concentrated on descriptions of changes at a community or system level, with little research into the population processes generating such changes (O'Connor, 1985; O'Reagain and Turner, 1992). Moreover, the studies have usually focused on

Table 13.2. Intensity of sampling of response variables in long-term experiments (>10 years duration, *n* = 44) in southern African rangelands. (From O'Connor, 1985.)

Response variable	Number of census per study[a]						Total number of studies	Mean number of years per census
	1	2	3	4	5	>5		
Herbaceous composition	9	8	4	2	0	15	38	3.7
Total basal cover	6	6	2	2	2	9	27	3.9
Yield	1	0	0	0	0	6	7	1.1
Plant density	2	1	0	0	0	0	3	11.0
Soil properties	4	1	0	0	0	0	5	19.0

[a] Sampling intensity given as number of studies.

single components and processes, whereas it is becoming increasingly apparent that it is the interactiveness of components that is important.

A case in point is fire experiments which are conducted in the absence of grazing. Long-term fire experiments were generally set up to evaluate the effects of frequency and seasonal timing of fire, or its exclusion, on woody plant density, size structure and species composition, as well as on herbaceous layer properties (Trapnell, 1959; Barnes, 1965; van Wyk, 1971; Kennan, 1972; Geldenhuys, 1977; Gertenbach and Potgieter, 1979; Trollope, 1980; Calvert, 1986; Chidumayo, 1988). Grazing may be hypothesized to reduce fuel loads, thereby lowering fire intensities and reducing the damage to coppice growth and woody plant seedlings and in some (but by no means all) cases the studies have also included investigations of the interactive effects of fire and herbivory, or fire and tree removal (Trapnell, 1959; Barnes, 1965; Trollope, 1980; Boultwood and Rodel, 1981; Chidumayo, 1988). The importance of looking at these interactive effects is seen in the results of Boultwood and Rodel (1981) who showed that tree densities declined on plots that were both lightly grazed and burnt annually over a 15-year period. Under a biennial fire regime, only the dominant species, *Brachystegia boehmii*, increased in numbers but the impact on the other species relative to annual burning was progressively less as fire frequencies declined and grazing pressure increased (Fig. 13.5).

A good example of the way that many of the above factors (time scale, the choice of variables for measurement, the importance of including interactive effects as well as main treatments, etc.) come together to influence or even bias the perceived outcome of long-term experiments is to be found in relation to experiments on bush clearing. The suppression of grass growth beneath the

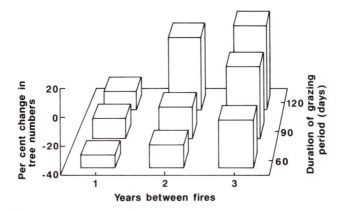

Fig. 13.5. Percentage change in number of trees in relation to fire frequency and grazing pressure at Henderson Research Station, Zimbabwe, 1961–75 (data from Boultwood and Rodel, 1981). On the unburnt, ungrazed control plot tree numbers increased by 87% in 15 years.

canopies of trees in central African woodlands has been well documented (Ward and Cleghorn, 1964; Dye and Spear, 1982). The effect depends on the size of the woody individuals, with small trees and shrubs having a greater suppressive effect than larger trees (Kelly *et al.*, 1978a,b). Grass yield can be higher in open canopy woodland than in either open grassland or closed canopy woodland (Kennard and Walker, 1973). The degree of moisture stress is also important, with the difference in grass yield between open and woodland areas generally being greatest in years of below average rainfall (Ward and Cleghorn, 1964; Dye and Spear, 1982).

These conclusions are derived from bush-clearing experiments in which all woody plants were cut down or ring-barked on the cleared plots and left intact on the woodland (control) plots. The most widely cited of these studies is that of Dye and Spear (1982) in which the results of bush clearing at four separate localities over periods ranging from 15 to 19 years were reported. At three of the four localities (Nyamandhlovu and Matopos – two sites: Hazelside and Two Tree Kop) grass yield on both wooded and cleared plots increased linearly with increasing annual rainfall, with an almost constant increase in yield across the range of annual rainfall amounts (and thus a decline in the relative increase in yield with increasing rainfall). At the fourth locality (Tuli) grass yield on the wooded plot increased linearly with increasing rainfall but on the cleared plot the increase was curvilinear, reaching an asymptote above 600 mm annual rainfall per annum.

Despite the extended duration of this study, the issue of sustainability of increased yield has not been addressed. Aside from the problem of a gradual re-establishment of woody plants (which did not occur on these plots), are the increased yields maintained through time? A plot of the difference in yield between cleared and wooded areas per unit rainfall for each of the four localities in Dye and Spear's (1982) study shows a long-term decline in the added yield, although at two of the sites relative yield initially increased before declining.

A number of explanations are possible. The increase in yield per unit of rainfall declines with increasing annual rainfall amounts. Thus the results could reflect the shift that occurred in average rainfall from a period of below-average rainfall at the start of the study to one of above-average rainfall towards the end of the study. Alternatively, the results could reflect increasing impoverishment of the soil nutrient stock in the absence of trees. Soil carbon and some of the cations were significantly lower on cleared compared to wooded plots at Matopos. The cleared plots on sandy soils (Hazelside) contained significantly lower levels of K, Ca and Mg than adjacent wooded plots. In contrast, in the cleared plots on more clayey soils (Two Tree Kop) only K was significantly lower than adjacent wooded plots (Campbell *et al.*, unpublished).

Removing or ring-barking trees must also result in an increase in soil nutrients as the tree roots decay. This in turn could stimulate an increase in herbaceous production in the same way that adding fertilizer does (Donaldson *et al.*, 1984). Thus the initially high yields (somewhat delayed in the case of two of the sites, perhaps reflecting delayed decomposition of the roots) could be due to a pulse of nutrients, followed by a decline as the nutrients are re-incorporated in biomass, leached, or exported.

These explanations are not wholly mutually exclusive but they cannot now be tested. The focus on effects rather than processes that characterizes this and many other long-term experiments meant that neither soil nutrients nor any other physical environmental variable apart from rainfall were measured during the study. Thus the nutrient pulse hypothesis cannot be evaluated. The hypothesis that the declines are mere artefacts of a shift in mean annual rainfall could be rejected if the declines continued despite a downturn in rainfall as the next dry cycle occurred. Regrettably the experiment was terminated just as this happened. Finally, interpreting the observation that nutrient and organic matter stocks are significantly lower in the cleared plots than in neighbouring wooded plots is complicated by the absence of any pretreatment measurements; by the small size of the plots (10 × 10 m), and by the presence of surrounding 1 m deep trenches, which are likely to have affected moisture dynamics.

Conclusion: Future Needs for Long-term Experiments in Africa

We have made an analysis of a sample of current and past experiments which may seem highly critical. This is not our intention. Our purpose is rather to see what lessons must be learnt before embarking on the very considerable investment that would be entailed in a new round of long-term experimentation in a continent which is already suffering a critical shortage of resources for agricultural and ecological research.

Selecting any individual or group of experiments for detailed review, using criteria which were not built into the initial design is of course unfair, but it makes the important, if obvious, point that an experiment set up with a limited design to achieve a limited objective will yield limited results, no matter how excellent in its own terms. It is often maintained that one of the values of long-term experiments lies simply in their longevity, for this means that insights can be gained, with hindsight, that were not envisaged when the experiment was started. But this is only possible if the design and the implementation procedure are sufficiently flexible. Long-term experiments or monitoring sites are usually set up to serve a specific purpose, to resolve a particular conundrum, to provide more substantial evidence for the difference between treatments where current evidence is only short term. During the progress of the experiment through time however new purposes may be discovered. Insights are gained which could not have been foreseen. Under such circumstances of changing expectation it is all too easy to be wise with hindsight and say 'if only they had monitored X and Y, how much easier it would be to now interpret these differences'. We labour this point to emphasize the critical importance of adopting very rigorous criteria in the selection and establishment of experiments to serve our future and long-term requirements.

It is clear from our earlier analysis that very significant action is required to establish a secure baseline of long-term data for both national and international research needs in Africa. The action, outlined here will involve support to rehabilitate extant experiments in a few places, to establish experiments in a number of new sites and to ensure the synthesis and publication of information from a very wide range of sites. All of this will be impossible without substantial international support from both the scientific and the donor communities. We are aware that this plea comes at a time of diminished support for agricultural research in general and in Africa in particular. But this is also a time of unprecedented interest in the long-term future of our environment. UNCED 2 has awakened interest for research into human impact on the natural environment at a global scale; within the CGIAR system, FAO, the World Bank and related institutions there is a significant reappraisal of the whole structure, function and priorities of international agricultural research. Is it inconceivable that this is a time when a significant investment could be made in long-term research on the sustainability of agricultural systems in Africa?

References

Barnes, D.L. (1965) The effects of frequency of burning and mattocking on the control of coppice in the marandellas sandveld. *Rhodesian Journal of Agricultural Research* 3, 55-56.

Boultwood, J.N. and Rodel, M.G.W. (1981) Effects of stocking rate and burning frequency on Brachystegia/Julbernardia veld in Zimbabwe. *Proceedings Grassland Society of South Africa* 16, 111-115.

Calvert, G.M. (1986) Fire effects in *Baikiaea* woodland, Gwaai forest. In: Piearce, G.D. (ed.) *The Zambezi Teak Forests*. The Forestry Department, Ndola, Zambia, pp. 319-325.

Chidumayo, E.N. (1988) A re-assessment of effects of fire on miombo regeneration in the Zambian Copperbelt. *Journal of Tropical Ecology* 4, 361-372.

Coppock, D.L., Ellis, J.E. and Swift, D.M. (1986) Livestock feeding ecology and resource utilization in a nomadic pastoral ecosystem. *Journal of Applied Ecology* 23, 573-583.

Donaldson, C.H., Rootman, G.T. and Grossmann, D. (1984) Long term nitrogen and phosphorus application to veld. *Journal of the Grassland Society of South Africa* 1, 27-32.

Dye, P.J. and Spear, P.T. (1982) The effects of bush clearing and rainfall variability on grass yield and composition in south-west Zimbabwe. *Zimbabwe Journal of Agricultural Research* 20, 103-118.

Dye, P.J. and Walker, B.H. (1987) Patterns of shoot growth in a semi-arid grassland in Zimbabwe. *Journal of Applied Ecology* 24, 633-644.

Edwards, K.A. and Blackie, J.R. (eds) (1979) Special Issue. *East African Agriculture and Forestry Journal* 45, 1-313.

Geldenhuys, C.J. (1977) The effect of different regimes of annual burning on two woodland communities in Kavango. *South African Forestry Journal* 103, 32-42.

Gertenbach W.P.D. and Potgieter, A.L.F. (1979) Veldbrandnavorsing in die struikmopanieveld van die Nationale Krugerwildtuin. *Koedoe* 22, 1-28.

Izac, A.-M. and Swift, M.J. (1993) On agricultural sustainability and its measurement in smallscale farming systems in sub-Sahara Africa. *Ecological Economics* (in press).

Jenkinson, D.S., Harkness, D.D., Vance, E.D., Adams, D.E. and Harrison, A.F. (1992) Calculating net primary production and annual input of organic matter to soil from the amount and radiocarbon content of soil organic matter. *Soil Biology and Biochemistry* 24, 295-308.

Kelly, R.D., Schwimm, W.F. and Barnes, D.L. (1978a) Effects of ringbarking larger trees in lowveld gneiss woodland on cattle production and on the botanical composition of the veld. *Annual Report of Division of Livestock and Pastures*, Department of Research and Specialist Services, Harare, Zimbabwe.

Kelly, R.D., Schwimm, W.F. and Barnes, D.L. (1978b) Effects of selective thinning of trees and shrubs in lowveld gneiss woodland on the productivity and botanical composition of the herbaceous layer. *Annual Report of Division of Livestock and Pastures*, Department of Research and Specialist Services, Harare, Zimbabwe.

Kennan, T.C.D. (1972) The effects of fire on two vegetation types at Matopos, Rhodesia. *Proceedings Annual Tall Timbers Fire Ecology Conference* 11, 53-98.

Kennard, D.G. and Walker, B.H. (1973) Relationships between tree canopy cover and *Panicum maximum* in the vicinity of Fort Victoria. *Rhodesian Journal of Agricultural Research* 11, 145-153.

Lynam, J.K. and Herdt, R.W. (1989) Sense and sustainability: sustainability as an objective in International Agricultural Research. *Agricultural Economics* 3, 381-398.

Martin, W.S. and Biggs, C.E.J. (1937) Experiments on the maintenance of soil fertility in Uganda. *East African Agricultural and Forestry Journal*, 371-378.

McNaughton, S.J. (1985) The ecology of a grazing system: the Serengeti. *Ecology Monograph* 55, 259-294.

McNaughton, S.J. and Campbell, K.L.I. (1991) Long-term ecological research in African Ecosystems. In: Risser, P.G. (ed.) *Long-term Ecological Research: An International Perspective*. SCOPE Report No. 47. Wiley, Chichester, UK.

Mentis, M.T. (1991) Are multi-paddock grazing systems economically justifiable? *Journal of the Grassland Society of South Africa* 8, 29-34.

Nix, H.A. (1985) Agriculture. In: Kates, R.W., Asubel, J.H. and Berberian, M. (eds) *Climate Impact Assessment*. SCOPE, Wiley, Chichester, UK, pp. 105-130.

O'Connor, T.G. (1985) A synthesis of field experiments concerning the grass layer in the savanna regions of southern Africa. *South African Natural Science Report* 114, 1-119.

O'Reagain, P.J. and Turner, J.R. (1992) An evaluation of the empirical basis for grazing management recommendations for rangeland in southern Africa. *Journal of the Grassland Society of South Africa* 9, 38-49.

Parry, M. (1992) The potential effect of climate changes on agriculture and land use. *Advances in Ecological Research* 22, 63-91.

Parton W.J., Schimel, D.S., Cole, C.V. and Ojima, D.S. (1987) Analysis of factors controlling soil organic matter levels in Great Plains Grasslands. *Soil Science Society of America Journal* 51, 1173–1179.

Savory, A. (1988) *Holistic Resource Management*. Island Press, Covalo, California, USA.

Scoones, I. (1989) Economic and ecological carrying capacity: implications for livestock development in Zimbabwe's communal areas. *Pastoral Development Network* 27b, ODI, London.

Spencer, D.S.C. and Swift, M.J. (1992) Sustainable agriculture: definition and measurement. In: Mulongoy, K., Gueye, M. and Spenser, D.S.C. (eds.) *Biological Nitrogen Fixation and Sustainability of Tropical Agriculture*. Wiley, Chichester, UK, pp. 15–24.

Tainton, N.M. (ed.) (1981) *Veld and Pasture Management in South Africa*. Shuter and Shooter, Pietermaritzburg.

Trapnell, C.G. (1959) Ecological results of woodland burning experiments in Northern Rhodesia. *Journal of Ecology* 47, 129–168.

Trollope, W.S.W. (1980) Controlling bush encroachment with fire in the savanna areas of South Africa. *Proceedings Grassland Society of South Africa* 15, 173–177.

van Wyk, P. (1971) Veld burning in the Kruger National Park, an interim report of some aspects of research. *Proceedings Annual Tall Timbers Fire Ecology Conference* 11, 9–32.

Walker, B.H., Matthews, D.A. and Dye, P.J. (1986) Management of grazing systems – existing versus an event-orientated approach. *South African Journal of Science* 82, 172.

Ward, H.K. and Cleghorn, W.B. (1964) The effects of ring-barking trees in Brachystegia woodland on the yield of veld grasses. *Rhodesian Agricultural Journal* 61, 98–105.

14

The Management of Long-term Agricultural Field Experiments: Procedures and Policies Evolved from the Rothamsted Classical Experiments

R.A. Leigh, R.D. Prew and A.E. Johnston
Institute of Arable Crops Research, Rothamsted Experimental Station, Harpenden, Hertfordshire AL5 2JQ, UK.

Introduction

By their very nature, long-term agricultural field experiments must remain in place for many years and, if they are to contribute useful results throughout their life, they must be well managed. At its simplest this management can be a set of policies and procedures that ensure that the application of treatments, methods of cultivation and harvesting techniques are consistent from year to year. However, in many cases, management will include the imposition of changes that will be needed from time to time for a number of reasons, such as: (i) ensuring that the experiment does not deteriorate in ways detrimental to its longevity; (ii) that it incorporates appropriate changes in agricultural practice or (iii) that it addresses newly emerging issues.

This chapter considers ways in which long-term agricultural experiments can be managed using the Rothamsted Classical experiments as examples (see Johnston, Chapter 2). Although the management practices adopted for the Rothamsted experiments represent a particular case and are probably not the only approach that can be used, the success and continuing usefulness of these experiments suggests that there may be something inherently useful in the procedures that have evolved.

©CAB INTERNATIONAL, 1994. From R.A. Leigh and A.E. Johnston (eds),
Long-term Experiments in Agricultural and Ecological Sciences.

Establishing Long-term Experiments

The establishment of a long-term experiment requires some initial decisions which can affect the subsequent usefulness of the experiment. The following are some of the important factors that need to be taken in to account. Others were considered by Johnston and Powlson (1994).

The question or hypothesis to be addressed

This is the starting point for all experimental work in science and applies equally to field and laboratory experiments. In many cases the question may not require a long-term experiment to provide the answer but the experiment itself may subsequently raise new questions or hypotheses that can be addressed by its continuation. Thus the Rothamsted Classical experiments were established by Lawes and Gilbert simply to determine the relative importance of the different chemical elements known to be present in plants and added to the soil in farmyard manure (FYM). The work was possible because these elements were available as simple, soluble inorganic salts so that they could be tested singly and in various combinations and so compared for efficacy with FYM. Each experiment addressed this for one of the crops grown in the traditional rotation of arable crops and for the herbage of grassland. This initial question was answered relatively quickly (Johnston, Chapter 2) but the experiments opened up other questions. In particular, the debate that developed with Liebig over the source of nitrogen used by crops was an important factor in Lawes' decision to continue the experiments beyond their original aim (Johnston, 1991 and Chapter 2). Subsequently Lawes and Gilbert also recognized that the experiments had become more valuable with time and that future generations of scientists, either with better insight or new techniques, might see uses for them that they could not (Lawes and Gilbert, 1895).

Treatments and design

These are determined by the question that is set. However, for experiments that are deliberately intended to be long term there should be relatively few treatments and thus the design should be simple but still incorporate the appropriate statistical requirements. The suggestion that there should be few treatments may, at first, seem illogical as more treatments provide a greater amount of information and therefore apparently increase the usefulness of the experiment. However, this may only be true in the short term. When, as inevitably happens, the output from, or the apparent usefulness of, the experiment declines, future generations of scientists may well question the need to maintain a complicated experiment. This happened in the University of Illinois Morrow Plots started in 1876 – a number of the original treatments being abandoned in the last century (T. Peck, personal communication). Also, the plots should be large to allow subsequent modification of the experiment if this becomes necessary.

The advantage of a simple design with large plots is exemplified by the Broadbalk experiment. The original experiment had relatively few treatments applied to 20 plots on 4.4 ha. This design has allowed subsequent subdivision of the plots first in 1925 and again in 1968 (Fig. 14.1) so that the experiment now incorporates a five-year crop rotation, including two years without wheat, no herbicide and no other pesticide treatments to continuous wheat, together with the original continuous wheat grown year after year. With some minor modifications in the amounts of nitrogen, the original fertilizer and manurial treatments are still applied to the whole of the plots as laid out initially. The sections growing continuous wheat and receiving these treatments are a continuation of the initial treatments giving an unbroken sequence back to 1843. One disadvantage of the design of the Rothamsted experiments is that they do not incorporate replication. However, they were instrumental in the development of more rigorous experimental design because, following early attempts to determine the significance of the results, Sir John Russell appointed R.A. Fisher who developed statistical analysis as known today (Russell, 1966; see Barnett, Chapter 10).

Uniformity of site

Choice of site is clearly important and factors to be considered are the uniformity of the soil, lack of shading, and slope. The soil on the site must be as uniform as possible. If it is not, another site should be chosen or the differences must be removed, for example, by differential fertilizer or lime additions. There should be no shading of plots by trees because this could differentially affect yields on some plots. Finally, the site should be relatively flat unless the effects of slope are to be a factor in the experiment, for instance, in studies of erosion.

Multidisciplinary approach

The availability of staff with a range of disciplines is important because as questions posed by one set of scientists are answered, ones of interest to others might then evolve. Thus the questions addressed in the later years of a long-term experiment may be quite different from those initially investigated. This is exemplified by some of the outputs from the Broadbalk experiment (Table 14.1). Results from this experiment have had impacts in the fields of soil science, crop nutrition, crop husbandry, statistics, plant pathology, entomology and weed biology. Table 14.1 also indicates the way new technological developments can influence the usefulness of long-term experiments. For example, work on nitrogen cycling (Jenkinson, 1990; Powlson, Chapter 6; Jenkinson *et al.*, Chapter 7) was only possible because of the availability of ^{15}N-labelled fertilizers and sensitive mass spectrometers.

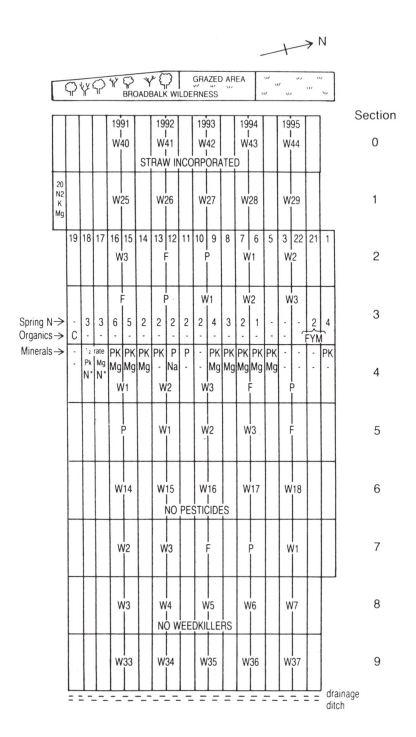

Table 14.1. Some achievements from the Broadbalk experiment at Rothamsted.

Period	Achievement
1843–67	Value of nitrogen for the yield of wheat demonstrated
1867–85	Discovery that P and K are not readily leached from soils
1885–1903	Long-term benefits of P and K established
1903–21	Statistical analysis pioneered using results from Broadbalk
1921–39	Take-all disease recorded and first identification of eyespot in UK
1939–57	Pattern of Take-all decline described Effect of fallow on wheat bulb fly populations studied
1957–75	Relationships between K fertilization and powdery mildew established Archived samples used to show accumulation of pollutants in soils
1975-93	Studies of soil-crop N cycle using ^{15}N Take-all incidence shown to be related to P deficiency

Maintaining the Integrity of Long-term Experiments

Once an experiment is established it is important to maintain the integrity of the experimental plots to make sure that they are not changed other than by the imposed treatments. If unwanted changes are detected they need to be to rectified by appropriate action. The following factors have been important in the Rothamsted experiments.

Accurate location of plot boundaries

Being able to find a plot following ploughing or other cultivations is obviously important because it ensures that treatments such as fertilizers and pesticides are applied in the way dictated by the design and that crop yields are taken from the same part of each plot each year. At Rothamsted this is achieved by a system of posts and sunken pegs. For each experiment there is a series of white

---

Fig. 14.1. The current plan of the Broadbalk winter wheat experiment. The original plots were in strips which ran the length of the field and have subsequently been subdivided into different sections which run across the original plots. The original manure and fertilizer treatments are still applied to the original strips. Sections 0, 1, 6, 8 and 9: wheat grown continuously; section 0: straw incorporated, section 6: no pesticides or fungicides; and section 8: no weedkiller. Sections 2, 3, 4, 5 and 7 have a rotation of fallow, potatoes, wheat, wheat, wheat.

posts in the field boundaries. Triangulation from these posts gives the location of wooden oak pegs, which are buried in the ground below the plough layer. The edges of the plots are located at accurately known distances from these buried pegs. However, even when the plots are located, care must still be taken to ensure that mistakes are not made in the application of treatments or the taking of yields. At Rothamsted all such field operations are supervised by specially designated members of the farm staff – the Recorders.

The Recorders are responsible for the maintenance of a very high standard of field operations, calibration of equipment and the documentation of all field operations. When basal operations (i.e. those applied uniformly to the whole experiment) are done, one Recorder is always present to ensure a high standard of work. When plot-specific treatments are applied, then two Recorders must be present so that the matching of treatments to plots can be double checked. Drilling is always done using Recorders with marker poles to ensure accuracy of siting, based on the permanent marker system described above. The Recorders are also responsible for weighing the harvested produce and taking samples for dry matter determinations or chemical analyses.

Prevention of soil movement across plot boundaries

Cultivation inevitably leads to the movement of soil and steps must be taken to minimize this so that it does not result in the blurring of plot boundaries. Two aspects in particular need to be considered. First, there should be no cross cultivation between plots as this can lead to the movement of soils of different nutrient status between the plots. Thus, on Broadbalk, ploughing is always along the length of the original plots (Fig. 14.1). Second, the direction of ploughing should be reversed annually to prevent unidirectional movement of soils across plots. The extent to which such techniques can maintain plot integrity is shown in Fig. 14.2 which shows the total P concentration in soils in a transect taken across neighbouring plots of Broadbalk. These plots receive different amounts or forms of P fertilizer each year and the long-term applications have resulted in different soil P levels on each plot. On each plot there are stable plateaux of soil P concentration separated by gradients of P concentrations. The very large differences between adjacent plots are obvious. This can be compared with a transect across plots on a lighter textured soil at Woburn Experimental Farm where there was much cross cultivation (Fig. 14.2). Here it is not possible to clearly differentiate the adjacent plots because there has been substantial soil movement. The importance of knowing that this soil movement had occurred on the Woburn plots was demonstrated by McGrath and Lane (1989) who showed that nearly all the heavy metals added to some of these plots in sewage sludge were retained in the topsoil if sideways movement of soil was taken into account. This substantially increased estimates of residence times of heavy metals in soils compared with previous estimates.

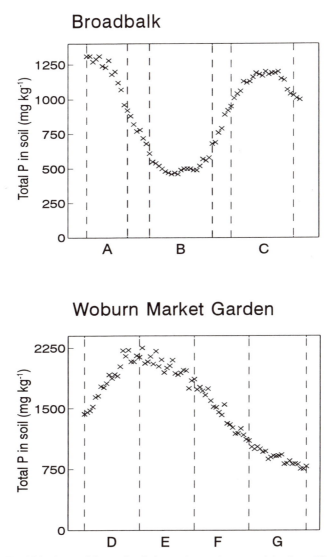

Fig. 14.2. Total P in the top 23 cm of soil along a transect across plots given different P treatments. Broadbalk: A, FYM 35 t ha⁻¹ year⁻¹ containing 40 kg P; B, unmanured; C, PK fertilizers supplying 33 kg P ha⁻¹ year⁻¹, all since 1843. Woburn Market Garden: total amounts of P, kg ha⁻¹, applied between 1942 and 1967, E, 13,130 kg in sewage sludge; F, 7720 kg in FYM, in addition to the 1000 kg P ha⁻¹ as superphosphate added to all four plots between 1942 and 1984.

Periodic checks for uniformity

It is essential that soil samples, taken periodically to follow the effects of treatments under test, are also used to monitor for unwanted side-effects so that remedial action can be taken when needed. For example, applying nitrogen as ammonium sulphate and sodium nitrate in the Classical experiments was intended as a comparison of two forms of nitrogen fertilizer not as a test of the acidifying effect of ammonium sulphate. Once this adverse side-effect of using ammonium sulphate was recognized, differences in soil acidity had to be rectified by applying lime ($CaCO_3$) and a liming policy adopted to prevent further acidification on the Classical experiments growing arable crops. On the Park Grass experiment soil acidification from ammonium sulphate has large effects on botanical composition and is monitored as a treatment effect.

Some changes may result from sources other than the treatments, such as the changes in soil pH from aerial inputs of acid rain. Although these changes also need to be rectified, recognizing their source can be important as it may lead to new information about the relationships between soil and the atmosphere, and between agriculture, industrial activity and the wider environment. Studies of pollutant inputs are an example (see Jones *et al.*, Chapter 9).

Soil variability can also occur as a result of treatments applied before the experiment was started and needs to be rectified because it will lead to unwanted differences between plots. An example at Rothamsted was the differential change in soil pH that occurred on the Exhaustion Land (Johnston and Poulton, 1977). This resulted from uneven liming of the field in the 18th century, and by 1954 had led to large differences in pH between one end of the experiment and the other. This was rectified by differential liming of the plots (Johnston and Poulton, 1977).

Keeping Records

Maintaining records needs to go beyond a database of yields or other outputs from an experiment. It needs also to include details of fertilizer additions, other treatments, anomalies of particular years, general observations of pests and diseases, and any other factors considered relevant to future interpretation of the results. Financial aspects are equally important and should include costs of treatments and value of outputs so that long-term trends in output/input ratios can be assessed in financial terms. At Rothamsted, the primary records of all field operations are maintained by the Recorders in handwritten books. They also record yields but dry matter determinations on all crops are the responsibility of the Crop Management Department. The Statistics Department is responsible for the computation of all yields and their statistical analysis. At the end of each year all relevant data are brought together in a booklet, *Yields of the Field Experiments*, published by Rothamsted.

All this information is now being assembled on to the Electronic Rothamsted Archive. This will allow faster access to the data as well as the application of computer-based techniques of statistical or other analyses to help further interpret the data. Whatever the means of storage, the final database is only as good as the original observations so the integrity of the latter must be ensured and this is helped at Rothamsted by the system of supervision of field operations described above. Secondary observations like chemical analysis of the crops, disease assessments, weed counts, soil sampling and analyses are the responsibility of individual scientists working on each experiment. The data collected are often processed by the Statistics Department but are not recorded in *Yields of the Field Experiments* although they are published in scientific papers.

A factor that can limit the usefulness of such databases to future generations is inappropriate or insufficient checks on the effects of changes in methodology that may be introduced. For instance, changes in harvesting technique can affect the yields and new cultivation methods can lead to deeper ploughing diluting the original topsoil with subsoil so changing soil nutrient concentrations. The Park Grass experiment provides an example where a change in harvesting technique was not sufficiently checked at the time of its introduction and which is now confounding the interpretation of some results. Originally the yields from this experiment were taken by cutting all the herbage on the plot and leaving it to dry on the plot before weighing it as hay and taking a sample for dry matter determination. In 1960 this was changed to one where a flail harvester is used to take a narrow sample cut of green herbage which is weighed immediately, and then a subsample is oven-dried to obtain percentage dry matter and dry matter yield. The rest of the herbage is then cut and made into hay on the plot to maintain the return of seed. This change led to an apparent increase in the yield of dry matter because when the herbage was left to dry in the field before being sampled there was shedding of seeds, some loss of small leaves and some respiratory losses which resulted in smaller dry matter yields than measured by direct cutting and immediate weighing (Table 14.2). This change in itself would not have been a problem if the two methods had been compared side-by-side for a number of years so that the new was properly calibrated against the old. Unfortunately, this was only done for one year when there was, by chance, little difference. Further comparisons in 1992 indicated a large difference between methods but in 1993 there were large differences only on the unmanured plots. Big differences between years indicates that any confidence shown in the earlier comparison was unfounded. This insufficient comparison of the two methods has confounded attempts to determine whether changes in atmospheric carbon dioxide levels have affected yields (Jenkinson *et al.*, 1994).

Table 14.2. Mean yields from plots of the Park Grass Experiment for periods before and after the change in harvesting method made in 1960.

Plot	Mean yield of dry matter (t ha^{-1})		Percentage increase after change in harvest method
	1891–1958 Weighed as hay	1960–92 Weighed green	
3d	1.47	2.76	88
12d	1.90	3.26	72
2d	1.71	2.79	63
16d	4.54	6.07	34
14d	5.92	7.66	29

The Importance of Archives of Soil and Crop Samples

A very important aspect of the Rothamsted experiments, which has added immensely to their value, is the existence of an archive of soil and crop samples taken from the experiments from their inception. Crop samples have been kept every year, initially from every plot, now from selected treatments, whereas soil samples are taken periodically. These samples have been used for variety of purposes including nutrient balances (e.g. Johnston and Penny, 1972), determining changes in soil organic matter (e.g. Jenkinson and John-ston, 1977), measuring soil P and K status (e.g. Johnston and Poulton, 1977) and following the accumulation of industrial inorganic and organic pollutants in soils (Jones *et al.*, Chapter 9). Because such changes have been determined with accuracy and can be related to the known history of the sites it is possible to construct good models for the changes. For example for the turnover of soil organic matter (Jenkinson *et al.*, Chapter 7).

   Today the soil and crop sample archive at Rothamsted is as important as the experiments from which it is derived. The uniqueness and comprehen-siveness of the archive distinguishes the Rothamsted experiments from all other long-term agricultural experiments. It is essential that all existing and new long-term experiments establish such archives. The samples in such an archive must be carefully taken and recorded to maximize their benefit and have to be appropriately labelled so that their provenance is known. Thus soil samples should be labelled with details of the field, plot number, date and depth of sampling. Additional details should be entered in a written record and appropriate use should now be made of electronic record keeping and retrieval (e.g. bar coding). In addition, it is important to remember that the number of samples will increase rapidly and that adequate space must be avail-able for storing them safely. The Rothamsted archive which dates back to 1843 now contains many thousands of samples and occupies floor-to-ceiling shelves in a substantial building.

The Ways Long-term Experiments Can Be Changed

It is unlikely that any experiment that lasts for many tens of years will remain in exactly its original form throughout that time. Alterations in farming practice may make the original treatments obsolete or impractical whereas new questions may dictate changes or additions to the treatments. At Rothamsted changes have been made to all of the long-term experiments although the degree of change ranges from minimal alterations that are required to maintain the longevity of the experiment to drastic changes in design that ultimately led to the demise of the experiment. The following describe five different scenarios of change that have been applied to individual long-term experiments at Rothamsted.

Minimum changes to treatments or husbandry to ensure longevity

Park Grass is an experiment that has changed minimally over its lifetime. In part this is because it is on permanent grassland, herbage is harvested only twice each year and no sprays are used to control non-grass species, pests or diseases. Originally, the soil pH was 5.7–5.8 and no lime ($CaCO_3$) was applied. As a result, there was a substantial decrease in pH on plots receiving ammonium sulphate. To test the effects of treatments on yield and botanical composition of the sward, the plots were divided in two in 1903 with one half receiving lime, the other not (Warren and Johnston, 1964). Subsequently, in 1965, it was decided to increase the range of soil pH values for each treatment by splitting the limed and unlimed plots in half thus giving four subplots on each of the original plots (Warren *et al.*, 1965). The aim now is to maintain a target of pH 5, 6 or 7 in three of the subplots with the fourth remaining unlimed as a continuation of the original treatment. No other substantial changes have been imposed so the experiment remains as otherwise unchanged from the original design of Lawes and Gilbert. This means that the experiment is probably not very relevant to modern grassland management but, because the treatments have resulted in major changes to the flora on different plots, it is of great ecological interest (see Tilman *et al.*, Chapter 16). It is also used in studies of soil acidification and testing the effects of liming (Goulding *et al.*, 1989).

Retain the major objectives but incorporate innovations in agricultural practice

Broadbalk provides a very good example of this. The cultivar is changed periodically; harvesting has changed from scything to reaper binder to combine harvester; fallowing was introduced to control weeds in 1925; herbicides were introduced in 1957 (although some plots have still never received herbicide and are now invaluable for studies on weed biology and herbicide resistance); a crop rotation was introduced from 1968 so that yields of wheat grown following a two-year break could be compared with continuous wheat and the effects of take-all (caused by the soil-borne fungus *Gaeumannomyces graminis*) estimated; larger N rates have been introduced on two occasions (1968 and 1985); and pesticides have been used as they have been developed and are needed for pest control. These changes have ensured that Broadbalk remains relevant to modern wheat growing in the UK.

Change the objectives and continue the experiment in a different form

The experiment on Exhaustion Land has undergone many changes in objectives and hence design during its lifetime (Johnston and Poulton, 1977). These changes are summarized in Table 14.3. Each change has been made to address agricultural questions current at the time. The history of this experiment reflects the flexibility that can be achieved with long-term experiments and how they can provide a test-bed for a range of ideas over long periods of time (Johnston and Warren, 1970).

Change the experiment to obtain more information in the short term but sacrificing longevity

This occurred on Agdell which was started in 1848 and in its original design had six very large plots to examine manurial treatments and the effect of crop rotation on yield (top plan, Fig. 14.3). At various intervals, especially after 1958, the plots were subdivided to answer new questions but this finally resulted in a very complicated design with very many subplots replacing each of the large original plots (bottom plan, Fig. 14.3). The purpose of this final design was to explore the effects of fresh P and K, added to soils with different levels of P and K residues and soil organic matter, on the yields of three arable crops grown in rotation (see Johnston, Chapter 2). Even this phase of the experiment lasted more than ten years – much longer than many field experiments. Once the plots had been so subdivided it was impossible to return to a larger plot design by differential fertilization and the experiment was stopped in 1990 and Agdell is now cropped in the normal farm rotation. Nonetheless these changes were necessary to answer major scientific problems and the

Table 14.3. Changes to the Exhaustion Land experiment.

Period	Purpose
1852–55	The 'Lois-Weedon' plots testing methods of husbandry for winter wheat
1856–75	Winter wheat experiment
1876–1901	Potato experiment
1902–48	Unmanured spring barley
1949–85	Spring barley grown with basal N to test effects of residues of P and K accumulated during 1856–1901. Test four rates of N on spring barley (1976–85)
1986–	Half of the plots used to test 4 levels of P
1986–	Half of the plots continued to test 4 levels of N (to be changed to 4 levels of K from 1994)

subdivision in this way was considered worthwhile even though it was recognized that it would eventually end the experiment.

Mothballing

This option recognizes that an experiment may still have a useful role to play in the future but that its immediate relevance is low. Therefore, rather than stopping the experiment, its treatments are retained under an easily managed crop until new uses are found for it. This is now the state of the Barnfield experiment started in 1843 (Johnston, Chapter 2). Originally it grew root crops continuously but this phase of the experiment ceased in 1959. Arable crops were then grown in rotation until 1973 but in 1975 the site was sown to grass with the original fertilizer treatments still applied. However, even in this form, Barnfield has proved useful. The response of grass to four amounts of fertilizer N indicated that on the fertilizer-treated plots with about 1.7% organic matter some nitrogen was being directed to increase soil organic matter. This did not appear to happen on the FYM treated plots with 4.2% organic matter (Johnston, Chapter 2). Also, the wide range of soil K levels resulting from past treatments and cropping have allowed the relationship between soil and crop K concentrations to be established (Barraclough and Leigh, 1993). This built on work previously done with barley on the Hoosfield experiment but where the restricted range of soil K levels prevented a detailed relationship from being established (Leigh and Johnston, 1983).

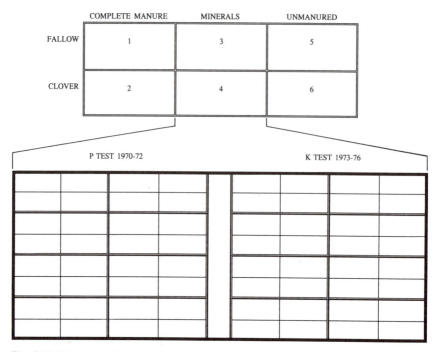

Fig. 14.3. Changes to the design of the Agdell experiment. The original experiment, started in 1848, consisted of six large plots (upper plan). By 1973, successive modifications had resulted in each of these large plots being subdivided into 64 subplots as shown for one of the plots in the lower plan. The 384 subplots in this final design examined the effects of fresh P and K on the yields of arable crops growing on soils with different residues of P, K and organic matter which had resulted from earlier treatments. When this phase of the experiment came to an end in 1985 it was not possible to return to a simpler design and the experiment was terminated in 1990.

Management Structures for Long-term Experiments

Long-term experiments must be the responsibility of a committee which can exist in perpetuity although its members will change. But committees cannot manage on a day-to-day basis and this responsibility needs to be delegated. At Rothamsted this delegation is to 'Sponsors'; two or three scientists from different disciplines who have both immediate and ongoing interests in the experiment. Thus Rothamsted has a two-tier system of management which has proved especially appropriate for considering changes.

Minor changes are considered by the Sponsors, who pass on their recommendations to 'Commodity Groups' (committees with responsibility for experiments on particular crops) and to the Working Party for Field Experiments, the senior committee at Rothamsted which oversees the field experimental programme. If major changes are to be undertaken they must have the approval of the Lawes Agricultural Trust, the governing body for Rothamsted, and a legally constituted body that owns the land at Rothamsted and to whom Lawes entrusted the Classical experiments. The Trust has a long-term responsibility in ensuring that the Classical experiments are not changed in a way that is likely to be detrimental to them, or is only for short-term interest, unless there is a good scientific reason for doing so. This system prevents strong-minded individuals from imposing changes on the experiments without full consideration of both the short- and long-term consequences.

Finally, Rothamsted has recently introduced a Standing Committee on Unpublished Data because there are now so many requests for information, particularly for access to the large bodies of unpublished primary data from the Classical experiments. To ensure control over the way these data are used, particularly that interpretations are made with the full knowledge of the previous treatments, all potential users of the data must agree to a set of conditions before access is given.

Concluding Remarks

The management practices established for the Rothamsted Classical experiments are ones which have withstood the tests of time and so provide a good blueprint which can be adapted for use in other situations. The approach adopted is summarized by two words – care and commitment. Care in the choice of site and the design of the experiment; care in maintaining the integrity of the experimental plots; care in taking yields and samples; care in managing change; and care in recording results so that future generations can know that they have integrity. Commitment is required both in maintaining experiments in times when there are troughs in the output and in ensuring that the results are interpreted and published in a timely way. The latter is particularly important because, as in any scientific endeavour, the results from long-term agricultural experiments must be submitted to peer review and are ultimately of most significance if they influence the work of others, be they scientists or farmers. Today, perhaps more than in the past, interpretation of the results in relation to environmental concerns is of vital importance so that the decisions of policy makers rest not only on the best possible data but also on good interpretation.

References

Barraclough, P.B. and Leigh, R.A. (1993) Grass yield in relation to potassium supply and the concentration of cations in tissue water. *Journal of Agricultural Science, Cambridge* 121, 157–168.

Goulding, K.W.T., McGrath, S.P. and Johnston, A.E. (1989) Predicting the lime requirement of soils under permanent grassland and arable crops. *Soil Use and Management* 5, 54-57.

Jenkinson, D.S. (1990) The turnover of organic carbon and nitrogen in soil. *Philosophical Transactions of the Royal Society, London* B 329 361-368.

Jenkinson, D.S. and Johnston, A.E. (1977) Soil organic matter in the Hoosfield Continuous Barley Experiment. *Rothamsted Experimental Station Annual Report for 1976, Part 2*, 87-101.

Jenkinson, D.S., Potts J.M., Perry, J.N., Barnett, V., Coleman, K. and Johnston, A.E. (1994) Trends in herbage yields over the last century on the Rothamsted Long-term Continuous Hay Experiment. *Journal of Agricultural Science, Cambridge* 122, 365-374.

Johnston, A.E. (1991) Liebig and the Rothamsted experiments. In: Judel, G.K. and Winnewisser, M. (eds) *Symposium "150 Jahre Agrikulturchemie"*, Justus Liebig-Gesellschaft zu Giessen, pp. 37-64.

Johnston, A.E. and Penny, A. (1972) The Agdell Experiment, 1848-1970. *Rothamsted Experimental Station Report for 1971*, Part 2, 38-67.

Johnston, A.E. and Poulton, P.R. (1977) Yields on the Exhaustion Land and changes in the NPK contents of the soils due to cropping and manuring, 1852-1975. *Rothamsted Experimental Station Report for 1976*, Part 2, 53-85.

Johnston, A.E. and Powlson, D.S. (1994) The setting up, conduct and applicability of long-term, continuing field experiments in agricultural research. In: Greenland, D.J. and Szabolcs, I. (eds) *Soil Resilience and Sustainable Land Use*. CAB International, Wallingford, pp. 395-421.

Johnston, A.E. and Warren, R.G. (1970) The value of residues from long period manuring at Rothamsted and Woburn. III. The experiments made from 1957 to 1962 the soils and histories of the sites on which they were made. *Rothamsted Experimental Station Report for 1969*, Part 2, 22-38.

Lawes, J.B. and Gilbert, J.H. (1895) The Rothamsted experiments over fifty years. *Transactions of the Highland and Agricultural Society of Scotland Fifth Series* 7, 11-354.

Leigh, R.A. and Johnston, A.E. (1983) Concentrations of potassium in the dry matter and tissue water of field grown spring barley and their relationships to grain yield. *Journal of Agricultural Science, Cambridge* 101, 675-685.

McGrath, S.P. and Lane, P.W. (1989) An explanation for the apparent losses of metals in a long-term field experiment with sewage sludge. *Environmental Pollution* 60, 235-256.

Russell, E.J. (1966) *A History of Agricultural Science in Great Britain*. George Allen and Unwin, London, pp. 493.

Warren, R.G. and Johnston, A.E. (1964) The Park Grass experiment. *Rothamsted Experimental Station Report for 1963*, 240-262.

Warren, R. G., Johnston, A.E. and Cooke, G.W. (1965) Changes in the Park Grass experiment. *Rothamsted Experimental Station Report for 1964*, 224-228.

IV MONITORING LONG-TERM ECOSYSTEMS, POPULATION DYNAMICS AND ENVIRONMENTAL CHANGE

15 Climate–Vegetation Relationships in the Bibury Road Verge Experiments

J.P. Grime, A.J. Willis, R. Hunt and N.P. Dunnett

NERC Unit of Comparative Plant Ecology, Department of Animal and Plant Sciences, The University of Sheffield, Sheffield S10 2TN, UK.

Introduction

Current attempts to predict the impacts of future climate changes on ecosystems involve reference to many different kinds of information. These sources vary in scale and focus from short-term studies of the responses of particular organisms, to long-term monitoring of field plots. If this research is to graduate into rigorous science it will be essential to incorporate a strong element of model building and hypothesis-testing. This is relatively easy in the case of individual processes or populations but much more difficult at the levels of the community and the ecosystem.

The objective in this chapter is to illustrate the role of one particular monitoring study, that conducted by Professor A.J. Willis on road verges near Bibury in Gloucestershire (Fig. 15.1), in a programme of research on the impacts of climate change on herbaceous vegetation. Our purpose is not to examine the data from Bibury in exhaustive detail but rather to place this study within a robust scientific protocol. In consequence, we are concerned not merely with the question as to how the monitored data should be analysed but also with the identification of the other kinds of research which are required in order to extract maximum value from such a monitoring study.

©CAB INTERNATIONAL, 1994. From R.A. Leigh and A.E. Johnston (eds), *Long-term Experiments in Agricultural and Ecological Sciences.*

271

Fig. 15.1. Professor A.J. Willis at Bibury in 1993.

Research Protocol

Figure 15.2 summarizes the main operations in a programme of research concerned with the impacts of climate change on British vegetation and involving the Bibury study. From this scheme, it is apparent that the protocol is iterative with each successive cycle involving both the development and testing of predictions. Two paths of hypothesis testing are suggested (left and right), only one of which utilizes the Bibury records directly. Each of the main components in the protocol will now be considered briefly.

Screening of plant attributes

Following a very common strategy in biological research, the protocol develops predictions at the system level (the plant community) by reference to information at one step below this level of organization (the functional characteristics of the component plant populations). In the case of Bibury, the source of standardized information on the component plants is the Integrated

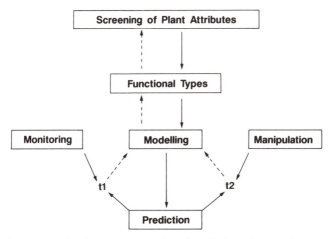

Fig. 15.2. Protocol for development and testing of predictions of vegetation response to climate change. Discrepancies revealed in the tests at t1 and t2 initiate further modelling cycles, each of which may necessitate refinement of the functional types or even additional screening.

Screening Programme (ISP) which has run continuously in Sheffield since 1987 and the methods of which are described in a laboratory manual (Hendry and Grime, 1993). Later in this chapter, examples are considered of plant attributes available from the ISP and useful in devising predictions of either the rate or direction of vegetation responses to climate change. Here, it will suffice to record that such attributes include aspects of life history, resource capture and utilization and regenerative biology as well as those more obviously related to climate change (responsiveness to carbon dioxide concentration, temperature and moisture supply). In closed, relatively productive vegetation, such as that at Bibury, it is likely that the structure and dynamics are strongly mediated through competitive interactions; accordingly, much of the input to modelling is related to the scale and precision (Campbell *et al.*, 1991) with which the leaf canopies and root systems of particular species forage for resources. Considerable effort is also devoted to the development of predictions of the resistance and resilience of component species when exposed to extreme climatic events such as severe summer drought or late frost following a mild winter.

Modelling and prediction of vegetation responses

Until recently, both conceptual and technical problems have limited the extent to which ecologists have been able to model the dynamics of multispecies systems. In order to accommodate the variety of plant life present in most communities it has been necessary to rely on relatively simple functional classifications in which predictions of response to changes in land use or climate are based on single attributes (Noble and Slatyer, 1979) or sets of attributes (Grime, 1974). An example of this approach is the expert system TRISTAR (Hunt *et al.*, 1991b) in which structural changes in vegetation are predicted from characteristics of both the established and regenerative phases of the life cycles of component populations.

A further stage in the development of predictions of vegetation responses to climate change is now in progress as greater computing power is allied to techniques in which plant behaviour in multispecies assemblages is represented by cellular automata (Wolfram, 1984). Already, cellular automata have been devised which reflect differences in life-span, clonal architecture, seed dispersal and gap exploitation (Barkham and Hance, 1982; Crawley and May, 1987; Green, 1989; Inghe, 1989) and, more recently, features of resource capture, utilization and release have been incorporated into this approach (Colasanti and Grime, 1993).

Further refinements will be required before cellular automata can provide detailed predictions of vegetation responses to specific climate scenarios. It is becoming apparent, however, that future progress is unlikely to be limited by the lack of sophisticated models. As suggested by Keddy (1992) a more serious problem may arise from the lack of data with which to parameterize the models.

Model testing using monitoring studies

Eventually, some of the monitoring studies now in progress may be expected to provide the first direct evidence of vegetation response to global climate change. For the present, however, their main use is to test the predictive value of existing models. Year-to-year variation in climate at most locations is considerable and it seems reasonable to judge our capacity to extrapolate into the future from the accuracy with which recent or contemporary fluctuations in vegetation composition can be predicted from the climate record.

In order for a monitoring study to provide a reliable basis for hypothesis testing with respect to the impacts of climate on vegetation, it is essential that the effects of climate are not confounded with observer errors or impacts of other undetected environmental or management changes (e.g. variation in atmospheric depositions or fluctuations in grazing pressure). At Bibury, it is particularly fortunate that the entire (and on-going) 35-year record is largely the work of a single observer (A.J. Willis) and that data have been collected from the permanent roadside plots on Cotswold Oolite, examined at the same time each year (the third week in July) and managed by a standard regime (autumn 'topping'). Recording methods have been the same throughout (Yemm and Willis, 1962; Willis, 1988). It is also important that, at intervals during the study, the non-destructive estimates of biomass used to provide the records of species composition have been calibrated against samples which were harvested, then dried and weighed in the laboratory. The Bibury plots included treatments with the selective herbicide, 2,4-dichlorophenoxyacetic acid (2,4-D), and the growth regulator, maleic hydrazide (MH); the present analysis concerns the unsprayed (control) plots of these treatments. Even in the rural, apparently unchanging conditions of Bibury, however, it has been necessary to take account of background changes. Table 15.1 shows that there have been gradual but consistent changes in species composition over the period of observation. These changes are unrelated to climatic variation and may at least partly be a response to eutrophication from air-borne deposition arising from progressive intensification of arable farming in the neighbouring landscape. This conclusion is prompted by the evidence (Table 15.1) that there has been a tendency for species, known to have inherently rapid potential relative growth rate (Grime and Hunt, 1975) and to be associated with fertile soils (Grime *et al.*, 1988), to expand at the expense of species normally restricted to unproductive vegetation. This complication has not invalidated the Bibury records as a basis for hypothesis testing but it has necessitated the use of statistical techniques whereby effects of suspected eutrophication are 'detrended' prior to analysis of climate–vegetation relationships (Hunt *et al.*, 1993b).

Model testing using manipulative experiments

There are two main reasons why we should suspect that tests involving monitoring studies will not provide a complete validation of models predicting vegetation response to climate change. First, it is clear that because weather variables frequently vary in concert (for example, high summer temperatures often coincide with low rainfall), different models may have similar predictive accuracy. In such cases, additional information may be required in order to achieve a reliable mechanistic interpretation. A second problem in the use of monitoring results is that, although year-to-year fluctuations in vegetation composition provide interesting insights into the homeostatic mechanisms which allow plants of different climatic tolerances to coexist in communities, they may be of limited relevance to the circumstances in which the global climate could initiate irreversible vegetation changes involving local extinctions and invasions. Many ecologists agree that such radical changes are most likely to follow extreme events such as a sequence of hot, dry summers. As shown later in the case of Bibury (Fig. 15.5), monitoring studies can assist our understanding of the potential of extreme events to initiate vegetation change. However, it is in the nature of extreme events that they are relatively rare occurrences making them unsatisfactory subjects for statistical analysis of plant–weather relationships. For this reason, additional approaches are now required to allow reliable inferences with respect to the impacts of extreme events. In various parts of the world considerable effort is now being expended to develop climate engineering techniques which will provide realistic simulations of extreme climatic events both in the laboratory and in

Table 15.1. Mean annual rates of change in shoot biomass in major component species of the Bibury plots. Dashes indicate that the species is absent from the series. Nomenclature follows Stace (1991).

Taxon	Change per year (%)	
	MH series	2,4–D series
Bare ground	−5.0	−3.2
Litter	−7.4	−7.9
Agrostis stolonifera	−4.3	−4.4
Brachypodium pinnatum	—	−15.8
Bromopsis erecta	−3.3	−9.0
Cirsium arvense	−4.6	—
Convolvulus arvensis	−4.2	—
Cruciata laevipes	−2.4	—
Dactylis glomerata	−2.4	—
Elytrigia repens	−10.1	−8.3
Festuca arundinacea	6.8	—
Festuca rubra	−4.0	−2.8
Glechoma hederacea	−3.3	−4.5
Heracleum sphondylium	—	−4.4
Hypericum spp.	—	−13.7
Knautia arvensis	3.7	—
Lolium perenne	3.5	—
Odontites vernus	2.0	3.7
Poa pratensis	−6.1	−5.8
Ranunculus repens	7.3	—
Stachys sylvatica	−6.6	—
Taraxacum officinale agg.	—	7.2
Tragopogon pratensis	—	7.2
Trifolium pratense	—	6.9
Trifolium repens	1.5	—
Urtica dioica	−10.5	−7.3
Veronica chamaedrys	—	−4.5
Vicia spp.	6.4	—
Viola hirta	2.3	−2.1
Brachythecium rutabulum	−2.2	—

The following are not shown because changes were insignificant: *Achillea millefolium, Anisantha sterilis, Anthriscus sylvestris, Arrhenatherum elatius, Centaurea nigra, Galium aparine, G. verum, Helictotrichon pubescens, Phleum pratense, Plantago lanceolata, Poa trivialis, Potentilla reptans, Rumex* spp., *Trisetum flavescens, Ulmus glabra* and *Eurhynchium praelongum.*

natural habitats. Some examples of this new technology, applied to Bibury, are provided in the final section of this chapter.

Tests Using the Floristic Records of A.J. Willis

Multivariate analyses have been conducted to compare year-to-year variation in floristic composition with the coincident climate, the climate of the preceding year and that recorded 24–12 months before (Hunt *et al.*, 1993b). The number of statistically significant relationships established between fluctuations in shoot biomass of particular species and specific weather variables exceeds those expected by chance by a considerable margin. These analyses confirm the existence within the Bibury vegetation of common constituents with quite contrasted responses to climatic variation. In the case of unusually warm summers, for example, growth is promoted in species such as *Achillea millefolium, Brachypodium pinnatum* and *Centaurea nigra* but suppressed in *Bromopsis erecta, Ranunculus repens* and *Trifolium repens*. Correlations of this kind can form a legitimate starting point for mechanistic interpretation. Here, however, with particular concern for the need to develop and test generalizing principles, we shall follow the protocol of Fig. 15.2 and examine three specific hypotheses.

Plant functional types and weather at Bibury

What has become known as the C–S–R model (Grime, 1974, 1988) describes various types of strategy or functional type in the established (vegetative) phase of plant life cycles (Fig. 15.3). The model recognizes two external groups of factors, both of which vitally affect the evolution of plants and the structure and dynamics of vegetation. The first group consists of factors such as shortages of light, water and mineral nutrients or suboptimal temperatures, all of which place prior restrictions on plant production. The second includes a variety of phenomena which cause the partial or total destruction of plant biomass: they include management factors such as grazing, trampling, mowing and ploughing and also events such as wind damage, frosting, droughting, soil erosion and fire.

The permutations of high and low productivity with high and low vegetation damage support a triangular matrix of distinct functional types (Grime, 1974). These are principally *competitors* in the case of high productivity and minimum damage, *stress tolerators* in the case of low productivity and minimal damage and *ruderals* in the case of high productivity and severe and frequent vegetation damage. The initials of these three give the C–S–R model its name; many intermediate types also exist.

With this theoretical framework it is possible to predict how climate variation should influence the composition of the Bibury vegetation with respect

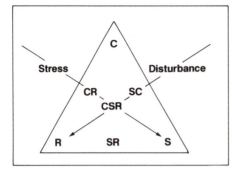

Fig. 15.3. A triangular array of primary functional types of plants. C, S and R are competitive, stress tolerant and ruderal plant species and the array summarizes their expected responses to stress or disturbance.

to plant strategies. First, it can be suggested that mild winters, by lengthening the growing season, will encourage an expansion by competitors. At the same time, stress-tolerators and ruderals may be expected to decline. This is because the slow growth-rates of the former will limit their ability to respond and the dependence of the latter upon vegetation gaps renders them susceptible to competitive suppression in years which support high biomass. A reversal of this pattern is forecast for years in which a hot, dry summer shortens the growing season.

Six of the seven main functional types and their intermediates identified in Fig. 15.3 exist within the Bibury flora. The results of the correlation analyses relating variation in climate to that of the aggregated shoot biomasses of the species of each functional type (Table 15.2) are in close agreement with the C–S–R model predictions.

Genome size and weather at Bibury

From a large number of published investigations (e.g. Stebbins, 1956; Hartsema, 1961; Bennett, 1971, 1976; Bennett and Smith, 1976; Levin and Funderburg, 1979; Grime and Mowforth, 1982; Grime, 1983; Grime *et al.*, 1985) it has been established that there is more than a thousandfold variation in nuclear DNA amount in vascular plants and that differences in genome size in cool temperate regions coincide with differences in the timing of shoot growth. A mechanism explaining variation in DNA amount, cell size and length of the cell cycle (the three attributes are inextricably linked) as a consequence of climatic selection has been proposed (Grime and Mowforth, 1982). In the context of Bibury, the most relevant aspects of this hypothesis are the suggestions: (i) that, in low DNA plants, growth is more dependent on current cell divisions (in high DNA plants growth appears to depend much more upon expansion of cells divided and stored in a preceding phase); and (ii) that the potentially short cell cycle associated with small genomes confers a greater capacity to respond to warmer conditions.

Unfortunately, these hypotheses are not easily transformed into predictions of plant responses to the temperatures experienced at Bibury. This is

Table 15.2. Correlation analysis relating variation in shoot biomass of species aggregated into primary functional types (see Fig. 15.3) to variation in climate at Bibury. Climate variation has been simplified by using temperature and rainfall data to define anthropocentrically 'good' (hot/dry) winters and 'bad' (cold/wet) summers and 'good' (mild/dry) summers and 'bad' (cold/wet) winters. * indicates a correlation significant at $P<0.05$. The SR functional type is not shown because it does not exist in the Bibury flora.

Functional type	'Good' summers		'Good' winters		Few or no effects
	Promoted	Retarded	Promoted	Retarded	
C		*	*		
S	*				
R	*				
CR					*
SC			*		
CSR			*		

because the annual July records of plant biomass are also likely to be affected by the greater sensitivity of many small genomes to low spring temperatures and summer drought. However, these complications do not completely exclude the possibility of hypothesis tests related to nuclear DNA amount. In particular we may explore the prediction that the shorter cell cycle and indeterminate growth of small genome species will make them more sharply responsive to year-to-year variations in climate. In Fig. 15.4, this hypothesis is tested by measuring the coefficient of variation in total shoot biomass over the duration of the Bibury experiment in three groups of species classified by genome size. The results show a progressive increase in coefficient in shoot yield with decreasing DNA amount.

Responses to extreme events at Bibury

Extreme climatic events are of particular relevance to studies of the impact of climate because they are likely to provide the 'engine' of vegetation change. This may be especially relevant to the circumstances prevailing at Bibury where the vegetation provides a continuous, fairly dense cover and, one suspects, may be susceptible to invasion and transformation only in conditions where the established dominants (*Arrhenatherum elatius* and *Dactylis glomerata*) are debilitated by some form of vegetation disturbance. Desiccation such as that experienced widely in Europe during the droughts of 1975 and 1976 provides a familiar example. Using a monitoring study conducted over this period in two neighbouring fields of contrasted productivity in Czechoslo-

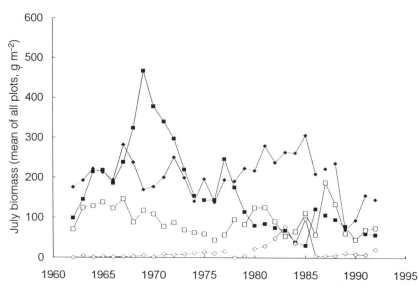

Fig. 15.4. Variation in shoot biomass in components of the Bibury vegetation from 1958 to 1992, classified with respect to nuclear DNA amount: ■, up to 5.0 pg (cv = 0.67); □, 5.1 to 10.0 pg (cv = 0.38); ♦, above 10.1 pg (cv = 0.26); ◇, unclassified (cv = 1.43).

vakia, Leps *et al.* (1982) were able to confirm the value of C–S–R theory in predicting both the resistance (initial damage) and resilience (subsequent recovery) of component species. Using the ISP database, a similar test will soon be conducted on the Bibury record. It is also intended that more detailed predictions will be developed using standardized information relating to root morphology, xylem conductivity and desiccation tolerance.

An important but elusive problem at Bibury is to test the hypothesis that the vegetation becomes temporarily invasible following extreme events. Although new species did not establish in significant numbers following the 1975–76 droughts, consistent structural changes were detected both immediately and subsequently. The most striking effect was a spectacular increase in the abundance of the competitive–ruderal (*sensu* Grime *et al.*, 1988), *Anthriscus sylvestris*, which rose to peak biomass in 1977 and did not return to its background values until 1980 (Fig. 15.5). A tentative conclusion from this observation is that drought damage to the dominant species did temporarily 'open the door' to robust, large-seeded invaders but that no effective colonization by new invaders took place in the monitored plots. Instead, it appears that one of the resident components, notable for its consistent ability to produce a substantial crop of large seeds and to colonize vegetation gaps (Thompson and Baster, 1992), benefited from a brief relaxation of dominance.

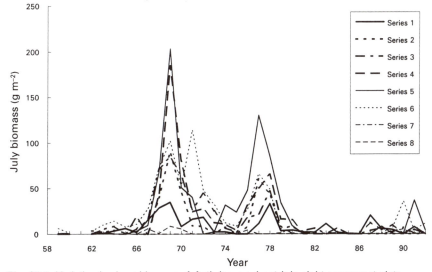

Fig. 15.5. Variation in shoot biomass of *Anthriscus sylvestris* in eight permanent plots (series) at Bibury between 1958 and 1992.

Tests Using Manipulative Experiments

Several aspects of vegetation response to climate at Bibury seem destined to remain beyond the reach of hypothesis-testing based upon the Willis records and will necessitate ancillary experiments, following the protocol of the right-hand side of Fig. 15.2 above. First, we may wish to consider whether the progressive changes in species composition which have occurred at Bibury over the period of the study (Table 15.1) are entirely independent of climate change. Although there may be little doubt that the site has, in common with much of lowland Britain, experienced agricultural eutrophication, it may be significant that atmospheric concentrations of CO_2 have also risen in parallel. Moreover, comparative laboratory studies of plant response to elevated CO_2 (Hunt *et al.*, 1991a, 1993a; Díaz *et al.*, 1993) reveal that the species expanding at Bibury include representatives of functional groups with a high potential for response. At Bibury and elsewhere (Körner, 1993) further research is needed to determine whether an interaction is taking place between mineral enrichment and rising CO_2 concentration.

Second, experiments are required to resolve questions arising from the many climate–vegetation correlations in the Willis records. There is a particular need to distinguish between effects of temperature and moisture supply in both the positive and negative responses to warm summers. It is also desirable to discriminate between direct effects of these factors on plant performance and those mediated by competitive interactions. An ideal approach to these problems would be to conduct climate engineering experiments on the Bibury verges. Practical and road safety considerations make this impossible.

As an alternative, key species have been transplanted to large containers in an experimental garden (Fig. 15.6) and, as monocultures and additive mixtures,

Fig. 15.6. A garden experiment in which monocultures and mixtures of species transplanted from Bibury are grown in large mobile containers which can be subjected to climate manipulations.

these are being subjected to climate manipulations involving mild winters, late frosts and summer drought.

A third topic requiring experimental development concerns the relationship between extreme events and receptivity of the Bibury vegetation to invasion by southern, perhaps Mediterranean, species. In order to pursue this line of investigation, garden experiments are in progress to examine the potential of imposed droughts to create circumstances conducive to the establishment from seed of *Anthriscus sylvestris* and several more exotic invaders.

Acknowledgements

For many years, the late Professor E.W. Yemm was an active collaborator in the Bibury experiments. Mrs D.P. Willis is an essential member of the recording team. We are grateful to Mr F. Sutton, Mr S.R. Band and Mr A.M. Neal for assistance in processing field data. This research has been supported by the Natural Environment Research Council, the Esmée Fairburn Trust, the Lindeth Trust, Rhône-Poulenc plc and Nuclear Electric plc.

References

Barkham, J.P. and Hance, C.E. (1982) Population dynamics of the wild daffodil (*Narcissus pseudonarcissus*). III. Implications of a computer model of 1000 years of population change. *Journal of Ecology* 70, 323-344.

Bennett, M.D. (1971) The duration of meiosis. *Proceedings of the Royal Society of London B, 178,* 277-299.

Bennett, M.D. (1976) DNA amount, latitude and crop plant distribution. *Environmental and Experimental Botany* 16, 93-108.

Bennett, M.D. and Smith, J.P. (1976) Nuclear DNA amounts in angiosperms. *Philosophical Transactions of the Royal Society B* 274, 227-274.

Campbell, B.D., Grime, J.P. and Mackey, J.M.L. (1991) A trade-off between scale and precision in resource foraging. *Oecologia* 87, 532-538.

Colasanti, R.L. and Grime, J.P. (1993) Resource dynamics and vegetation processes: a deterministic model using two-dimensional cellular automata. *Functional Ecology* 7, 169-176.

Crawley, M.J. and May, R.M. (1987) Population dynamics and plant community structure: competition between annuals and perennials. *Journal of Theoretical Biology* 125, 475-489.

Díaz, S., Grime, J.P., Harris, J. and McPherson, E. (1993) Effects of elevated atmospheric carbon dioxide on plant communities during the initial stages of secondary succession. *Nature* 364, 616-617.

Green, D.G. (1989) Simulated effects of fire, dispersal and spatial pattern on competition within forest mosaics. *Vegetatio* 82, 139-153.

Grime, J.P. (1974) Vegetation classification by reference to strategies. *Nature* 250, 26-31.

Grime, J.P. (1983) Prediction of weed and crop response to climate based upon measurements of nuclear DNA content. *Aspects of Applied Biology* 4, 87-98.

Grime, J.P. (1988) The C-S-R model of primary plant strategies - origins, implications and tests. In: Gottlieb, L.D. and Jain, S.K. (eds) *Plant Evolutionary Biology*. Chapman and Hall, London, pp. 371-393.

Grime, J.P. and Hunt, R. (1975) Relative growth rate; its range and adaptive significance in a local flora. *Journal of Ecology* 63, 393-422.

Grime, J.P. and Mowforth, M.A. (1982) Variation in genome size - an ecological interpretation. *Nature* 299, 151-153.

Grime, J.P., Hodgson, J.G. and Hunt, R. (1988) *Comparative Plant Ecology: A Functional Approach to Common British Species*. Unwin Hyman, London.

Grime, J.P., Shacklock, J.M.L. and Band, S.R. (1985) Nuclear DNA contents, shoot phenology and species coexistence in a limestone grassland community. *New Phytologist* 100, 435-444.

Hartsema, A.M. (1961) Influence of temperature on flower formation and flowering of bulbous and tuberous plants. In: Ruhland, W. (ed.) *Handbuch der Pflanzenphysiologie 16. Ansenfaktoren in Wachstum und Entwicklung*. Springer-Verlag, Berlin, pp. 123-167.

Hendry, G.A.F. and Grime, J.P. (eds) (1993) *Comparative Plant Ecology - A Laboratory Manual*. Chapman and Hall, London.

Hunt, R., Hand, D.W., Hannah, M.A. and Neal, A.M. (1991a) Response to CO_2 enrichment in 27 herbaceous species. *Functional Ecology* 5, 410-421.

Hunt, R., Middleton, D.A.J., Grime, J.P. and Hodgson, J.G. (1991b) TRISTAR: an expert system for vegetation processes. *Expert Systems* 8, 219-226.

Hunt, R., Hand, D.W., Hannah, M.A. and Neal, A.M. (1993a) Further responses to CO_2 enrichment in British herbaceous species. *Functional Ecology* 7, 661-668.

Hunt, R., Willis, A.J., Ward, L.K., Grime, J.P., Hodgson, J.G., Dunnett, N.P., Sutton, F., Band, S. and Neal, A.M. (1993b) A thirty-five year study of vegetation and climate in road verges at Bibury, Gloucestershire, and a twenty-one year study in chalk grassland at Aston Rowant, Oxfordshire. *The NERC Unit of Comparative Plant Ecology, Terrestrial Ecology Research on Global Warming: Phase 2. Contract report to Nuclear Electric plc*.

Inghe, O. (1989) Genet and ramet survivorship under different mortality regimes - a cellular automata model. *Journal of Theoretical Biology* 138, 257-270.

Keddy, P. (1992) A pragmatic approach to functional ecology. *Functional Ecology* 6, 621-626.

Körner, C. (1993) CO_2 fertilization: the great uncertainty in future vegetation development. In: Solomon, A.M. and Shugart, H.H. (eds) *Vegetation Dynamics and Global Change*. Chapman and Hall, London, pp. 53-70.

Leps, J., Osbornova-Kosinova, J. and Rejmanek, M. (1982) Community stability, complexity and species life-history strategies. *Vegetatio* 50, 53-63.

Levin, D.A. and Funderburg, S.W. (1979) Genome size in angiosperms; temperate versus tropical species. *American Naturalist* 114, 784-795.

Noble, I.R. and Slatyer, R.O. (1979) The use of vital attributes to predict successional changes in plant communities subject to recurrent disturbances. *Vegetatio* 43, 5-21.

Stace, C. (1991) *New Flora of the British Isles*. Cambridge University Press, Cambridge.

Stebbins, G.L. (1956) Cytogenetics and evolution of the grass family. *American Journal of Botany* 43, 890-905.

Thompson, K. and Baster, K. (1992) Establishment from seed of selected Umbelliferae in unmanaged grassland. *Functional Ecology* 6, 346-352.

Willis, A.J. (1988) The effects of growth retardant and selective herbicide on roadside verges at Bibury, Gloucestershire, over a thirty-year period. *Aspects of Applied Biology* 16, 19-26.

Wolfram, S. (1984) Cellular automata as models of complexity. *Nature* 311, 419-424.

Yemm, E.W. and Willis, A.J. (1962) The effects of maleic hydrazide and 2,4-dichloro-phenoxyacetic acid on roadside vegetation. *Weed Research* 2, 24-40.

16

The Park Grass Experiment: Insights from the Most Long-term Ecological Study

D. Tilman[1], M.E. Dodd[2], J. Silvertown[3], P.R. Poulton[4], A.E. Johnston[4] and M.J. Crawley[5]
[1]University of Minnesota, Department of Ecology, Evolution and Behaviour, 100 Ecology Building, 1987 Upper Buford Circle, St Paul, Minnesota 55108-6097, USA; [2]The Open University, Biology Department, Walton Hall, Milton Keynes MK7 6AA, UK; [3]The Open University, Department of Statistics, Walton Hall, Milton Keynes MK7 6AA, UK; [4]Rothamsted Experimental Station, Harpenden, Herts AL5 2JQ, UK; [5]Imperial College at Silwood Park, Centre for Population Biology, Department of Biology, Ascot, Berks SL5 7PY, UK.

Introduction

In 1856 J.B. Lawes and J.H. Gilbert started an experiment still in progress, which is now, undoubtedly, the most long-term ecological study in the world. The Park Grass Experiment was initially designed to determine the effects of different amounts and combinations of mineral fertilizers and organic manures on the productivity of permanent grassland. However, there were such dramatic effects of the treatments on plant species composition and plant diversity that Lawes and Gilbert (1880) quickly concluded that the experiment was of greater interest to the 'botanist, vegetable physiologist, and the chemist than to the farmer'. And this has proved to be so.

As demonstrated below, the treatments within the experiment have continued to provide insights into issues in evolutionary, population, community and ecosystem ecology. The long-term data have become increasingly powerful with each additional sampling, despite the lack of replication and randomization. However, Lawes and Gilbert can hardly be faulted for their experimental design. Important contrasts were placed adjacent to each other so that treatment effects could be directly observed all along the 80 m boundary that the two plots shared. Moreover, it was during R.A. Fisher's tenure at Rothamsted that he articulated, more than 70 years after the establishment of Park Grass, the concepts upon which modern experimental design are based (Barnett, Chapter 10).

Design of the Park Grass Experiment

The Park Grass Experiment was laid out in a four hectare area that had been maintained as grassland by grazing or mowing for hay for several hundred years before the study began (Lawes and Gilbert, 1863; see also Johnston, Chapter 2). It had also been periodically fertilized with farmyard manure. Lawes and Gilbert originally established 13 plots and later added seven more; they ranged in size from 0.05 to 0.2 ha. Initially each plot received either no nutrient addition (the control plots, Plots 3 and 12), farmyard manure (Plot 2), or various combinations and rates of annual addition of mineral nutrients (the remaining 17 plots). Some treatments were altered (Warren and Johnston, 1964) and most plots were halved first in 1903 and then in 1965 into four subplots that received different rates of liming. The most recent subdivision was to give four different levels of soil pH. A summary of the major changes in manurial history and management was given by Warren and Johnston (1964). These authors, together with Thurston *et al.* (1976) summarized both chemical and botanical results and Thurston *et al.* (1976) discussed initial effects of the liming treatments introduced in 1965. The layout of the plots, and the treatments applied to each plot, are summarized in Fig. 16.1. These treatments provide 89 plots in which the responses of plants, of herbivores, and of soil organisms, properties and processes to soil nutrients can be directly observed.

The Ecology of Park Grass: a Brief Review

The first significant papers published on the Park Grass Experiment were those of Lawes and Gilbert (1863, 1880) and Lawes *et al.* (1882). These presented results of surveys in which the relative abundances of plant species in each plot were estimated in 1862, 1867, 1872, and 1877. These papers also included soil chemistry and hay yields. Brenchley and Warington (1958) published the next major compilation of botanical data and included data from 1903, 1914, 1919, 1920, 1926, 1936, 1948 and 1949. Some plots were not sampled in some of these years and a few were sampled more frequently. Thurston (1969) and Williams (1978) have published more recent data on the botanical composition of these plots, and Crawley and co-workers have collected

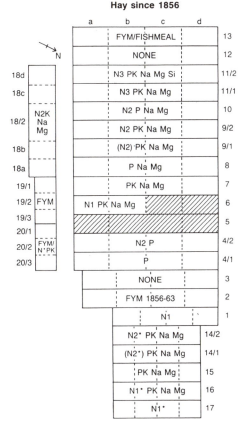

PARK GRASS
Hay since 1856

Fig. 16.1. Treatments (every year except as indicated). *Nitrogen* (applied in spring): N1, N2, N3 sulphate of ammonia supplying 48, 96, 144 kg N ha⁻¹; N1*, N2* nitrate of soda supplying 48, 96 kg N ha⁻¹; (N2), (N2*) last applied 1989. *Minerals* (applied in winter): P 35 kg; K 225 kg; Na 15 kg; Mg 10 kg ha⁻¹; Si Silicate of soda at 450 kg ha⁻¹ of water soluble powder; Plot 20, rates of manuring in years when FYM not applied: 30 kg N*, 15 kg P, 45 kg K ha⁻¹. *Organics* (each applied every fourth year since 1905): FYM, 35 t ha⁻¹ farmyard manure (bullocks) (1989, 1993); fish meal (about 6.5% N) to supply 63 kg N ha⁻¹ (1991, 1995). *Lime*: a,b,c, lime applied as needed to maintain pH 7, 6 and 5 respectively; d, no lime applied (pH range 3.5 (plot 11–1) to 5.7 (plot 17)).

the most recent data (1991–93). In addition, Cashen (1947), Smith (1960), Warren and Johnston (1964), Warren *et al.* (1965), Silvertown (1980, 1987), Tilman (1982, 1986), and others have presented analyses of both yield and botanical data. The above data and analyses, and the analyses presented below, demonstrate that:

1. In the absence of experimental perturbation, plant species composition, diversity and productivity are quasi-stable, but are influenced by climatic variation. Some usually rare species have brief periods of dominance, and a few usually dominant species have brief periods of rarity.

2. Plant community composition is highly dependent on both the rate and ratio of supply of limiting soil resources. Nitrogen, phosphorus and potassium are the major limiting soil resources for Park Grass species.

3. Soil pH, which is modified via liming and fertilization, greatly affects plant species composition and diversity.

4. Different rates of nutrient supply act as a selective force on plant populations, causing measurable evolutionary change.

These four points are discussed in the following sections.

Compositional Dynamics

Because few ecological communities have been sampled over periods of even a decade, it is difficult to know the extent to which present compositional patterns are representative of the long-term state of most plant communities. The 130-year record of plant species abundances in the control plots of Park Grass provides insight into this issue and illustrates the difficulty of establishing true experimental 'controls'. Consider, for instance, the dynamics in Plot 12 of eight of the most abundant plant species in Park Grass (Fig. 16.2). Three perennial grasses, *Festuca rubra*, *Dactylis glomerata*, and *Agrostis capillaris* generally remained the co-dominant species during this period, but their absolute and relative abundances changed from one sampling to the next. *Agrostis*, for instance, was the most abundant species at the first sampling in 1862, was the fourth most abundant species in 1949, and the second most abundant species in 1991. The least abundant of the eight species in 1862, *Arrhenatherum*, generally remained less abundant throughout the 130-year period. The shifts in the absolute and relative abundances of these species were also shown by other species during the 130 years. *Plantago lanceolata* was the most abundant species in 1919, but never in any other year; in 1877 it was the 17th most abundant species of the 29 in this plot. *Leontodon hispidus* was recorded as having 'trace' abundance through 1877, but was 10% of biomass in 1949. Its low initial abundance may have resulted from the periodic grazing of the field before the experiment began (see later discussion). *Leontodon* is a preferred forage species for cattle and would have been freed from such selective herbivory once the experiment began. Whereas many species tended to remain dominants, subdominants or rare throughout this 130-year

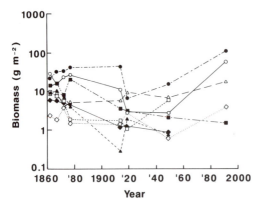

Fig. 16.2. The dynamics of eight common species in Plot 12, an unfertilized control plot. Reasons for some of the initial changes are discussed in the text. ○ *Agrostis*; □ *Alopecurus*; ◇ *Arrhenatherum*; △ *Dactylis*; ● *Festuca*; ■ *Holcus*; ◆ *Lathyrus*; ▲ *Rumex*.

period (Silvertown, 1987), occasionally a species which was rare in most years sporadically became a dominant. We do not know what caused these periodic outbreaks of usually rare plant species.

The other control, Plot 3, was sampled more frequently, and shows apparently greater variation (Fig. 16.3). Two grasses, *Festuca* and *Agrostis*, were the co-dominants throughout most of the period. In 1903 *Agrostis* seemingly declined and another grass, *Briza*, increased; but this may be a case of misidentification because these two species are difficult to distinguish vegetatively. Many species had order-of-magnitude variations in estimated biomass, and some, such as *Leontodon* and *Poterium* showed a marked pattern of increase during the initial 50–70 years of the experiment (Fig. 16.3). When viewed on a coarser scale by summing together all grasses, all legumes, or all forbs, species composition shows a long-term pattern of change on Plot 3. Grasses were initially the most dominant group, but grasses declined and forbs (non-legume, dicotyledonous species) increased until the early 1900s (Fig. 16.4). Legumes remained rare throughout and showed little directional change in relative abundances. The data in Fig. 16.4 come from visual estimates of abundances, so that some of the variation may be caused by differences in technique between observers over the 130 years. However, the major trends seem too great to be accounted for in this way. The marked increases in

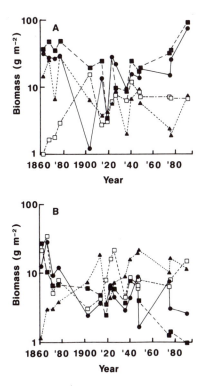

Fig. 16.3. The dynamics of eight species in Plot 3, another unfertilized control plot which was sampled more frequently than Plot 12 (Fig. 16.2). (A) ● *Agrostis*; ■ *Festuca*; ▲ *Holcus*; □ *Poterium*; (B) ● *Anthoxanthum*; ■ *Helictotrichon*; ▲ *Leontodon*; □ *Plantago*. Note that most variation is between 1 and 100 in both plots.

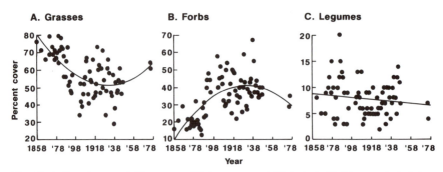

Fig. 16.4. Abundances of (A) grasses, (C) legumes, and (B) forbs in Plot 3, a control plot; estimated visually (Williams, 1978). (The fitted curves were those which gave the highest R^2 values.)

the abundances of particular forb species (Fig. 16.3) and in the relative abundances of forbs as a group (Fig. 16.4) during the first half century of the experiment has several possible explanations. The field had been both mown for hay and grazed and periodically fertilized with farmyard manure prior to 1856 after which the occasional addition of nutrients ceased. From 1856 the plots were cut for hay each year with the aftermath being grazed by sheep until 1874, from 1875 the aftermath was cut and weighed green. The changes in vegetational composition might be caused by either the cessation of manuring or herbivory or both. The changes during the first 50–70 years in Plots 3 and 12 probably reflect such effects. The dynamics since then provide a better estimate of the stability of unmanipulated grassland.

Climate and Grassland Productivity

The effect of the climate, or 'season' in agricultural terms, on aboveground net primary productivity, which is reasonably estimated by the size of the hay crop, has been frequently discussed since the early days of the experiment. In general the hay yield of English meadows tends to be strongly correlated with actual transpiration, and rainfall tends to be the most important determinant of this (Smith, 1960). This relationship is usually strongest in grasslands not receiving nitrogen fertilizer. Cashen (1947) statistically demonstrated both of these aspects of the relationship between hay yield and rainfall for the Park Grass plots in a study directed by R.A. Fisher. More recently, Jenkinson *et al.* (1994) found that meteorological variables, including rainfall, could account for only between 12% and 21% of the variance in total annual hay yield on the unlimed parts of Plots 2, 3, 12, 14 and 16 over the period from 1891 to 1958. This improved to between 45% and 63% for 1960 to 1992, during which time yields of herbage at the traditional hay-making stage of growth were estimated by cutting a part of each plot with a forage harvester. This provides a better estimate of the dry matter of the standing crop than traditional haying methods which were continued on the remainder of the plot, although these hay yields were not determined.

The relationship between climate and hay yield is important from an ecological point of view because it demonstrates climatic effects on the plant communities. Climatic variation may affect plant community composition: (i) directly by differentially influencing the growth and death of individual plant species; or (ii) indirectly by affecting comparative growth rates between species which in turn affects composition through competitive effects among species; or (iii) indirectly through effects on nutrient dynamics, pathogen densities, litter decomposition, etc. Silvertown *et al.* (unpublished) have looked at different combinations of the relationships between rainfall, hay yield and composition (the ratio of grasses/legumes/other species) and found stronger relationships between yield variation and variation in composition than between rainfall and composition. Of twelve regression models used, the most successful was a model which explained variation in composition based on variation in hay yield in the year before composition was measured. This may mean that direct effects of rainfall were less important than indirect ones and may indicate the importance of interspecific competition or of accumulated plant litter in the response of the Park Grass communities to climatic perturbation.

Nutrient Supply and Species Composition

The addition of inorganic N, of P, and of various combinations and ratios of N, P, K, Ca, and other nutrients (Fig. 16.1 above) caused dramatic shifts in the abundances of plant species, as predicted by a theory of resource competition (Tilman, 1982). Consider, for instance, Plot 11-1, which received complete mineral fertilizer, including 144 kg N ha^{-1} as ammonium sulphate. A perennial grass, *Holcus lanatus* (Yorkshire fog) became dominant by 1950 (Fig. 16.5A) and virtually all other species were displaced from the unlimed subplot by 1976. In contrast, Plot 14 received complete mineral fertilizer, including 96 kg N ha^{-1} as sodium nitrate. It was dominated by *Alopecurus* and *Arrhenatherum*, but *Holcus* was extremely rare (Fig. 16.6). The difference between these two N treatments seems to be caused by the acidifying effects of N when added as ammonium sulphate. The top 23 cm of soil on the unlimed part of Plot 11-1 had a pH less than 4 in 1959, whereas that on the unlimed part of Plot 14 had a pH of 6 (Warren and Johnston, 1964). The part of Plot 11-1 which received lime every fourth year from 1903 (Fig. 16.5B) had a composition much more similar to that of Plot 14 than to Plot 11-1, from 1959 onwards. This supports the hypothesis that the dramatically different effects of the two forms of N addition came from the effects on soil pH. Further effects of liming and soil pH are discussed below.

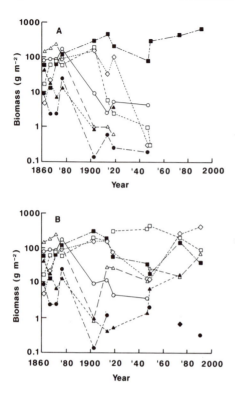

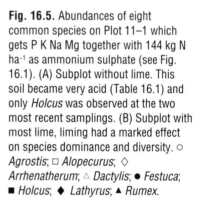

Fig. 16.5. Abundances of eight common species on Plot 11–1 which gets P K Na Mg together with 144 kg N ha⁻¹ as ammonium sulphate (see Fig. 16.1). (A) Subplot without lime. This soil became very acid (Table 16.1) and only *Holcus* was observed at the two most recent samplings. (B) Subplot with most lime, liming had a marked effect on species dominance and diversity. ○ *Agrostis*; □ *Alopecurus*; ◇ *Arrhenatherum*; △ *Dactylis*; ● *Festuca*; ■ *Holcus*; ◆ *Lathyrus*; ▲ *Rumex*.

Another critical determinant of the species composition of the plots is the N:P ratio (Tilman, 1982). Again considering the limed sections of the plots, low

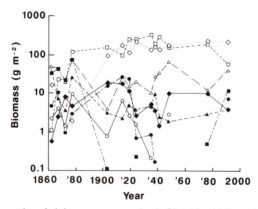

Fig. 16.6. The dynamics of eight common species in Plot 14, which received the same amounts of P, K, Na and Mg as Plot 11–1 (Fig. 16.5), but 96 kg N ha⁻¹ as sodium nitrate. The unlimed soil had pH 6.0 in 1959. ○ *Agrostis*; □ *Alopecurus*; ◇ *Arrhenatherum*; △ *Dactylis*; ● *Festuca*; ■ *Holcus*; ◆ *Lathyrus*; ▲ *Rumex*.

N:P ratios, such as created by the addition of just P (Plot 4-1) or of P, Na and Mg (Plot 8) or of P, K, Na, and Mg (Plot 15) led to markedly higher relative abundances of the legumes *Lathyrus, Lotus* and *Trifolium* and lower relative abundances of the grasses *Festuca rubra, Arrhenatherum,* and *Dactylis glomerata.* In contrast, the addition of just N (Plots 1 and 17) led to about fivefold lower relative abundances of *Lathyrus* and *Lotus* and threefold greater relative abundances of the grasses *Dactylis glomerata* and *Festuca rubra.* In general, increased productivity led to greater grass dominance and to lower legume and forb relative abundances. However, within limed plots with similar pH values, the abundances and identity of the dominant grass species depended on the ratio of supply of N and P. For instance, Plot 16, which received N and P together with K, Na, Mg, had more than 10-times the *Arrhenatherum* abundance of Plots 1 and 17, which received the same rate of N addition, but no P or K, Na, Mg. Such effects of N:P ratios on species composition and relative abundances are consistent with the theory of competition for multiple resources presented in Tilman (1982).

There are also effects of nutrients other than N and P. Consider *Taraxacum officinale,* the dandelion. Its relative abundance in Park Grass ranges from 0 in some plots to over 20% of aboveground living biomass in others. Using the data summarized in Williams (1978), the nine limed subplots receiving K had 67 times greater relative abundance of dandelions than the unlimed subplots receiving no K. The unlimed part of Plot 4-2, which received only 96 kg N ha^{-1} as ammonium sulphate together with P, was 100% grass (*Agrostis* and *Anthoxanthum*) and the part limed to pH 5 was 99.8% grass (*Festuca, Agrostis* and *Poa*). Both have experienced a rain of dandelion seed from neighbouring plots, but have virtually no dandelions, possibly because they receive no K. This suggests that *Taraxacum* may be a poor competitor for K and cannot persist in soils in which K is limiting.

Productivity and Diversity

The addition of various amounts and combinations of nutrients also influenced the number of plant species occurring in these plots. Although there are some interesting exceptions, in general, the greater the productivity caused by a given fertilizer treatment, the lower the species diversity (Fig. 16.7). This effect was evident in 1862 and every year since (Tilman, 1982). However, the effect of productivity on species richness depended on soil pH. At a given level of productivity, more acidic plots had fewer species (see Table 16.1). This experimentally demonstrated inverse relationship between productivity and species richness is the opposite of the commonly reported pattern for habitats along latitudinal productivity gradients (e.g. Pianka, 1966; Tilman and Pacala, 1993). However, results similar to those observed in Park Grass have been observed in a wide variety of terrestrial and aquatic habitats that have had their

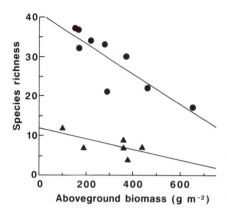

Fig. 16.7. Relationship between species richness (number of plant species recorded during a given sampling period) and productivity on soils, above (●) and below (▲) pH 4.9 (Redrawn from Tilman, 1982, using 1948 and 1949 data).

productivity increased via nutrient addition (Huston, 1979; Tilman, 1982). An explanation for this effect is that nutrient addition eliminates spatial heterogeneity and causes a large number of plant species to become limited by the same resource (often light), with the best competitor for that resource displacing all other species limited by it from the habitat (Tilman, 1982).

The Effects of Liming

Most fertilized plots were halved in 1903 to comprehensively test the effects of liming after some early tests between 1881 and 1897 (Warren and Johnston, 1964). Half of each plot was limed and half left alone. In 1965, these half plots were again halved, with the intention of achieving pH values of 7, 6, 5 on subplots a, b, c (Warren *et al.*, 1965) by periodic additions of lime. Some effects on soil pH of lime treatments introduced in 1965 were discussed by Johnston (1972) especially the slow change in mineral soil pH on plots with a 'mat' of partially decomposed plant debris.

The liming treatments started in 1903 and 1965 both dramatically illustrate the effects of an abiotic environmental factor, soil acidity, on plant species composition and diversity. Soil pH can influence the way in which other soil resources are utilized and can control the number of plant species capable of persisting on a subplot. Declining pH increases the ionic concentration of potentially toxic metals like iron and aluminium. Here we describe the effects of liming on five plots: the two unfertilized controls (Plots 3 and 12) to give an impression of the degree of background variation between plots, a plot receiving 144 kg N ha^{-1} as ammonium sulphate (the largest rate of nitrogen tested on Park Grass; Plot 11-1), a plot receiving potassium and phosphorus but no nitrogen (Plot 7), and a plot receiving phosphorus only (Plot 8). (Plots 7, 8 and 11-1 also receive small amounts of Na and Mg which almost certainly do not affect the results reported here.) Soil pH, plant species richness and hay yields from these plots are shown in Table 16.1 for the most recent sampling (1991–93).

Table 16.1. Effects of lime treatment on soil pH, species richness and biomass at the first cut on five plots on Park Grass.[a]

Plot	Soil pH in H_2O (0–23 cm) in 1991			
	(a)[b]	(b)	(c)	(d)
3	6.4	6.4	5.0	4.8
12	7.0	6.1	4.8	4.8
7	6.5	6.2	5.2	4.8
8	6.7	6.4	5.1	4.9
11-1	5.3	5.4	4.1	3.4
	Plant species richness (total species per 10 m x 15 m area)			
3	34	33	35	35
12	44	42	34	41
7	22	23	25	25
8	32	32	28	29
11-1	11	13	9	2
	Biomass (t dry matter ha^{-1}) from the early June hay crop; 1991–93			
3	2.77	3.28	2.00	2.16
12	2.31	2.50	1.85	2.15
7	5.39	5.47	4.59	4.35
8	2.96	3.43	3.66	3.56
11-1	5.84	5.76	5.30	5.59

[a] Species richness was estimated in a 10 x 15 m rectangle in the centre of each subplot so that the area used for the estimates was the same throughout.
[b] Subplot (a) limed to achieve a pH of 7 since 1976 but limed since 1903; (b) limed to achieve a pH of 6 since 1965, but limed since 1903; (c) limed to achieve a pH of 5 since 1965, unlimed from 1856 to 1964; (d) unlimed since 1856.

Liming and soil pH

The aim of the liming regime introduced in 1965 was to produce three different soil pH values for each fertilizer treatment (subplots a,b,c), as discussed above, whilst subplot d would remain unlimed and its pH would depend on the various acidifying inputs. As yet it has been difficult to maintain the target pH values of 7, 6 and 5 on subplots a,b and c respectively. The liming treatments have, however, created a sequence of plots that differ in soil pH in the desired directions (Table 16.1). Soil pH has fallen below 4 only where ammonium sulphate has been applied, and liming has not raised and maintained the pH above 6 on the plot receiving 144 kg N ha^{-1} as ammonium sulphate (Plot 11-1a). The difference in pH between subplots 3a and 12a of the unfertilized controls is probably because of the different amounts of lime applied between 1965 and 1991.

Lime and plant species richness

There are at least two important diversity-reducing processes at work on Park Grass: (i) increasing soil acidity may reduce the number of species capable of tolerating the abiotic conditions; and (ii) increasing productivity may lead to competitive exclusion of small-stature plant species by taller species that are presumably better competitors for light (which is likely to be a major limiting factor in productive plots). Given these processes, we expect diversity to be least on acid plots with tall individual species and high biomass, and most on limed plots with small individuals and low total biomass. Comparisons of species richness, based on 1991–93 visual censuses, for Plots 11-1 and 12, and Plots 7 and 8 (Table 16.1) support this. However, this does not explain the differences between the control plots, Plot 12 is seemingly more diverse than Plot 3. The effect of acid-tolerance on the size of the species-pool is suggested by the steeply declining trend of species richness with increasing acidity on the plot (11-1) receiving most ammonium nitrogen (see also Fig. 16.5A). Of the 126 species that have been recorded at some time on Park Grass, *Holcus lanatus* was the dominant grass in 1947 when the pH was 3.8 and it now thrives below pH 3.5 with large inputs of ammonium sulphate together with P and K. This is likely to be a direct pH effect, but it may also be caused by some other factor, such as the dense layer of litter (thatch or mat) which has developed gradually on the unlimed plots receiving most ammonium nitrogen. Clearly, although the Park Grass Experiment provides evidence that is consistent with these hypothesized processes, additional studies are required to determine whether there are actually direct effects of pH and productivity on diversity, or if the observed patterns are mediated through other correlated processes.

The effect of pH on species biomass

On Plot 11-1 there is no clear trend in hay yield with pH, but there is a major impact on botanical composition (Table 16.2). The 6 t ha^{-1} of dry matter on subplot 11-1a is made up of *Arrhenatherum*, *Alopecurus* and *Dactylis* and several tall herbs, whereas the 6 t ha^{-1} of dry matter on subplot 11-1d is made up entirely of *Holcus lanatus*. There is no trend in hay yield with liming and hence pH on the control plots, but the more recently limed 'c' subplots have lower yields than the more long-term stable subplots. The declining pH over the course of the experiment on the unlimed control subplots suggests an effect of acid deposition. It is estimated that the soil pH of the original meadow was about 6.0. Since the industrial revolution, the pH on the unlimed sub-plots has declined to 4.8 on both 12d and 3d. This is associated with greater relative abundances of the grasses *Agrostis capillaris* and *Anthoxanthum odoratum*,

Table 16.2. Average dry matter yield (g m^{-2}) for 1991–93 based on quadrat sampling just prior to the first hay crop, of four grass species showing a range of responses to the application of lime. Note (a), (b), (c) and (d) refer to liming treatments; for pH values see Table 16.1.

Plot	(a)	(b)	(c)	(d)
	Arrhenatherum elatius			
3	0.6	0	0	3.3
12	2.1	0	3.4	1.5
7	123.0	134.7	12.7	3.3
8	9.8	27.1	1.3	8.2
11-1	397.7	320.0	118.2	0
	Alopecurus pratensis			
3	0.6	0	0	0
12	0	0.6	1.0	0
7	19.4	35.4	11.6	5.0
8	0	0	0.8	0.2
11-1	67.7	50.6	71.7	0
	Holcus lanatus			
3	2.6	4.2	0.7	3.8
12	6.6	2.5	1.5	0.9
7	39.3	43.4	27.1	19.4
8	19.9	14.6	13.3	30.6
11-1	66.7	91.2	280.2	457.6
	Agrostis capillaris			
3	32.0	38.4	55.2	72.8
12	55.2	50.4	54.4	65.6
7	4.8	10.4	131.2	154.4
8	32.8	16.8	80.0	72.8
11-1	0.8	0.2	37.6	0

and thus with lower relative abundances of herbs and legumes species, but not yet with changes in species richness.

Liming and abundances of selected plant species

The average dry weight at first harvest for 1991–93 of four of the more characteristic grass species are shown in Table 16.2. The species were chosen to exemplify contrasting responses to the application of lime.

Arrhenatherum elatius is the characteristic tall grass of the limed subplots receiving both P and K; in the absence of K it is rare. It grows more than 1 m high in the crop cut in June and forms the major component of the regrowth cut in October–November. It flowers in both periods. The effects of lime are shown on high yielding plots with nitrogen (Plot 11-1) and without (Plot 7) (Table 16.2). Plot 7 yields well because of the legume component of the sward.

Alopecurus pratensis has a very similar distribution to *Arrhenatherum* but differs in its ecology in important ways. It reaches only about one-fifth of the peak biomass shown by *Arrhenatherum* in the first crop (Table 16.2) and is virtually absent from the second. It flowers only in the first crop and is a strongly early-season grass. Like *Arrhenatherum* it disappears from the sward in the absence of K when P is applied.

Holcus lanatus is ubiquitous on Park Grass, but rarely achieves high biomass with the spectacular exception of the most acid subplots of Plot 11-1 (Table 16.2). On Plot 11-1 it declines with increasing lime from a monoculture of 458 g m^{-2} on subplot 11-1d to a modest 66 g m^{-2} on the limed subplot 11-1a.

Agrostis capillaris has a general increase in absolute abundance with declining pH (Table 2). On control Plot 3 it more than doubled in the standing crop from subplot a to d; the trend on Plot 12 was similar but less pronounced. The increase was especially pronounced where no nitrogen but both P and K were applied; Plot 7 subplots c and d had 131 and 154 g m^{-2}. *Agrostis* has less absolute biomass on Plot 8 without potassium. *Agrostis* is rare on all but subplot c of Plot 11-1 (144 kg N ha^{-1}, limed to attain pH 5 since 1965). Here *Agrostis* has increased markedly since 1965.

Microevolution in the Park Grass Experiment

Because the various rates and patterns of nutrient addition to the Park Grass plots have resulted in changes in the abundances of plant species, it seems reasonable that the treatments might also act as selective forces on individual species. However, the rapidity of evolutionary change in response to such selective forces at Park Grass has been surprising. Genetic differentiation has been demonstrated in *Anthoxanthum odoratum* (sweet vernal grass) which occurs across a broad range of nutrient and pH differences on many of the Park Grass plots. Snaydon (1970) sampled tillers and collected seeds of *Anthoxanthum odoratum* in limed and unlimed parts of Plots 1, 10 and 18 and grew these in boxes of acid and calcareous soils. Plants from acid plots did better on acid soil than on calcareous soil, whereas the reverse was the case for plants from less acid plots. The results were similar for plants grown from vegetative material or seed, suggesting that differences were genetic rather than due to phenotypic carryover (e.g. Bullock *et al.*, 1993). Liming was begun in Plots 1 and 10 in 1903 and in Plot 18 in 1920. Thus, genetic differentiation occurred in less than 40 years. Following the introduction of new liming treatments on Park Grass in 1965, there were significant morphological differences between populations that had received different liming treatments for only six years.

In another experiment, Snaydon and Davies (1972) collected plants from the limed and unlimed parts of Plots 1, 3, 4, 9, 10, 17 and 18, propagated the samples vegetatively, and then grew these in a common garden. They observed correlations between disease resistance and plot treatments. Mildew (*Erisyphe graminis*) resistance of plants in the common garden was positively correlated ($r^2 = 0.83$) with total soil nitrogen in source plots. Rust (*Puccinia poae-nemoralis*) resistance was positively correlated with vegetation height ($r^2 = 0.69$). Nitrogen fertilizer increases susceptibility to mildew, and humid atmospheric conditions (as found in tall vegetation) aids rust infection, so there appears to have been selection for resistance in those populations most exposed to each disease. However, these experiments were carried out with vegetative samples, so a genetic basis was not fully proven.

The decisive evidence that natural selection has operated on *Anthoxanthum* in Park Grass comes from reciprocal transplants between contrasting plots (Plot 3 unlimed and Plot 9 limed, Plot 1 limed and Plot 1 unlimed, Plot 8 limed and Plot 7 unlimed). Davies and Snaydon (1976) collected plants from each of the six populations, propagated them vegetatively in a garden bed and then transplanted tillers back into source plots and contrasting ones. All alien transplants had lower survival and lower tillering rates than native transplants. Selection coefficients against aliens (calculated as 1 – alien/native performance) 18 months after transplanting were in the range 0.09–0.77 for survival and 0.23–0.57 for tillering rate.

The rapid microevolution of *Anthoxanthum* probably results from its short generation time and the large selection pressures in Park Grass. If an outcrossing (self-incompatible) species like *Anthoxanthum* can adapt to local conditions so rapidly, then it would be surprising if there was not genetic differentiation among many of the other species found in the Park Grass plots. As yet, no other species have been investigated, but Park Grass provides a unique opportunity to conduct a comparative study of microevolution among species with a variety of life histories and mating systems growing in a common set of environments with a well-documented and long-term history.

Conclusions

The experiment started by Lawes and Gilbert to determine the effects of nutrient additions on hay yield has provided a large number of unanticipated ecological and evolutionary insights. However, there are many more such insights to be gained from further studies on Park Grass. Effects of treatments on herbivorous insects, for instance, should provide insights into the validity of hypothesized relationships between productivity and the intensity of herbivory (Hairston *et al.*, 1960; Oksanen, 1990). The effects of treatments on both aboveground and belowground foodweb structure have likewise not yet been studied in the Park Grass ecosystems. Published analyses of Park Grass data have suggested a host of mechanistic explanations for the observed effects of nutrient ratios and productivity on plant species composition and diversity (e.g. Tilman, 1982), but these hypotheses have not yet been tested. The Park Grass Experiment provides a richer and longer history than for any other site of interest to experimental ecologists. This bestows on Park Grass a unique potential to contribute to the central issues of ecology, evolution and environmental science.

Acknowledgements

We thank Roger Leigh for inviting us to prepare this chapter and David Jenkinson for his support in our efforts to resample Park Grass and to collate and organize the historical data.

References

Brenchley, W.E. and Warington, K. (1958) *The Park Grass plots at Rothamsted 1856-1949*. Reprinted 1969. Rothamsted Experimental Station, Harpenden, Herts.

Bullock, J.M., Mortimer, A.M. and Begon, M. (1993) Carryover effects on the clonal growth of the grass *Holcus lanatus* L. *New Phytologist* 124, 301-307.

Cashen, R.O. (1947) The influence of rainfall on the yield and botanical composition of permanent grass at Rothamsted. *Journal of Agricultural Science* 37, 1-10.

Davies, M.S. and Snaydon, R.W. (1976) Rapid population differentiation in a mosaic environment. III. Measurements of selection pressures. *Heredity* 36, 59-66.

Hairston, N.G., Smith, F.E. and Slobodkin, L.B. (1960) Community structure, population control, and competition. *American Naturalist* 94, 421-425.

Huston, M.A. (1979) A general hypothesis of species diversity. *American Naturalist* 113, 81-101.

Jenkinson, D.S., Potts, J.M., Perry, J.N., Barnett, V., Coleman, K. and Johnston, A.E. (1994) Trends in herbage yields over the last century on the Rothamsted Long-term Continuous Hay Experiment. *Journal of Agricultural Science* 122, 365-374.

Johnston, A.E. (1972) Changes in soil properties caused by the new liming scheme on Park Grass. *Rothamsted Experimental Station Report for 1971*, Part 2, 177-180.

Lawes, J.B. and Gilbert, J.H. (1863) The effect of different manures on the mixed herbage of grassland. *Journal of the Royal Agricultural Society of England* 24, Part I, 1-36.

Lawes, J.B. and Gilbert, J.H. (1880) Agricultural, botanical and chemical results of experiments on the mixed herbage of permanent meadow, conducted for more than twenty years in succession on the same land. Part I. The agricultural results. *Philosophical Transactions of the Royal Society* 171, 289-415.

Lawes, J.B., Gilbert, J.H. and Masters, M.T. (1882) Agricultural, botanical and chemical results of experiments on the mixed herbage of permanent meadow, conducted for more than twenty years in succession on the same land. Part II. The botanical results. *Philosophical Transactions of the Royal Society (A & B)* 173, 1181-1413.

Oksanen, L. (1990) Predation, herbivory, and plant strategies along gradients of primary productivity. In: Grace, J.B. and Tilman, D. (eds) *Perspectives on Plant Competition*. Academic Press, San Diego, CA, pp. 445-474.

Pianka, E. (1966) Latitudinal gradients in species diversity: a review of concepts. *American Naturalist* 100, 33-46.

Silvertown, J.W. (1980) The dynamics of a grassland ecosystem: botanical equilibrium in the Park Grass Experiment. *Journal of Applied Ecology* 17, 491-504.

Silvertown, J. (1987) Ecological stability: a test case. *American Naturalist* 130, 807-810.

Smith, L.P. (1960) The relation between weather and meadow hay yields in England. *Journal of the British Grassland Society* 15, 203-208.

Snaydon, R.W. (1970) Rapid population differentiation in a mosaic environment. I. Response of *Anthoxanthum odoratum* to soils. *Evolution* 24, 257-269.

Snaydon, R.W. and Davies, M.S. (1972) Rapid population differentiation in a mosaic environment. II. Morphological variation in *Anthoxanthum odoratum* L. *Evolution* 26, 390-405.

Thurston, J. (1969) The effect of liming and fertilizers on the botanical composition of permanent grassland, and on the yield of hay. In: Rorison, I. (ed.) *Ecological Aspects of the Mineral Nutrition of Plants*. Blackwell Scientific Publications, Oxford, pp. 3-10.

Thurston, J.M., Williams, E.D. and Johnston, A.E. (1976) Modern developments in an experiment on permanent grassland started in 1856: effects of fertilizers and lime on botanical composition and crop and soil analyses. *Annales Agronomiques* 27, 1043-1082.

Tilman, D. (1982) *Resource Competition and Community Structure*. Princeton University Press, Princeton, New Jersey.

Tilman, D. (1986) Resources, competition and the dynamics of plant communities. In: Crawley, M.J. (ed.) *Plant Ecology*. Blackwell Scientific Publications, Oxford, pp. 51-75.

Tilman, D. and Pacala, S. (1993) The maintenance of species richness in plant communities. In: Ricklefs, R. and Schluter, D. (eds) *Species Diversity*. University of Chicago Press, Chicago.

Warren, R.G. and Johnston, A.E. (1964) The Park Grass experiment. *Rothamsted Experimental Station, Report for 1963*, 240-262.

Warren, R.G., Johnston, A.E. and Cooke, G.W. (1965) Changes in the Park Grass experiment. *Rothamsted Experimental Station, Report for 1964*, 224-228.

Williams, E.D. (1978) Botanical composition of the Park Grass plots at Rothamsted 1856-1976. Harpenden, Rothamsted Experimental Station.

17 Long-term Studies of Tropical Forest Dynamics

M.D. SWAINE
*Department of Plant and Soil Science, University of
Aberdeen, Aberdeen AB9 2UD, UK.*

Introduction

Long-term studies of tropical forests centre on tree populations, and mostly
have their origins in government forest departments charged with the management of extensive timber resources. The focus on trees has a number of implications which impose considerable demands on the temporal and spatial scale
of such research (see also Evans, Chapter 5). Trees are large plants, usually of
great longevity and rates of demographic processes are relatively slow. In the
tropics the difficulties are compounded by high species diversity and the lack
of practically useable annual growth rings by which to determine tree age and
growth rates.

Large plot sizes are needed to provide reasonable sample sizes for even the
commonest species, especially if interest focuses on the largest individuals
(the potential timber harvest for foresters). Trees >10 cm dbh (diameter at
1.3 m) in natural lowland tropical rain forest number between 500 and 600
ha^{-1}; those larger than 100 cm dbh, 10–15 ha^{-1}. Tree species richness varies
greatly in different areas of tropical rain forest, but there are commonly more
than 100 species ha^{-1} among trees >10 cm dbh. Of these only a small number
exceed 30 stems ha^{-1}. For example, in a West African forest (Swaine *et al.*,
1987) only five species exceeded 30 trees ha^{-1}, and of these four were small
understorey species. In Costa Rica (Lieberman and Lieberman, 1987) four species averaged densities >30 ha^{-1}, three of which were palms. In Malaysia
(Manokaran and Kochummen, 1987), the most numerous species on 2 ha
had a density of 20.5 ha^{-1}.

Rates of demographic change are slow in natural forest: mortality rates for
trees >10 cm dbh are typically between 1 and 2 % year^{-1}. The number of small
trees which are recruited into the population >10 cm dbh is equal, in mature

natural forest, to the number of trees (of any size >10 cm dbh) which die over the same period. For a mortality and recruitment rate of 1.5% year⁻¹, 46 years elapse before 50% of the trees are replaced, and 138 years for 75% replacement (Swaine and Lieberman, 1987). Growth rates of individual trees are highly variable, but the majority of changes in diameter increments are very slow. Typically, the modal diameter growth rate for a stand of trees, including all sizes, is less than 1 mm year⁻¹, but the distribution is highly skewed with few negative increments (most such trees soon die), but with a long positive tail which may extend to more than 15 mm year⁻¹ in the fastest-growing individuals of pioneer species (Fig. 17.1). Precise ages of trees in tropical forest are effectively unknown. Attempts have been made to estimate tree age from growth rate data. These estimates vary widely, depending on whether they are based on mean, modal or maximum rates (Swaine *et al.*, 1987). These constraints mean that an adequate sample to examine differences between even the commoner species must cover many hectares and span several decades. None of the existing permanent samples in tropical forest meets these requirements.

This chapter reviews existing sites where long-term permanent sample plots have been established, considers some of the results and offers some ideas for the future.

Long-term Data on Tropical Forest Tree Populations

Table 17.1 sets out some details of existing data from experiments I have been able to locate in Africa, Asia and South America. Considering the area of the vegetation type served by these data, the sample is extremely small, and most are within only one forest formation, lowland tropical forest, and are typically restricted to mature forest without recent human disturbance. It is likely that other enumerated plots exist which await a second enumeration, particularly in the records of the numerous government forestry authorities. However, valuable early records may have been lost, or the plots themselves may be

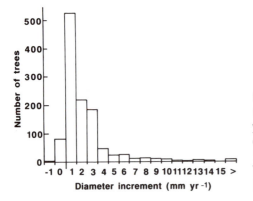

Fig. 17.1. Frequency distribution of mean diameter growth rate (mm year⁻¹) for all trees > 30 cm gbh (9.7 cm dbh) (*n*=1166) between 1970 and 1987 in 2 ha of forest at Kade, Ghana. Values on the abscissa are the upper limit for each class.

Table 17.1. Permanent sample plots of tropical forest tree populations.

Country	Site	Number of plots	Plot area (ha)	Minimum size (cm dbh)	Period of observation (dates)	Number of enumerations	Reference
Costa Rica	La Selva	3	4–4.4	10	1969–1985	2	Lieberman et al. (1990)
Panama	Barro Colorado	1	50	1	1982–1990	3	Hubbell and Foster (1990)
		8	1	10	1975–1980	2	Putz and Milton (1993)
Jamaica	Blue Mtns	1	1.5	10	1968–1978	2	Lang and Knight (1983)
Puerto Rico	El Verde	1	0.1	3.2	1974–1984	3	Bellingham et al. (1992)
	Luquillo	1	0.72	10	1943–1976	4	Crow (1980)
Venezuela	lowland forest	8	0.2–0.8	4	1946–1976		Brown et al. (1983)
		8	0.25	10	1962–1985	9–14	Carey et al. (in press)
	montane forest	9	0.25	10	1962–1984	13–20	Carey et al. (in press)
	San Carlos	1	1	10	1975–1980		Uhl et al. (1988)
		1	1	10	1980–1985		Uhl et al. (1988)
Brazil	Manaus	1	1	15	1974–1978	2	St John (in Rankin-de-Merona 1990)
		5	1	10	1981–1986	2	Rankin-de-Merona (1990)
				25	(5 yr)		Higuchi (1987)
Ghana	Kade	2	1	9.5	1968–1993	8	Swaine et al. (1987, 1990)
	Shai Hills	1	0.49	3.2	1979–1987	2	Swaine (1992)
Nigeria	Akure F.R.		23.5	—	1934–1970	2	Mervart (1972, 1974)
	Ilaro F.R.		97		1962–1969	2	Mervart (1972, 1974)
Gabon	M'passa	1	0.4	4.8	1972–1979	2	Hladik (1982)
		1	0.9	30	1972–1979	2	Hladik (1982)
Sumatra	G. Gadut	9	0.06–1	5.8	1981–1988	4–7	Hotta (1984, 1986, 1989)
Malaysia	Sungei Menyala	1	2.02	10.1	1947–1985	15	Manokaran and Kochummen (1987); Manokaran (1988)
	Bukit Lagong	1	2.02	10.1	1949–1985	15	Manokaran (1988)
	Pasoh F.R.	4	2	10	1971–1974	3	Manokaran (1988)
	Pasoh F.R.	1	50	1	1987–1990	2	Manokaran et al. (1990, 1993)
Sarawak	Bako		2.4	10	1965–1985	5	Hall (1991)
	Mersing		3	10	1966–1986	5	Hall (1991)
	Lambir		2.4	10	1964–1985	5	Hall (1991)
Sabah	Sepilok	1	1.81	9.7	1956–1962	4	Nicholson (1965)
	Danum Valley	2	4	10	1986–	1	Newbery et al. (1992)
New Guinea	Madang	4	0.5	10	1989–1993	2	Siaguru (1992)
Solomon Is.	Kolombangara	21	0.63	30.3(15)	1964–1986	12	Whitmore (1974, 1989)
Peru	Manu N.P.	2	0.94	10	1974–1985	2	Gentry and Terborgh (1990)
Bolivia	Beni	2	1	—	1987(8)–1990(1)	2	Dallmeier (1992)
Virgin Is.	St John	1	1	4	1992–	—	Dallmeier (1992)

difficult to find. Furthermore, foresters have particular objectives for such data, and may confine their studies to the larger individuals of species marketable at the time of plot establishment. Other samples are based on the 'leading desirable' idea in which one tree is selected in each subplot as the individual which should be nurtured to provide a future harvestable log of timber. These data can provide useful information on the performance of the better individuals of a species, but exclude information on stand density, meaningful mortality/recruitment data and species which have since become, or may in the future become important (commercially or otherwise).

The variety of the samples listed in Table 17.1 reflects the compromises that were necessary given the problems discussed in the Introduction. A common size for a plot is 1 ha, which is clearly inadequate for the analysis of any but the commonest species. A relatively small plot size should allow for replication, but this is rarely done; more often the second plot is contiguous with the first (Wyatt-Smith, 1966; Manokaran and Kochummen, 1987) or is chosen to represent a contrasting condition (Swaine *et al.*, 1987). Others have set up or adopted larger plots (Newbery *et al.*, 1992, 2 × 4 ha; Lieberman and Lieberman, 1987, *c.* 3 × 4 ha). The largest plots are 50 ha in Panama (Hubbell and Foster, 1990) and in Malaysia (Manokaran *et al.*, 1990). There are plans to create a network of such plots throughout the tropics (Anon., 1990) but apparently without replication.

The 50 ha plots include all woody plants >1 cm dbh, and thus include many shrub species as well as trees. The plot in Panama includes approximately half a million measured stems; the commonest species being the shrub *Hybanthus prunifolius* (Violaceae) with 41,106 individuals in 1985 (Hubbell *et al.*, 1990). Large plots thus provide large samples for many species, but rare taxa are still inadequately covered. Although sample sizes may be large, they may represent only a small fraction of the breeding population of a species. Fifty hectares of forest inevitably includes a variety of environmental conditions at various scales and offer intriguing correlations between species distributions and environment. Parallel information on even the most likely controlling environmental factors is never as extensive as the plant data, so that useful hypothesis generation by this means is limited.

The periods of observation for these samples (Table 17.1) is short relative to the life span of most of the tree species. The longest span is 1947–85 for a 2 ha plot in Malaysia (Manokaran, 1988) established by Wyatt-Smith (1966). During this period of 38 years, 40% of trees >10 cm dbh were replaced. As an example of much more extensive sampling undertaken by forestry authorities in the tropics, large numbers of 1 ha permanent sample plots (PSPs) were established in Ghana in the 1950s and 1960s. Their purpose was to provide data on tree growth in order to inform management and to allow the calculation of sustainable yield, and thus they were of limited ecological value for the reasons discussed in the Introduction. Since the late 1980s, however, a programme of re-establishing PSPs has been initiated in Ghana (Wong, 1989), aiming to sample the forest reserves of the country with 600 PSPs. This objective is close to being fulfilled and re-enumerations on a five-year cycle have begun. These plots, each of 1 ha, are located at random within Forest Reserves, include all species of trees >5 cm dbh and a subsample of the smaller size classes. Supported by Forestry Department staff experienced in tree identification and by a data processing unit, this initiative holds great prospects for understanding differences between species and for variation within species' ranges.

Interpretations of Observations in Permanent Sample Plots

The results of long-term observations in permanent sample plots in tropical forest concentrate on tree population dynamics and to some extent on the temporal changes in demographic processes and species composition. Other opportunities provided by the plots and their records will be considered later.

Forest dynamics

There is a considerable body of evidence that mortality rates do not differ among tree sizes classes >10 cm dbh, at least in mature lowland tropical forest (Lieberman and Lieberman, 1987; Swaine *et al.*, 1987; Manokaran, 1988). In other kinds of forest subject to cataclysmic events (Johns, 1986) such as fire (Goldammer, 1990; Swaine, 1992), cyclones/hurricanes (e.g. Whitmore, 1974, 1989; Bellingham *et al.*, 1992; Dallmeier, 1992) or severe drought, certain size classes are more prone to damage or death: fire kills more small trees, drought the larger.

In size classes below 10 cm dbh, mortality rises steeply, particularly below 5 cm dbh (Lieberman *et al.*, 1985; Swaine, 1990; Clark and Clark, 1992). Considered as part of the life of an individual, this trend of increasing mortality with decreasing age/size extends to include small seedlings, seeds and fertilized ovules on the parent trees. Losses of genetic individuals may be very high for a cohort of seeds between dispersal and germination (e.g. Sarukhán, 1980). One tree of *Heritiera utilis* (Sterculiaceae) in Ghana produced an estimated 4.5 million flowers in March and April 1976, which yielded an estimated 67,000 seeds; no seedlings could subsequently be found of this seed crop (M.D. Swaine and J.B. Hall, unpublished data).

The idea of tree life span sits uncomfortably with the logarithmic survivorship of tree populations (in such models the last tree in a cohort dies at infinite age). Rare, very large individuals can be found in all natural forests and may be of great age, but the average tree within a population dies well before maturity. A more robust measure of longevity is population half-life (Swaine and Lieberman, 1987), effectively the inverse of mortality rate.

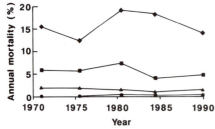

Fig. 17.2. Annual mortality (%) for four common tree species over a 23-year period at Kade, Ghana. *Greenwayodendron oliveri* (Annonacaeae), ♦, and *Aidia genipiflora* (Rubiaceae), ■, are small understorey trees, *Calpocalyx brevibracteatus* (Mimosaceae), ▲, is a medium-sized tree, and *Celtis mildbraedii* (Ulmaceae), ●, is a large emergent tree. The census intervals were: 1970–72, 1972–77, 1977–82, 1982–87 and 1987–93.

Mortality varies among species. Repeated estimates of mortality rates over successive periods show that these differences may be maintained over extended periods (Fig. 17.2), suggesting that they may be genetically determined. Mortality also varies with tree growth rate. This is evident from direct measurement (Fig. 17.3; Manokaran, 1988; Swaine, 1990), and indirectly in studies which show decreases in mortality with increasing crown illumination (position) class (e.g. Korsgaard, 1986). Fast-growing trees are more likely to reach maturity.

Tree growth rates are highly variable (Fig. 17.1), but the variance may be partitioned and attributed to various interrelated factors. Growth is correlated with crown illumination class (Korsgaard, 1986), canopy gap size (Brown, 1990) or 'microsite' (Clark and Clark, 1992), and incident photosynthetically active radiation (Oberbauer and Strain, 1984). It differs among species (Fig. 17.4) and ecological species groups (Table 17.2; Lieberman *et al.*, 1985; Manokaran and Kochummen, 1987; Swaine, 1990), and is related to past growth history (Swaine *et al.*, 1987; Swaine, 1990) especially in larger size classes whose light environment is less subject to change (Clark and Clark, 1992). It is thus possible to calculate the future growth of an individual tree with fairly precise confidence limits from an initial measurement of growth. This autocorrelation of growth augments the probability that fast-growing individuals will reach maturity. These relationships in mortality and growth are manifested in

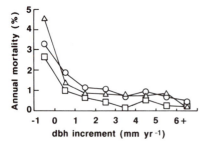

Fig. 17.3. Relationship between annual mortality (%) and prior diameter growth rate (mm year[-1]) for three forest samples (each 2 ha) in Malaysia at Bukit Lagong, △, and Sungei Menyala, ○, (Manokaran, 1988) and in Ghana at Kade, □ (Swaine, 1990).

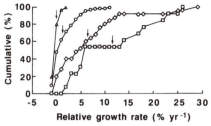

Fig. 17.4. Differences in relative diameter growth rate (% year⁻¹) among tree species at Kade, Ghana. The curve for each species shows the cumulative percentage increments with increasing growth rate. Species means are arrowed. The species represent the full range of shade tolerance: *Baphia pubescens* (Papilionaceae), △, is a small understorey tree, *Guarea cedrata* (Meliaceae), ○, is a large shade-tolerant timber species, *Ricinodendron heudelotii* (Euphorbiaceae), ◇, is a large pioneer, and *Musanga cecropioides* (Moraceae), □, is a medium-sized extreme pioneer.

the strong positive correlation of growth with increasing tree size (Swaine *et al.*, 1987).

Age–size relationships based on the growth of the fastest-growing individuals in size classes (those most likely to survive to maturity) suggest that trees may reach 100 cm dbh in 75–125 years, depending on species. These relationships may be unreliable for small understorey species where the selection for fast-growing individuals may be less strong, and all estimates are seriously confounded by high variance among growth rates. Few data are available for trees <10 cm dbh, the majority of which are in deep shade and very slow-growing. Figure 17.5 shows the distribution of relative growth rate in a sample of 416 tree seedlings and saplings between 50 cm tall and 30 cm gbh. As in Fig.

Table 17.2. Differences among ecological species groups in demographic rates for trees >30 cm gbh between 1970 and 1987 at Kade, Ghana. Mortality and recruitment do not differ significantly within any group, but all groups differ significantly in mortality. The groups are defined *a priori* from field experience: pioneers need full sunlight for establishment, non-pioneers can establish in forest shade; among non-pioneers there is a well-recognized but ill-defined difference in shade tolerance; small and large refer to size attained by mature individuals.

Species group	Number of trees in 1970	Mortality (% year⁻¹)	Recruitment (% year⁻¹)	Mean dbh increment (mm year⁻¹)	% increments > 5 mm year⁻¹
Pioneer					
All	67	5.08	4.48	4.25	28.1
Small	30	9.06	6.72		
Large	37	3.17	3.18		
Non-pioneer					
Shade intolerant	311	1.65	1.58	1.86	16.5
Shade tolerant					
Small	607	2.39	1.81	0.62	2.2
Large	335	1.10	1.03	1.24	8.3

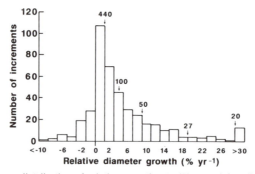

Fig. 17.5. Frequency distribution of relative growth rate (% year⁻¹) for all tree seedlings and saplings ⩾50 cm tall and <30 cm gbh (9.7 cm dbh) in 0.0875 ha of forest at Kade, Ghana (1979–87); mean = 5.4 % year⁻¹, n = 416. Abscissa values are the lower limit for each class. Arrowed values are the approximate number of years for a tree to grow from 50 cm tall to 10 cm dbh at the indicated growth rate.

17.1, the great majority of trees grow very slowly: at 5% year⁻¹ relative growth rate a tree would take *c.* 100 years to reach 10 cm dbh (Fig. 17.5). The fastest-growing individuals, however, can achieve the same result in less than 20 years. Trees growing in the forest understorey experience much more variation in light from year to year, so that they are unlikely to show the same growth rate over many years as seems to be the case for larger trees (Clark and Clark, 1992). Thus the time taken for a seedling to grow to enumerated size is very unpredictable.

The robust relationships emerging from such studies offer some optimism for the development of models (e.g. Alder, 1990) to predict forest growth and composition which will greatly facilitate the management of natural tropical forest (e.g. for timber) and will perhaps strengthen arguments against the conversion of tropical forest into plantations which are currently seen as more predictable and manageable.

Differences among species

Species-specific differences in demographic processes clearly exist, but our understanding of them is limited. We do not know if clear differences among species at one site will be preserved at other sites where the species occur. Since recruitment, growth and mortality are all strongly moderated by the environment, disentangling genetic and environmental controls will be an important task to be addressed by these long-term studies.

Support for a significant genetic component is provided by calculation of demographic rates for guilds of species with similar ecology. The results for a classification of species based on West Africa (Table 17.2) are paralleled in Malaysia (Manokaran and Kochummen, 1987) and in central America (Lieberman and Lieberman, 1987). A pantropical classification (Swaine and Whitmore, 1988) of sufficient detail and robustness for meaningful application in ecology and forestry remains elusive, but progress is being made towards objective methods for the delimitation of groups on life-history traits. Clark and Clark (1992) used 6-year demographic data for selected species in relation to microsite and life stage to define four classes of non-pioneer species. Welden *et al.* (1991) used BCI demographic data for 1982–85 on growth, sapling recruitment and mortality to recognize three categories of species: pioneers, understorey specialists and generalists.

Changes in forest composition

In mature forests, mortality and recruitment are approximately equal both for all species combined and for the great majority of individual species (Swaine and Lieberman, 1987; Manokaran, 1988), implying a degree of equilibrium in forest composition. Change, or lack of it, in the floristic composition in tropical forests is the focus of some attention among theoretical ecologists. Explanations are sought for the high species richness of tropical forests – differences in species ability to establish in gaps of different size have been proposed and tested as a mechanism whereby otherwise similar species may coexist (Denslow, 1980; Whitmore, 1984; Brokaw, 1987; Brown and Whitmore, 1992; Kennedy and Swaine, 1992). Others have looked for differences between common and rare species in demographic processes to find mechanisms which maintain an equilibrium in forest composition (see above). An alternative approach is to ask how forest composition is determined (Swaine and Hall, 1988).

It has been argued elsewhere (Swaine, 1989) that to ask if forest is in equilibrium or non-equilibrium cannot be practicably answered. In all forests studied, significant changes in the abundance of some species may be detected over time, often gradual over many years (Manokaran and Kochummen, 1987; Manokaran, 1988). Figure 17.6 (compiled from Manokaran, 1988) shows the changes in abundance of those species whose total mortality and recruitment were significantly imbalanced ($X^2 > 6.64$, df=1, $P < 0.01$) over the 36–38 years of observation. With about 250 species in each of the 2 ha plots, we expect about two or three to exceed the 1% limit by chance, so these examples offer only slight evidence for floristic non-equilibrium. Watching these changes happen tells us nothing of their causes, about which we can only speculate.

Although the trends in Fig. 17.6 show persuasively persistent changes in species abundance over many years, they have little influence on the nature of the community. The three species represent <2% of the species in the 2 ha plot, and their mortality and recruitment a similarly small proportion of all changes in tree numbers. Multivariate analysis of changes in forest composition at Kade, Ghana, over 17 years, during which time 25% of the trees were replaced, showed (Fig. 17.7) that floristic changes had no clear direction and were small compared to local differences in forest composition.

M.D. Swaine

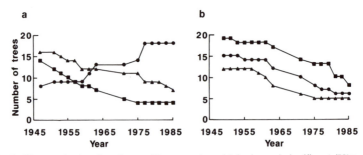

Fig. 17.6. Changes in the abundance of tree species which showed significant ($X^2 > 6.64$, df=1, $P < 0.01$) inequality in mortality and recruitment in 2 ha plots in Peninsular Malaysia. Adapted from Manokaran (1988). (a) During 36 years at Bukit Lagong, *Pimeleodendron*, ●; *Shorea*, ▲; *Dacryodes*, ■. (b) During 38 years at Sungei Menyala, *Lindera*, ■; *Castanopsis*, ●; *Endospermum*, ▲.

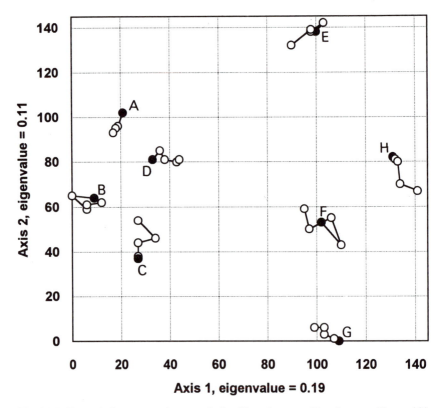

Fig. 17.7. Detrended correspondence analysis of forest composition (trees > 30 cm gbh) in eight 0.25 ha samples at Kade, Ghana. The four samples A, B, C, D are contiguous, forming a 100 m × 100 m plot, and are about 200 m distant from the samples E, F, G, H which form a second 1 ha plot. Points, representing five enumerations in 1970 (●), 1972, 1977, 1982 and 1987, are joined in chronological order.

The kinds of minor drifts in forest composition which are likely to be seen in any such enumeration data are insignificant in the face of effects of major disturbances such as rare fires, and windstorms. These events need occur only once in two or more centuries to have a permanent influence of forest composition (Swaine and Hall, 1988; Swaine, 1989). It should also be recognized that in small plots of 1 ha, the fall of a large emergent tree (which may be 60 m tall with a canopy 50 m wide) can affect as much as a quarter of the whole plot, and will thus appear as catastrophic in the data.

Conclusions

Some consideration of sampling is useful. The lack of replication in most cases is cause for concern. Whether to use a few large plots or more smaller plots is a decision normally determined by the objective for the study. However, experience shows that objectives may change, certainly over the several decades spanned by the most useful records. It is tempting to adopt a comprehensive approach to cover all possibilities, but this is scarcely feasible. The compromise is probably parallel series of samples incorporating large plots and a larger number of small plots. Large plots are undoubtedly valuable for the spatial information they contain on species distribution and the interactions between individuals. Small, 1 ha plots, appropriately distributed over large areas will give us a better grasp of genetic variability and improve the predictive power of demographic analyses.

We also need to include in our studies forest other than that in lowland tropical areas. Montane forests are likely to become an increasing proportion of the surviving natural forests as the more accessible and productive forests of the lowlands are used more intensively. Rare or unusual forest types such as heath, peat swamp or ultrabasic also need to be understood. By far the most extensive forests in the tropics are, and will continue to be, managed or exploited forests – those with various degrees of human disturbance. These are an important resource and though some are in the care of foresters, we need studies to understand how processes in natural forests are affected by exploitation. Natural forest ecologists, therefore, should collaborate with foresters, whose manpower resources are often substantial, but who have different objectives. Foresters often take an experimental approach, apparently lacking in the studies reviewed here.

A persistent problem with PSP data is its voyeurism: we watch natural processes occurring but have no equipment but speculation to determine their causes. An experimental approach is lacking in tropical forest PSP studies. No doubt the rarity of these resources deters their use in such a way. Foresters, however, have often invested heavily in large-scale experiments to examine the effects of logging or silvicultural treatments (e.g. Maitre, 1987). It would be sensible if ecologists and foresters collaborated more often in these endeavours especially as the two groups have very close objectives.

A final point: many of the examples of long-term studies are closely guarded personal property. We need to share our data because more progress will be made if the same questions can be asked of several data sets.

References

Alder, D. (1990) *GHAFOSIM: a projection system for natural forest growth and yield in Ghana.* Final Report to Government of Ghana Ministry of Lands and Natural Resources, Department of Forestry, Accra.

Anon. (1990) *The Center for Tropical Forest Science.* Smithsonian Tropical Research Institute, Panama City.

Bellingham, P.J., Kapos, V., Varty, N., Healey, J.R., Tanner, E.V.J., Kelly, D.L., Dalling, J.W., Burns, L.S., Lee, D. and Sidrak, G. (1992) Hurricanes need not cause high mortality: the effects of Hurricane Gilbert on forests in Jamaica. *Journal of Tropical Ecology* 8, 217.

Brokaw, N.V.L. (1987) Gap phase regeneration of three pioneer tree species in a tropical forest. *Journal of Ecology* 75, 9-19.

Brown, N.D. (1990) Dipterocarp regeneration in tropical rain forest gaps of different sizes. Unpublished DPhil Thesis, University of Oxford.

Brown, N.D. and Whitmore, T.C. (1992) Do dipterocarp seedlings really partition tropical rain forest gaps? *Philosophical Transactions of the Royal Society of London B* 335, 369-378.

Brown, S., Lugo, A.E., Silander, S. and Liegel, L. (1983) Research history and opportunities in the Luquillo Experimental Forest. *USDA Forest Service General Technical Report* SO-44.

Carey, E.V., Brown, S., Gillespie, A.J.R. and Lugo, A.E. (1994) Tree mortality in mature lowland and montane moist forests of Venezuela. *Biotropica* (in press).

Clark, D.A. and Clark, D.B. (1992) Life history diversity of canopy and emergent trees in a neotropical rain forest. *Ecological Monographs* 62, 315-344.

Crow, T.R. (1980) A rainforest chronicle: a 30-year record of change in structure and composition at El Verde, Puerto Rico. *Biotropica* 12, 42-45.

Dallmeier, F. (1992) Long-term monitoring of biological diversity in tropical forest areas. Methods for establishment and inventory of permanent plots. *MAB Digest* 11, UNESCO, Paris.

Denslow, J.S. (1980) Gap partitioning among tropical rainforest trees. *Biotropica* 12 (supplement), 47-55.

Gentry, A.H. and Terborgh, J. (1990) Composition and dynamics of the Cocha Cashu 'mature' floodplain forest. In: Gentry, A.H. (ed.) *Four Neotropical Rainforests.* Yale University Press, New Haven, pp. 542-564.

Goldammer, J.G. (ed.) (1990) *Fire in the Tropical Biota.* Springer-Verlag, Berlin.

Hall, P. (1991) Structure, stand dynamics and species compositional change in three mixed dipterocarp forests of northwest Borneo. PhD Thesis, Boston University.

Higuchi, N. (1987) Short-term growth of an undisturbed tropical moist forest in the Brazilian Amazon. PhD Thesis, Michigan State University.

Hladik, A. (1982) Dynamique d'une forêt équatoriale africaine: mesures en temps réel et comparaison du potentiel de croissance des différentes espèces. *Acta Oecologia* 3, 373-392.

Hotta, M. (ed.) (1984) *Forest Ecology and Flora of Gunong Gadut, West Sumatra.* Sumatra Nature Study (Botany), Yoshida College, Kyoto University, Kyoto.

Hotta, M. (ed.) (1986) *Diversity and Dynamics of Plant Life in Sumatra.* Sumatra Nature Study (Botany), Yoshida College, Kyoto University, Kyoto.

Hotta, M. (ed.) (1989) *Diversity and Plant-Animal Interactions in Equatorial Rain Forests.* Sumatra Nature Study (Botany), Kagoshima University, Kagoshima.

Hubbell, S.P. and Foster, R.B. (1990) Structure, dynamics and equilibrium status of old-growth forest on Barro Colorado Island. In: Gentry, A.H. (ed.) *Four Neotropical Rainforests.* Yale University Press, New Haven, pp. 522-541.

Hubbell, S.P., Condit, R. and Foster, R.B. (1990) Presence and absence of density dependence in a neotropical tree community. *Philosophical Transactions of the Royal Society of London B* 330, 269-281.

Johns, R.J. (1986) The instability of the tropical ecosystem in New Guinea. *Blumea* 31, 341-371.

Kennedy, D.N. and Swaine, M.D. (1992) Germination and growth of colonizing species in artificial gaps of different sizes in dipterocarp rain forest. *Philosophical Transactions of the Royal Society of London B* 335, 357-367.

Korsgaard, S. (1986) An analysis of the potential timber production under conservation management in the tropical rain forest of south-east Asia. Unpublished PhD Thesis, Royal Danish Veterinary and Agricultural University.

Lang, G.E. and Knight, D.H. (1983) Tree growth, mortality and recruitment, and canopy gap formation during a 10-year period in a tropical moist forest. *Ecology* 64, 1075-1080.

Lieberman, D. and Lieberman, M. (1987) Forest tree growth and dynamics at La Selva, Costa Rica (1969-82). *Journal of Tropical Ecology* 3, 347-358.

Lieberman, D., Lieberman, M., Hartshorn, G.S. and Peralta, R. (1985) Growth rates and age-size relationships of tropical wet forest trees in Costa Rica. *Journal of Tropical Ecology* 1, 97-109.

Lieberman, D., Hartshorn, G.S., Lieberman, M. and Peralta, R. (1990) Forest dynamics at La Selva Biological Station, 1969-85. In: Gentry, A.H. (ed.) *Four Neotropical Rainforests.* Yale University Press, New Haven, pp. 509-521.

Maitre, H.F. (1987) Natural forest management in Côte d'Ivoire. *Unasylva* 30, 53-60.

Manokaran, N. (1988) Population dynamics of tropical forest trees. Unpublished PhD Thesis, University of Aberdeen.

Manokaran, N. and Kochummen, K.M. (1987) Recruitment, growth and mortality of tree species in a lowland dipterocarp forest in Peninsular Malaysia. *Journal of Tropical Ecology* 3, 315-330.

Manokaran, N., Lafrankie, J.V., Kochummen, K.M., Quah, E.S., Klahn, J.E., Ashton, P.S. and Hubbell, S.P. (1990) Methodology for the fifty-hectare research plot at Pasoh Forest Reserve. *Research Pamphlet No. 104.* Forest Research Institute Malaysia, Kepong.

Manokaran, N., Kassim, A.R., Hassan, A., Quah, E.S. and Chong, P.F. (1993) Short-term population dynamics of dipterocarp trees in a lowland rain forest in Peninsular Malaysia. *Journal of Tropical Forest Science* 5, 97-112.

Mervart, J. (1972) Growth and mortality rates in the natural high forest of western Nigeria. *Nigeria Forestry Information Bulletin* (n.s.) No. 22.

Mervart, J. (1974) Appendix to the paper on growth and mortality rates in the natural high forest of western Nigeria. *Nigeria Forestry Information Bulletin* (n.s.) No. 28.

Newbery, D.McC., Campbell, E.J.F., Ridsdale, C.E. and Still, M.J. (1992) Primary lowland dipterocarp forest at Danum Valley, Sabah, Malaysia: structure, relative abundance and family composition. *Philosophical Transactions of the Royal Society B* 335, 341-356.

Nicholson, D.I. (1965) A study of virgin rain forest near Sandakan, North Borneo. In: *Proceedings of the Symposium on Ecological Research into Humid Tropics Vegetation, Kuching.* UNESCO, Paris, pp. 67-87.

Oberbauer, S.F. and Strain, B.R. (1984) Photosynthesis and successional status of Costa Rican rain forest trees. *Photosynthesis Research* 5, 227-232.

Putz, F.E. and Milton, K. (1983) Tree mortality rates on Barro Colorado Island. In: Leigh, E.G., Rand A.S. and Windsor, O.M. (eds) *The Ecology of a Tropical Forest.* Oxford University Press, Oxford, pp. 95-108.

Rankin-de-Merona, J.M., Hutchings, R.W. and Lovejoy, T.E. (1990) Tree mortality and recruitment over a five-year period in undisturbed upland rainforest of the central Amazon. In: Gentry, A.H. (ed.) *Four Neotropical Rainforests.* Yale University Press, New Haven, pp. 573-584.

Sarukhán, J. (1980) Demographic problems in tropical systems. In: Solbrig, O.T. (ed.) *Demography and Evolution in Plant Populations.* Blackwell Scientific Publications, Oxford, pp. 161-188.

Siaguru, P. (1992) Effect of shade on growth of lowland forest tree seedlings in Papua New Guinea. Unpublished PhD Thesis, University of Aberdeen.

Swaine, M.D. (1989). Population dynamics of tree species in tropical forest. In: Holm-Nielsen, L.B., Nielsen, I. and Balslev, H. (eds) *Tropical Forests: Botanical Dynamics, Speciation and Diversity.* Academic Press, London, pp. 101-110.

Swaine, M.D. (1990) Population dynamics of moist tropical forest at Kade, Ghana. In: Maitre, H.F and Puig, H. (eds) *Actes de l'Atelier sur l'Aménagement de l'Ecosystème Forestier Tropical Humide,* Cayenne 1990. MAB/UNESCO/FAO, pp. 40-61.

Swaine, M.D. (1992) Characteristics of dry forest in West Africa and the influence of fire. *Journal of Vegetation Science* 3, 365-374.

Swaine, M.D. and Hall, J.B. (1988) The mosaic theory of forest regeneration and the determination of forest composition in Ghana. *Journal of Tropical Ecology* 4, 253-269.

Swaine, M.D. and Lieberman, D. (eds) (1987) The dynamics of tree populations in tropical forest. *Journal of Tropical Ecology* 3, Special Issue (4), Cambridge University Press, Cambridge.

Swaine, M.D. and Whitmore, T.C. (1988) On the definition of ecological species groups in tropical rain forests. *Vegetatio* 75, 81-86.

Swaine, M.D., Hall, J.B. and Alexander, I.J. (1987) Tree population dynamics at Kade, Ghana (1968-82). *Journal of Tropical Ecology* 3, 331-345.

Swaine, M.D., Lieberman, D. and Hall, J.B. (1990) Structure and dynamics of a tropical dry forest in Ghana. *Vegetatio* 53, 31-51.

Uhl, C., Clark, K., Dezzeo, N. and Maquirino, P. (1988) Vegetation dynamics in Amazonian treefall gaps. *Ecology* 69, 751-763.

Welden, C.W., Hewett,S.W., Hubbell, S.P. and Foster, R.B. (1991) Sapling survival, growth, and recruitment: relationship to canopy height in a neotropical forest. *Ecology* 72, 35–50.

Whitmore, T.C. (1974) Change with time and the role of cyclones in tropical rain forest on Kolombangara, Solomon Islands. *Commonwealth Forestry Institute Paper* 46, 1–78.

Whitmore, T.C. (1984). Gap size and species richness in tropical rain forests. *Biotropica* 16, 239.

Whitmore, T.C. (1989) Changes over twenty-one years in the Kolombangara rain forests. *Journal of Ecology* 77, 469–483.

Wong, J. (ed.) (1989) *Ghana Forest Inventory Project*. Seminar Proceedings, March 1989, Accra. Overseas Development Administration, London.

Wyatt-Smith, J. (1966) Ecological studies on Malayan forest I. *Research Pamphlet No.52*, Forest Research Institute, Kepong.

18 Flying in the Face of Change: The Rothamsted Insect Survey

I.P. Woiwod and R. Harrington
Institute of Arable Crops Research, Rothamsted
Experimental Station, Harpenden, Herts AL5 2JQ, UK.

Introduction

Insects form a key component of most terrestrial ecosystems in terms of biomass and species richness. It is therefore important that we understand long-term fluctuations in insect populations to aid prediction of changes in pest status, to conserve beneficial and non-pest species, and to develop and test fundamental ideas of population biology and dynamics. Also, insects may provide very useful indicators of subtle environmental changes. For example, insect populations often react very quickly to changes in plant quality and may amplify and thereby indicate deleterious changes in ecosystems long before they become apparent by conventional botanical observations (Erhardt and Thomas, 1991; Woiwod and Thomas, 1993). However, the diversity of form and function and the highly aggregated nature of insect populations make them very difficult to sample over a sufficient range of species and at suitable synoptic spatial scales for long-term monitoring.

One approach to this problem, adopted by the Rothamsted Insect Survey (RIS), that of monitoring the flying populations of adults from two contrasting and diverse groups of insects, the larger moths and the aphids, owes much to the foresight of L.R.(Roy) Taylor. A brief history of the background to the work may be instructive, especially as interest in long-term ecological research has recently increased and includes attempts to set up new long-term studies such as at the Long Term Ecological Research sites in USA (Magnuson and Bowser, 1990) and the Environmental Change Network in Britain (Lewis, 1993; Tinker, Chapter 22). This interest has been accompanied by a much greater realization of the difficulties associated with such work (e.g. Strayer *et al.*, 1986; Likens, 1989) some of which have been highlighted by the experience gained through the operation of the Rothamsted Insect Survey (Taylor, 1989, 1991; Woiwod,

1991). The historical background to the Survey, problems of long-term monitoring and their solution, and a summary of the research results, emphasizing
the long-term nature of the data, form the basis of this review.

Historical Background to the Insect Survey

Two long-standing entomological interests at Rothamsted came together with
the foundation of the Insect Survey in the early 1960s. One of these was a
background of work on insect migration, including aphids (Davidson, 1924;
Johnson, 1954) and butterflies and moths (Williams, 1958). The other was an
expertise in sampling and the quantitative analysis of aerial populations of
insects.

   In one set of studies a Rothamsted light trap designed by C.B. Williams
(1948) had been operated at the edge of Barnfield, the long-term Rothamsted
root crop experiment, between 1933 and 1937 and again between 1946 and
1950. All macrolepidoptera, the larger moths, were counted and identified.
These data were used to study the effect of weather on insects and to develop
and quantify the concept of diversity based on the observed species frequency
distribution found in these samples (Williams, 1939, 1940, 1953; Fisher *et al.,*
1943). Suction trapping was also developed as a quantitative method for studying aerial insect populations (Johnson, 1950; Taylor, 1951, 1962; Johnson and
Taylor, 1955). These authors later ran such traps simultaneously at different
heights to examine the importance of wind-borne migration in insects and
establish the form of the relationship between insect density and height (Johnson, 1957; Taylor, 1974a).

   In the late 1950s, some of the first mathematical models of population
dynamics were developed and it soon became apparent at Rothamsted that
they were unrealistic when applied to very mobile species such as aphids
because no account was taken of migration behaviour except to equate it to
birth or death processes (e.g. Bartlett, 1960). A problem with making these
models more realistic was that virtually no quantitative information was available about the spatial variability or movement of insect populations and it was
the attempt to obtain such information that inspired Roy Taylor to establish
the Rothamsted Insect Survey.

   Initially, with few resources available, the only practical way to obtain the
required information was to enlist the expert taxonomic skills of the many
amateur entomologists in Britain, provide them with a cheap, simple sampling
device and collate the resulting population estimates at Rothamsted. The Rothamsted light trap was the obvious choice and the larger moths a suitable insect
group, with a large pool of species well known to many amateur lepidopter-
ists. The first step in this work was taken in 1960 when a trap was placed
beside Barnfield at the same site that was sampled by C.B.Williams in the 1930s
and 1940s. Although, initially, the main purpose of the Survey was to obtain

spatial data, this trap immediately provided temporal information on changes in the moth fauna on the Rothamsted estate since 1933 and, as will be shown later, established the importance of long-term monitoring of insect populations (Taylor, 1991).

Gradually a network of light traps, operated by volunteers, was set up from which all the larger moths were identified and counted daily so that, by 1968, enough sites were operating to begin mapping of populations on a national scale (Woiwod, 1986). Altogether, 370 different sites and 688 species have been sampled nationally with about 95 traps currently in operation in Britain (Fig. 18.1). As well as at the Barnfield site, three other traps have been in operation in different habitats on Rothamsted farm for over 20 years and in 1990 a new grant enabled a further 22 traps to be set up throughout the Rothamsted estate to study local population variability and movement at the farm scale. In addition, Rothamsted light traps have been operated at various sites throughout the world as part of more general studies into moth biodiversity. In Europe, this has involved sites in Denmark, Finland, France and Ireland, and elsewhere more exotic locations include Aldabra, Iraq, Malaysia, Seychelles, Sulawesi and Tenerife.

In 1963, the publication of *Silent Spring* (Carson, 1963) stimulated public awareness of the environmental hazards associated with pesticide use and funds became available for research leading to the reduction of unnecessary applications of these chemicals. Roy Taylor used the opportunity to design a new type of suction trap with the stated aim of investigating the prospect of developing a pest warning scheme for aphids, the main agricultural insect pest group in Britain (Taylor, 1989). The trap design was based on previous insect density height experiments and resulted in the choice of sampling height at 12.2 m so that the more mobile wind-borne species were selected and the samples were representative of a wide area (Macaulay *et al.*, 1988). The first '12-metre' trap was erected at Rothamsted in 1964 and gradually a network of 23 of these was set up throughout Britain. It was quickly established that with training, professional staff could operate such traps on a daily basis, that all the aphid species could be identified rapidly and timely information provided to the agricultural industry (Taylor and Palmer, 1972; Woiwod *et al.*, 1984; Tatchell, 1985). At the same time, the suction trap network began to provide complementary data to the light trap network on the spatial, and later temporal, population dynamics of insect populations. It was these two sampling networks that together became known as the Rothamsted Insect Survey (RIS). That title is perhaps unfortunate in that it implies that the main purpose of the work is purely to assess the presence of species, however the RIS has always been much more than that as it is based on standardized quantitative daily sampling so that insect populations can be studied comparatively and at scales of time and space not previously possible (Taylor, 1986).

Currently 15 suction traps operate in Britain including four in Scotland run by the Scottish Agricultural Science Agency. The success of the British trap

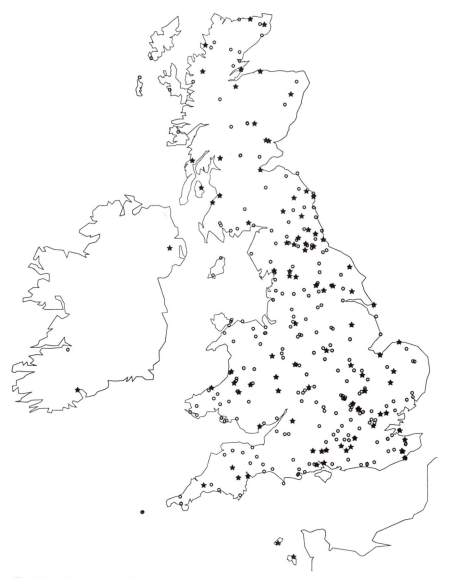

Fig. 18.1. Distribution of RIS light trap samples in the British Isles. Sites operating in 1993 (★); sites no longer operating (○).

network has inspired the operation '12-metre' suction traps in many other European countries including a network of 14 traps in France (Fig. 18.2). These traps are often operated for local and specific applied purposes and although attempts have been made to coordinate the sampling and data collection throughout Europe (Cavalloro, 1989), funding has not been forthcoming.

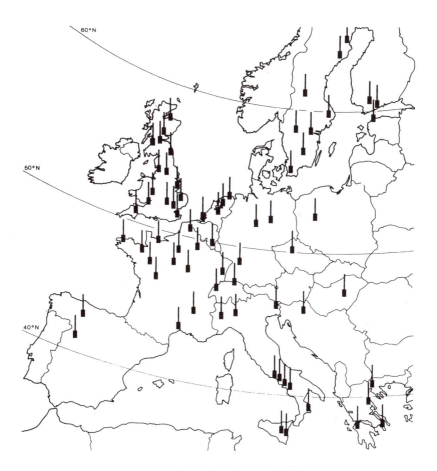

Fig. 18.2. Current distribution of '12-metre' suction traps in Europe.

However, pooling data across national boundaries has led to successful research collaboration, particularly in climatic studies (Harrington *et al.*, 1992).

Between the two insect sampling networks of the RIS over 20 million individuals, comprising more than one thousand species of aphids and moths, have been identified and counted since the first light trap was started in 1960, making this the largest database on insect populations anywhere in the world. The data have already been used in many studies ranging from the applied, such as pest forecasting, to the fundamental, for example, research into spatial and temporal population dynamics. A selection of the scientific results concentrating on the long-term nature of the data from the RIS is given below but as there are now over 850 publications relating to RIS data only an introduction to the main areas of interest is possible here. More details are to be found in previous reviews of the work (e.g. Taylor, 1974b, 1977, 1978, 1986; Taylor *et al.*, 1981; Tatchell, 1991; Woiwod, 1991).

Pest Status and Forecasting

As might be expected at Rothamsted, the applied agricultural aspects of the RIS's data have received considerable attention. Generally, macrolepidoptera are not very important pests in Britain, although this might alter in the future under some climate change scenarios. Therefore the main use of the light trap samples has been for fundamental ecological studies, although data from RIS catches have been used occasionally in applied contexts (Bowden and Jones, 1979; Bowden *et al.*, 1983; Gordon *et al.*, 1988). However, the aphids, monitored by the suction trap network, contain some of the most damaging agricultural pests in Britain. Their importance is partly due to the damage they can do by feeding but mainly to their role as efficient vectors of plant virus diseases (Tatchell, 1989). Applied studies have therefore been focused on this group.

Aphids usually have very complex life cycles, often involving alternating host plants and up to three migration cycles per year, so useful information provided by the samples has included basic data on the migration biology of the group (e.g. Taylor *et al.*, 1982; Cammell *et al.*, 1989). There are two other important inputs into agricultural pest control programmes, one is the provision of information on the current situation and status of pest species and the other is in the development of forecasting systems so that control measures can be taken at the best time, thereby improving control and preventing unnecessary precautionary spraying (Tatchell, 1991).

Current information on pest aphid abundance has been provided since 1968 in the form of the *Aphid Bulletin*, a table of weekly totals for the most important pest species from all the sample sites in Britain. It is possible to produce this information within a week of the samples being collected which is adequate in many cases for control measures or for individual field assessments to be made as necessary. As more information became available it was possible to supplement this information by means of an *Aphid Commentary* which put the data in the *Bulletin* into context with previous years and included any relevant forecasts (Woiwod *et al.*, 1984). In 1989 a change of Government policy led to the withdrawal of funding from 'near market' applied research resulting in a reduction in the RIS aphid sampling network, the cessation of *Commentary* production and a restriction in the circulation of the *Bulletin* (Woiwod *et al.*, 1990), thus emphasizing the continuing difficulty of maintaining long-term studies within current short-term funding objectives and erratic changes in political policy.

Successful forecasting systems have been developed, using RIS data, for aphid damage to hops, cereals, sugar beet, beans and potatoes. The objective of forecasting differs widely and depends on the crop–pest–disease combination under consideration and is linked to external economic variables. There are three important questions in aphid forecasting for which RIS suction trap data have been shown to be particularly useful. These are: will pests reach damage threshold level? when will these thresholds be reached or important migration events take place? and are there qualitative attributes, such as pesticide resistance or virus transmission potential that will make aphids difficult to control or affect their pest status? (Tatchell and Woiwod, 1989; Tatchell, 1991).

One of the most comprehensive forecasts for regional abundance has been developed for the black bean aphid, *Aphis fabae* (Scop.), a species for which there is a very full historical set of data on crop and overwintering abundance from field sampling, and aerial abundance from the RIS database. Crop infestation on spring-sown field beans, *Vicia faba* L., can be forecast most successfully from the size and timing of the spring migration measured by RIS suction traps, although earlier forecasts (Fig. 18.3) can be made but with less accuracy from autumn trap catches or counts of overwintering eggs (Way *et al.*, 1981).

In other successful aphid forecasts, temperature, often in the form of day-degrees above a threshold temperature, has been used to predict abundance (Tatchell, 1991). An example of this approach is that for the peach potato aphid, *Myzus persicae* (Sulzer), an important virus vector species for many crops. In this species and several others, particularly anholocyclic species (ones that do not overwinter as eggs), there is a strong statistical relationship (Fig. 18.4) between winter temperatures and subsequent abundance (A'Brook, 1983; Harrington *et al.*, 1990).

For many aphid pest problems it is as important to be able to predict the timing of migration into crops as the subsequent population size. For example, when aphids act as plant virus vectors the initial damage is sometimes done as soon as the first few aphids have arrived and transmitted the disease, this particularly applies in crops grown to provide seed stocks which have to have very low virus levels if future crops are to remain healthy (Turl, 1980). The date of the start of the migration of *M. persicae* is also closely related to the incidence of sugar beet virus yellows. The forecast of this date, based on temperature in January and February, is used routinely to advise beet growers on the likely value of insecticidal granules at planting. On the day that the first aphid is caught in the trap an updated forecast of virus incidence is issued, and RIS trap catches form an essential part of the season-long spray warning scheme issued by Broom's Barn Experimental Station to the sugar beet industry (Harrington *et al.*, 1989). In crops normally sprayed frequently it can be important to forecast the start and end of migration so that unnecessary spraying is reduced and the development of insecticide resistant clones limited. Successful forecasts of this type have used the same approach as for predicting abundance except that dates to particular stages in a migration are used. For example such a forecast has been successfully established for the spring migration of the hop

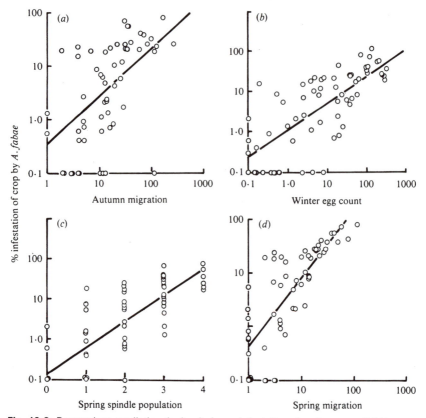

Fig. 18.3. Regressions predicting the level of crop infestation of spring-sown field beans by the black bean aphid, *Aphis fabae* from (a) autumn aerial migration, measured by RIS suction traps; (b) from winter-host egg counts; (c) from spring populations on winter-host plants; (d) from spring migration, measured by RIS suction traps. The forecasts improve from (a) to (d). (After Way *et al.*, 1981.)

aphid, *Phorodon humuli* (Schrank), into hop gardens, and for the spring migrations of several other pest species, mainly based on relationships with winter temperatures (Turl, 1980; Harrington *et al.*, 1990, 1992; Tatchell, 1991).

Several interesting features apply to nearly all the successful forecasts so far developed. They have all required long runs of data, partly because only a single point for any relationship is obtained each year (e.g. between winter temperature and spring abundance; Fig. 18.4), but also because adequate ranges of variables, such as temperature, are needed to establish the form of the relationships. Ideally data are then required to test and update the predictions periodically to ensure they are not based on some short-term cycle or artefact. The most successful forecasts so far have resulted from very simple robust statistical models, often simple linear regression (e.g. Figs 18.3 and

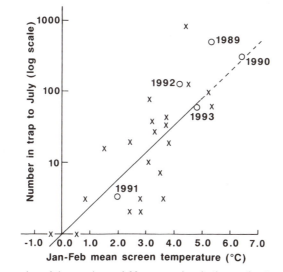

Fig. 18.4. Regression of the numbers of *Myzus persicae* in the suction trap at Rothamsted against mean temperature between January and February, based on data between 1968 and 1988. The relationship predicts well 1989–93 data and is accurate for 1990 despite being outside the previous temperature range.

18.4). More complicated multivariate regressions have been tried extensively but without any consistent improvement on the simpler relationships (Harrington *et al.*, 1991; Howling *et al.*, 1993). Detailed simulation models, although helpful in understanding the system, have not fulfilled their early promise and have been notably unsuccessful in forecasting real problems. In part, this is because they often include parameters which are difficult to measure routinely in the field, notably the population parameters of the many natural enemies of aphids (Carter, 1985).

Only recently has it been possible to start measuring some of the biological attributes of individual aphids from RIS samples. For *M. persicae* it is now possible to test the degree of resistance to insecticides of individual aphids from suction trap catches and discover which viruses each aphid is carrying, by collecting samples in a specially developed storage solution which preserves enzymes and viruses for later immunoassay (Tatchell *et al.*, 1988b). This is now done regularly at several sites to provide current information on resistance, but again, long runs of such data will be required before the dynamics of resistance can be understood and predicted with confidence. Suction trapping also forms the basis of the infectivity index system for assessing the risk to autumn sown crops from barley yellow dwarf virus (Plumb, 1983). The index is improved by distinguishing the morphologically identical gynoparae (which will not reproduce on cereals) and virginoparae (which will) by using living aphids in host plant choice tests (Tatchell *et al.*, 1988a). Such tests are time consuming, and methods are being developed to separate these morphs routinely, direct from suction trap samples.

Population Dynamics

One of the original purposes of the RIS was to obtain data on insect populations for studies of the spatial structure of population dynamics. In the 1960s very little attention was being paid to this subject. As soon as the coverage of sample sites permitted, dynamic mapping was found to be useful for understanding changes in population densities of insects at a national scale (Woiwod, 1979, 1986; Taylor, 1986). Later analyses established a species-specific power law relationship between the large-scale spatial variability (variance) of populations and mean density for 360 common species selected from the RIS database (Taylor *et al.*, 1980). A similar analysis later extended the finding to temporal variability, and the spatial and temporal parameters were compared (Taylor and Woiwod, 1980, 1982). The explanation for the ubiquity of the power law relationship has remained controversial although several models have now been proposed which can produce similar patterns (e.g. Taylor and Taylor, 1979; Perry, 1988; Hanski and Woiwod, 1993). However, the basic empirical pattern is now well established and can be used with confidence in designing sampling programmes (Taylor, 1984; Taylor *et al.*, 1988; Perry and Woiwod, 1992).

Over the last 30 years, ecological population dynamic theory has developed greatly. Many models are now available to predict how populations may fluctuate in time and recent developments in metapopulation dynamic theory specifically incorporate the spatial elements (Gilpin and Hanski, 1991). However, many controversies still remain and there is a more pressing need than ever for extensive spatial and temporal population data against which such theories can be tested (May and Hassell, 1990). The RIS has been gradually obtaining such data and we are now in a good position to start this important process of testing these ecological theories.

One long-running controversy in insect population dynamics has been that concerning the relative importance of density-dependent and density-independent processes in population regulation. Many ecologists believe that density dependence is a logical necessity but it has proved notoriously difficult to detect in real populations (e.g. Stiling, 1988). In a recent analysis of RIS data, density dependence was sought in over 5700 time series of longer than ten years using several commonly applied tests. There were two orders of magnitude more time series used in this analysis than any previously attempted. Appreciable levels of direct density dependence were found but the amount was very dependent on the length of time series analysed (Fig. 18.5). It was immediately apparent why previous attempts had been so unsuccessful as

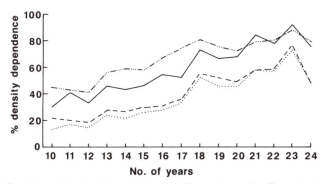

Fig. 18.5. The effect of length of time series on ability to detect significant density dependence using three commonly used tests on RIS moth data. Bulmer test (·····), Ricker test(———), Pollard test(– – –), all tests significant(···). (After Woiwod and Hanski, 1992).

they had usually dealt with shorter time series and often unrepresentative species, such as pests with outbreak dynamics (Woiwod and Hanski, 1992).

Although the density dependence question required resolution, a more important and interesting question is, what type of dynamic behaviour exists in real populations? Theoretical population models suggest there might be a range of possible patterns: stable equilibria, limit cycles, or chaos, yet the relative frequency of any of these dynamics in real populations is currently unknown. Of particular interest at the moment is whether chaotic dynamics are as widespread in population dynamics as theory suggests (Godfrey and Grenfell, 1993). The problem in answering such questions has been that standard methods of detecting chaos require very long time series of hundreds or even thousands of points. Such time series are almost completely unknown in ecological systems. However, techniques are starting to become available which are more suitable for ecological data where 20–30-year runs are considered long time series. One such method, the response-surface methodology, has recently been developed and tested on a range of species but the only set consistent with chaotic dynamics was a 19-year run of RIS data on the aphid *Phyllaphis fagi* from a suction trap at Dundee (Turchin and Taylor, 1992). A re-examination of these data using a more complete set covering 25 years and a further 27-year run of data on the same species at Rothamsted failed to confirm the result suggesting that length of time series is again of critical importance in such tests (Perry *et al.*, 1993). A more complete analysis is currently underway using only the longest time series available to answer some of these questions more thoroughly.

Environmental Change

Biodiversity

Biodiversity and its conservation has become an increasingly important area of study in recent years although measurement of biodiversity still remains difficult and controversial (e.g. Magurran, 1988; Stork, 1993). A big problem with many sampling programmes aimed at detecting change in diversity is that the most common measure, species richness, depends very directly on sample size. One of the first attempts to derive a quantitative index of diversity which could be used to compare samples of different size was based on the light-trap catches obtained at Rothamsted in the 1930s and 1940s (Fisher *et al.*, 1943; Williams, 1953). Since then, many alternative measures have been proposed and used with varying degrees of success.

Some of the most widely used methods were re-examined using the spatial and temporal replication available from the RIS light trap samples (Kempton and Taylor, 1974; Taylor *et al.*, 1976). The results from these analyses, rather surprisingly, confirmed that the original parameter, α from the log-series distribution, as proposed by Fisher *et al.* (1943) was better than any of the more recent indices. This parameter was found to exhibit a whole range of important properties including: independence from sample size, power to discriminate between sites, lack of sensitivity to unstable very abundant species or transient rare species, and perhaps most importantly responds to known environmental change (Taylor, 1978).

The structure of moth diversity can be mapped for Britain (Fig. 18.6) and exhibits a very consistent pattern at this scale (Taylor, 1986). The pattern is quite complex and has latitudinal and altitudinal trends, presumably related to climate. It has also been possible to detect urbanization effects (Taylor *et al.*, 1978) and relate the general pattern to the Institute of Terrestrial Ecology's ecological land class stratification of Britain (Luff and Woiwod, 1994). Detailed knowledge of the biodiversity of macrolepidoptera in Britain invites comparison with patterns in Europe and elsewhere in the world. In principle, it is now possible to obtain comparable data anywhere using a standard sampling technique, the Rothamsted light trap. In practice, operating traps and identifying samples from poorly studied areas is difficult and so far only a few such samples have been obtained from widely scattered places. As expected, the contrast between temperate and tropical samples is enormous (Barlow and Woiwod, 1989,1990), but we still have very little idea of tropical variability compared to Britain or even across Europe.

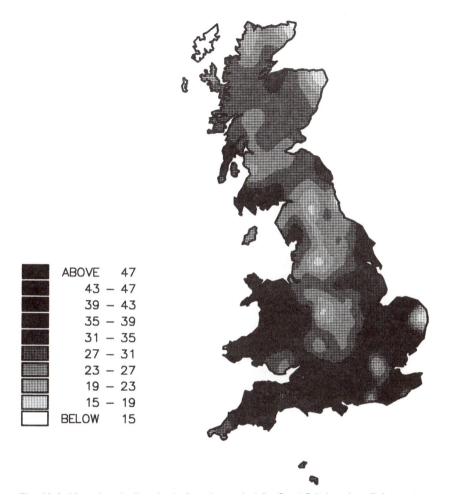

ABOVE   47
43 – 47
39 – 43
35 – 39
31 – 35
27 – 31
23 – 27
19 – 23
15 – 19
BELOW   15

Fig. 18.6. Map of moth diversity (α from log-series) for Great Britain using all data up to 1982.

Land use

When the light trap was placed beside Barnfield in 1960 at the site originally sampled by C.B.Williams it became immediately apparent that there had been a large drop in numbers of macrolepidoptera and their diversity during the 1950s. This reduced moth community has continued until the present time (Fig. 18.7). It was during the 1950s that many changes were taking place in British agriculture: fields were enlarged to accommodate mechanization, hedges were removed in areas where arable farming became dominant and the first insecticides and pesticides were introduced. Because sampling was not done while many of these changes took place and there is no replication of the

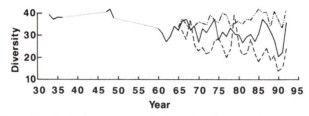

Fig. 18.7. Moth diversity (α from log-series) of the three long-running traps on the Rothamsted estate from 1933 to 1993. (————) Barnfield, (-···-) Geescroft Wilderness, (– –) Allotments.

sample at other sites it is not possible to quantify the relative importance of these effects formally but it seems likely that the introduction of herbicides and changes in land use and farm management would have had major effects.

Two other sites have also been sampled for long periods on the Rothamsted estate. In 1965 a trap was placed in Geescroft Wilderness, an area of relatively stable woodland on the farm resulting from successional reversion following the cessation of cultivation in 1888. This is one of Rothamsted's Classical sites. The following year, 1966, a light trap was placed in some mature allotments which were due to be grubbed up. From that time the site has experienced many changes of land use including fallow, arable and building development. As might be expected, the moth population at this 'Allotments' trap has fluctuated and the overall diversity has continued to decline (Fig. 18.7). In contrast, the Geescroft woodland site has maintained a diversity and general level of moth population not too dissimilar from that to be found at Barnfield during the 1930s and 1940s. It therefore seems likely that much of the original moth community that existed in open field situations until the 1950s remains on the Rothamsted estate but at a much reduced level following agricultural intensification. If this is so, it may be possible to re-establish some of the lost diversity and population abundance in the future if agriculture becomes less intensive.

With the establishment of a network of 26 light traps on the Rothamsted estate in 1990 it has been possible to study the variability of moth biodiversity in much more detail at the farm landscape scale (Woiwod and Thomas, 1993). By measuring the land use in the immediate vicinity of each trap and regressing diversity on this it is possible to explain up to 60% of the pattern of biodiversity shown in Fig. 18.8. The importance of land use change has become clear from such analyses. Areas of seminatural habitat on the farm such as hedges and woodland seem to be particularly important in promoting diversity and large moth populations whereas arable and grassland areas now provide very poor habitats for non-pest species (Woiwod and Thomas, 1993; Luff and Woiwod, 1994). This result goes a long way to explaining the important changes observed in the 60-year time series from the Barnfield trap and has important implications for conservation of insect populations on the Rothamsted farm.

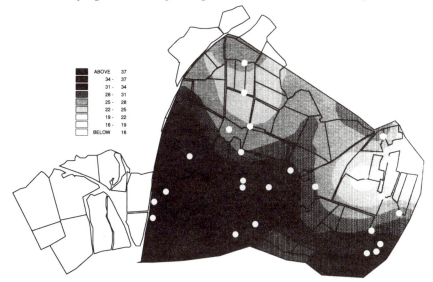

Fig. 18.8. Map of moth diversity (α from log-series) at Rothamsted in 1990 determined with data from the network of 26 light traps (indicated by white circles) on the estate.

Climate

There is widespread concern at the moment over possible anthropogenic climate change. Whether this concern is justified or not it has led to a growing area of research into the possible effects such changes might have on insect populations (Harrington and Stork, 1994). The RIS provides some of the few data of sufficient length to establish and test such predictions and this is an active area of current research. Fortunately, as already mentioned, many of the pest-forecasting systems developed for agricultural purposes are based on climatic variables such as mean temperature. These forecasts can be adapted almost immediately to provide predictions for climate change scenarios (Bale *et al.*, 1992; Howling *et al.*, 1992; Harrington *et al.*, 1994).

As an example, the population of *M. persicae* shows a strong linear relationship between abundance and mean screen temperatures in January and February (Fig. 18.4). This relationship is robust and has been shown to be capable of extrapolation beyond the data from which it was derived (Fig. 18.4; Harrington, 1991; Bale *et al.*, 1992). Commonly used climate change projections have mean winter temperatures that lie within the range already observed so we are fairly confident that the forecast of an average increase in abundance for this aphid of 15 times by the end of June is a realistic one (Fig. 18.9). Similar types of forecast of flight phenology also suggest that *M. persicae* will migrate into crops up to 6 weeks earlier than the current average. Such forecasts are well founded on real data obtained over the last 30 years but it

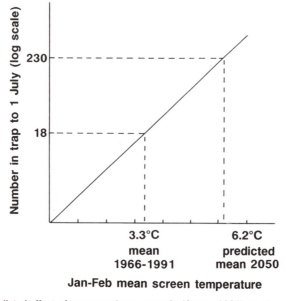

Fig. 18.9. Predicted effect of warmer winter scenario (Anon., 1991) on abundance of *Myzus persicae* at Rothamsted up to July 1, based on the relationship in Fig. 18.4.

could be argued that changing climate would change other important factors such as crop cultivation or the population dynamics of natural enemies. Some answers to such questions will come from analyses utilizing the spatial variation already available within the European network (Harrington *et al.*, 1992). For example, the trap at Elgin in Scotland to the one at Montpellier in France covers over 14° of latitude and forecasts can therefore be verified against existing climate patterns within Europe.

A question being asked with increasing frequency is whether actual changes are detectable yet in any of the RIS data. This is a much more difficult question as any changes are likely to be small over the period we have so far sampled. However, an analysis of flight phenology for five species of aphid at eight different sites does suggest that flight is beginning significantly earlier in more recent years (Fleming and Tatchell, 1993) and similar patterns are emerging from analyses of the moth data. It is clearly important to continue as much long-term monitoring as possible during the time when any change is predicted so that forecasts can be verified and we can follow changes as they occur. Only with such data will it be possible to adapt control strategies for pests and conserve the vulnerable non-pest fauna.

Conclusions

The original purpose of the RIS was to study spatial change and variability in insect populations but now, over 30 years later, the long-term nature of the data has become as important and changes in insect populations over spatial scales of a few metres to hundreds of kilometres and temporal scales ranging from days to decades can be studied in ways not previously possible. This underlines an important feature found in many long-term studies, including the Rothamsted 'Classical' experiments, that it is difficult to predict the future uses for the data or the experimental sites themselves (Taylor, 1989; Woiwod, 1991). However, we can be certain of one thing, there is no going back in time to fill in gaps in missing historical data. For example, we would like to know in much more detail what happened to the moth fauna in the 1950s when such large environmental changes occurred as it would help us to interpret future land-use change.

The difficulty of setting up and maintaining long-term monitoring exercises should not be ignored. Apart from the obvious problems of attracting long-term funds in a political and scientific climate that emphasizes short-term objectives (see also Tinker, Chapter 22), there are some more subtle problems. For example, the fundamental question about scientific methodology that is behind such damning phrases as 'mere monitoring'. The validity of this type of criticism has been thoroughly repudiated in several recent publications (Likens, 1989; Taylor, 1991; Woiwod, 1991; Tinker, Chapter 22) but attitudes are often deeply ingrained and arguments about the vital importance of monitoring are still not accepted by some scientists involved in more conventional experimental studies. Certainly in ecology, evolution and other sciences which have a definite historical component a temporal perspective is crucial though often lacking and can often only be obtained from regular quantitative observations over long periods of time (Taylor, 1989).

Interest in the effects of environmental change, particularly climate, has increased rapidly over the last few years. Here again data from the RIS and other long temporal data sets are beginning to come into their own in a way not originally envisaged, both to develop and, perhaps more crucially, test models and theories about future effects of both artificial and natural change. Change is inevitable in all biological systems and the flying insects monitored by the RIS are well adapted to tracking and reflecting such change. However, scientists trying to study these changes are perhaps not so well adapted to dealing with short-term changes in funding and administrative cycles. Perhaps we must try to fly in the face of these problems if we are to secure the survival of vital long-term studies.

References

A'Brook, J. (1983) Forecasting the incidence of aphids using weather data. *Bulletin OEPP/EPPO Bulletin* 13, 229-233.

Anon (1991) *The potential effects of climate change in the United Kingdom - First Report of the United Kingdom Climate Change Impacts Review Group* HMSO, London.

Bale, J.S., Harrington, R. and Howling, G.G. (1992) Aphids and winter weather I. Aphids and climate change. *Proceedings of the Fourth European Congress of Entomology and the XIII. Internationale Symposium für die Entomofaunistik Mitteleuropas, Volume 1,* Hungarian Natural History Museum, Budapest, pp. 139-143.

Barlow, H.S. and Woiwod, I.P. (1989) Moth diversity in lowland tropical forest in Peninsular Malaysia. *Journal of Tropical Ecology* 5, 37-50.

Barlow, H.S. and Woiwod, I.P. (1990) Seasonality and diversity of macrolepidoptera in two lowland sites in the Dumoga-Bone National Park, Sulawesi Utara. In: Knight, W.J. (ed.) *Insects and the Rain Forests of South East Asia (Wallacea),* The Royal Entomological Society of London, pp.167-172.

Bartlett, M.S. (1960) *Statistical Population Models in Ecology and Epidemiology,* Methuen, London.

Bowden, J. and Jones, M.G. (1979) Monitoring wheat bulb fly, *Delia coarctata* (Fallen) (Diptera: Anthomyiidae), with light-traps. *Bulletin of Entomological Research* 69, 129-139.

Bowden, J., Cochrane, J., Emmett, B.J., Minall, T.E. and Sherlock, P.L. (1983) A survey of cutworm attacks in England and Wales, and a descriptive population model for *Agrotis segetum* (Lepidoptera: Noctuidae). *Annals of Applied Biology* 102, 29-47.

Cammell, M.E., Tatchell, G.M. and Woiwod, I.P. (1989) Spatial pattern of abundance of the black bean aphid, *Aphis fabae,* in Britain. *Journal of Applied Ecology* 26, 463-472.

Carson, R. (1963) *Silent Spring.* Hamish Hamilton, London.

Carter, N. (1985) Simulation modelling of the population dynamics of cereal aphids. *Biosystems* 18, 111-119.

Cavalloro, R. (ed.) (1989) *'Euraphid' network: Trapping and aphid prognosis.* Commission of the European Communities, Luxembourg.

Davidson, J. (1924) The biological and ecological aspects of migration in aphids. *Science Progress,* London 22, 57-69.

Erhardt, A. and Thomas, J.A. (1991) Lepidoptera as indicators of change in the semi-natural grasslands of lowland and upland Europe. In: Collins, N.M. and Thomas, J.A. (eds) *The Conservation of Insects and Their Habitats: 15th Symposium of the Royal Entomological Society,* Academic Press, London, pp. 213-236.

Fisher, R.A., Corbet, A.S. and Williams, C.B. (1943) The relation between the number of species and the number of individuals in a random sample of an animal. *Journal of Animal Ecology* 12, 42-58.

Fleming, R.A. and Tatchell, G.M. (1994) Long term trends in aphid flight phenology consistent with global warming: methods and some preliminary results. In: Leather, S.R., Watt, A.D., Mills, N.J. and Walters, K.F.A. (eds) *Individuals, Populations and Patterns in Ecology.* Intercept, Andover, pp. 63-71.

Gilpin, M. and Hanski, I. (eds) (1991) *Metapopulation Dynamics: Empirical and Theoretical Investigations*. Academic Press, London.

Godfrey, H.C.J. and Grenfell, B.T. (1993) The continuing quest for chaos. *Trends in Ecology and Evolution* 8, 43–44.

Gordon, S.C., McKinlay, R.G., Riley, A.M. and Osborne, P. (1988) Observations on the biology and distribution of the double dart moth (*Graphiphora augur* (Fabricius)) in Scotland, and on the damage caused by the larvae to red raspberry. *Crop Research* 28, 157–167.

Hanski, I. and Woiwod, I.P. (1993) Mean-related stochasticity and population variability. *Oikos* 67, 29–39.

Harrington, R. (1991) Aphid borne viruses: what's the forecast? *Potato Review* 1, 32–34

Harrington, R. and Stork, N.E. (eds) (1994) *Insects in a Changing Environment: Proceedings of the 17th International Symposium of the Royal Entomological Society*. Academic Press, London (in press).

Harrington, R., Dewar, A.M. and George, B. (1989) Forecasting the incidence of virus yellows in sugar beet in England. *Annals of Applied Biology* 114, 459–469.

Harrington, R., Tatchell, G.M. and Bale, J.S. (1990) Weather, life cycle strategy and spring populations of aphids. *Acta Phytopathologica et Entomologica Hungarica* 25, 423–432.

Harrington, R., Howling, G.G., Bale, J.S. and Clark, S. (1991) A new approach to the use of meteorological and suction trap data in predicting aphid problems. *Bulletin OEPP/EPPO Bulletin 21*, 499–505.

Harrington, R., Hullé, M., Pickup, J. and Bale, J.S. (1992) Forecasting the need for early season aphid control: geographical variation in the relationship between winter temperature and early season flight. *Association of Applied Biologists 1992 Presidential Meeting*, Rennes, France 8–11.

Harrington, R., Bale, J.S. and Tatchell, G.M. (1994) Aphids in a changing climate. In: Harrington, R. and Stork, N.E. (eds) *Insects in a Changing Environment: Proceedings of the 17th International Symposium of the Royal Entomological Society*. Academic Press, London (in press).

Howling, G.G., Harrington, R. and Bale, J.S. (1992) Aphids and winter weather II. The analysis of meteorological and suction trap data. *Proceedings of the Fourth European Congress of Entomology and the XIII Internationale Symposium für die Entomofaunistik Mitteleuropas, Volume 1*, Hungarian Natural History Museum, Budapest, pp. 174–178.

Howling, G.G., Harrington, R., Clark, S.J. and Bale, J.S. (1994) The use of multiple regression via principal components in forecasting early season aphid (Hemiptera: Aphididae) flight. *Bulletin of Entomological Research* 83, 377–381.

Johnson, C.G. (1950) A suction trap for small airborne insects which automatically segregates the catch into successive hourly samples. *Journal of Animal Ecology* 26, 479–494.

Johnson, C.G. (1954) Aphid migration in relation to weather. *Biological Reviews* 29, 87–188.

Johnson, C.G. (1957) The distribution of insects in the air and the empirical relation of density to height. *Journal of Animal Ecology* 26, 479–494.

Johnson, C.G. and Taylor, L.R. (1955) The development of large suction traps for airborne insects. *Annals of Applied Biology* 43, 51–61.

Kempton, R.A. and Taylor, L.R. (1974) Log-series and log-normal parameters as diversity discriminants for the Lepidoptera. *Journal of Animal Ecology* 43, 381-399.

Lewis, T. (1993) The Environmental Change Network. *AFRC Institute For Arable Crops Research, Report for 1992,* The Lawes Agricultural Trust, Harpenden, pp. 51-52.

Likens, G.E. (ed.) (1989) *Long-term Studies in Ecology, Approaches and Alternatives.* Springer-Verlag, New York.

Luff, M. and Woiwod, I.P. (1994) Insects as indicators of land use change: a European perspective, focusing on moths and ground beetles. In: Harrington, R. and Stork, N.E. (eds) *Insects in a Changing Environment: Proceedings of the 17th International Symposium of the Royal Entomological Society.* Academic Press, London (in press).

Macaulay, E.D.M, Tatchell, G.M. and Taylor, L.R. (1988) The Rothamsted Insect Survey '12-metre' suction trap. *Bulletin of Entomological Research* 78, 121-129.

Magnuson, J.J. and Bowser, C.J. (1990) A network for long-term ecological research in the United States. *Freshwater Biology* 23, 137-143.

Magurran, A.E. (1988) *Ecological Diversity and Its Measurement.* Princeton University Press, Princeton.

May, R.M. and Hassell, M.P. (eds) (1990) *Regulation and Relative Abundance of Plants and Animals.* The Royal Society, London.

Perry, J.N. (1988) Some models for spatial variability of animal species. *Oikos* 51, 124-130.

Perry, J.N. and Woiwod, I.P. (1992) Fitting Taylor's power law. *Oikos* 65, 538-542.

Perry, J.N., Woiwod, I.P. and Hanski, I. (1993) Using response surface methodology to detect chaos. *Oikos* 68, 329-339.

Plumb, R.T. (1983) Barley yellow dwarf virus–a global problem. In: Plumb, R.T. and Thresh, J.M. (eds) *Plant Virus Epidemiology.* Blackwell Scientific Publications, Oxford, pp. 187-198.

Stiling, P. (1988) Density-dependent processes and key factors in insect populations. *Journal of Animal Ecology* 57, 581-593.

Stork, N.E. (1993) How many species are there? *Biodiversity and Conservation* 2, 215-232.

Strayer, D., Glitzenstein, J.S., Jones, C.G., Kolasa, J., Likens, G.E., McDonnell, M.J., Parker, G.S. and Pickett, S.T.A. (1986) *Long-term Ecological Studies.* Institute of Ecosystem Studies, New York.

Tatchell, G.M. (1985) Aphid control advice to farmers and the use of aphid monitoring data. *Crop Protection* 4, 39-50.

Tatchell, G.M. (1989) An estimate of the potential economic losses to some crops due to aphids in Britain. *Crop Protection* 8, 25-29.

Tatchell, G.M. (1991) Monitoring and forecasting aphid problems. In: Peters, D.C. and Webster, J.A. (eds) *Aphid–Plant Interactions: Populations to Molecules. Oklahoma Agricultural Experimental Station, Miscellaneous Publication* 132, 215-230.

Tatchell, G.M., Plumb, R.T. and Carter, N. (1988a) Migration of alate morphs of the bird cherry aphid (*Rhopalosiphum padi*) and implications for the epidemiology of barley yellow dwarf virus. *Annals of Applied Biology* 112, 1-11.

Tatchell, G.M., Thorn, M., Loxdale, H.D. and Devonshire, A.L. (1988b) Monitoring for insecticide resistance in migrant populations of *Myzus persicae*. *Proceedings of the Brighton Crop Protection Conference – Pests and Diseases* 1, 439-444.

Tatchell, G.M. and Woiwod, I.P. (1989) Aphid migration and forecasting. In: Cavalloro, R. (ed.) *'Euraphid' Network: Trapping and Aphid Prognosis*. Commission of the European Communities, Luxembourg, pp. 15–28.

Taylor, L.R. (1951) An improved suction trap for insects. *Annals of Applied Biology* 38, 582–591.

Taylor, L.R. (1962) The absolute efficiency of insect suction traps. *Annals of Applied Biology* 50, 405–421.

Taylor, L.R. (1974a) Insect migration, flight periodicity and the boundary layer. *Journal of Animal Ecology* 43, 225–238.

Taylor, L.R. (1974b) Monitoring change in the distribution and abundance of insects. *Rothamsted Experimental Station, Report for 1973*, Part 2, pp. 240–269.

Taylor, L.R. (1977) Aphid forecasting and the Rothamsted Insect Survey. *Journal of the Royal Agricultural Society* 138, 75–97.

Taylor, L.R. (1978) Bates, Williams, Hutchinson – a variety of diversities. In: Mound, L.A. and Waloff, N. (eds) *The Diversity of Insect Faunas*. Royal Entomological Society of London, pp. 1–18.

Taylor, L.R. (1984) Assessing and interpreting the spatial distributions of insect populations. *Annual Review of Entomology* 29, 321–357.

Taylor, L.R. (1986) Synoptic dynamics, migration and the Rothamsted Insect Survey. *Journal of Animal Ecology* 55, 1–38.

Taylor, L.R. (1989) Objective and experiment in long-term research. In: Likens, G.E. (ed.) *Long-term Studies in Ecology: Approaches and Alternatives*, Springer-Verlag, New York, pp. 20–70.

Taylor, L.R. (1991) Proper studies and the art of the soluble. *Ibis* 133 suppl. 1, 9–23.

Taylor, L.R. and Palmer, J.M.P. (1972) Aerial sampling. In: van Emden, H.F. (ed.) *Aphid Technology*. Academic Press, London, pp. 189–199.

Taylor, L.R. and Woiwod, I.P. (1980) Temporal stability as a density-dependent species characteristic. *Journal of Animal Ecology* 49, 209–224.

Taylor, L.R. and Woiwod, I.P. (1982) Comparative synoptic dynamics I. Relationship between inter- and intra-specific spatial and temporal variance/mean population parameters. *Journal of Animal Ecology* 51, 879–906.

Taylor, L.R., Kempton, R.A. and Woiwod, I.P. (1976) Diversity statistics and the log-series model. *Journal of Animal Ecology* 45, 255–272.

Taylor, L.R., French, R.A. and Woiwod, I.P. (1978) The Rothamsted Insect Survey and the urbanization of land in Great Britain. In: Frankie, G.W. and Koehler, C.S. (eds) *Perspectives in Urban Entomology*. Academic Press, New York, pp. 31–65.

Taylor, L.R., Woiwod, I.P. and Perry, J.N. (1980) Variance and the large-scale spatial stability of aphids, moths and birds. *Journal of Animal Ecology* 49, 831–854.

Taylor, L.R., French, R.A., Woiwod, I.P. Dupuch, M.J. and Nicklen, J. (1981) Synoptic monitoring for migrant insect pests in Great Britain and Western Europe. I. Establishing expected values for species content, population stability and phenology of aphids and moths. *Rothamsted Experimental Station, Report for 1980*, Part 2, 41–104.

Taylor, L.R., Woiwod, I.P., Tatchell, G.M., Dupuch, M.J. and Nicklen, J. (1982) Synoptic monitoring for migrant insect pests in Great Britain and Western Europe III. The seasonal distribution of pest aphids and the annual aphid aerofauna over Great Britain 1975–1980. *Rothamsted Experimental Station, Report for 1981*, Part 2, 129–157.

Taylor, L.R., Perry, J.N., Woiwod, I.P. and Taylor, R.A.J. (1988) Specificity of the spatial power-law exponent in ecology and agriculture. *Nature* 332, 721-722.

Taylor, R.A.J. and Taylor, L.R. (1979) A behavioural model for the evolution of spatial dynamics. In: Anderson, R.M., Turner, B.D. and Taylor, L.R. (eds) *Population Dynamics*. Blackwell Scientific Publications, Oxford, pp. 1-27.

Turchin, P. and Taylor, A. (1992) Complex dynamics in ecological time series. *Ecology* 73, 289-305.

Turl, L.A.D. (1980) An approach to forecasting the incidence of potato and cereal aphids in Scotland. *Bulletin OEPP/EPPO Bulletin 10*, 135-141.

Way, M.J., Cammell, M.E., Taylor, L.R. and Woiwod, I.P. (1981) The use of egg counts and suction trap samples to forecast the infestation of spring-sown beans, *Vicia faba*, by the black bean aphid, *Aphis fabae. Annals of Applied Biology* 98, 21-34.

Williams, C.B. (1939) An analysis of four years captures of insects in a light trap. Part 1. General survey; sex proportion; phenology; and time of flight. *Transactions of the Royal Entomological Society of London* 89, 79-132.

Williams, C.B. (1940) An analysis of four years captures of insects in a light trap. Part II. The effect of weather conditions on insect activity; and the estimation and forecasting of changes in the insect population. *Transactions of the Royal Entomological Society of London* 90, 227-306.

Williams, C.B. (1948) The Rothamsted light trap. *Proceedings of the Royal Entomological Society of London A.* 23, 80-85.

Williams, C.B. (1953) The relative abundance of different species in a wild animal population. *Journal of Animal Ecology* 22, 14-31.

Williams, C.B. (1958) *Insect Migration*. Collins, London

Woiwod, I.P. (1979) The role of spatial analysis in the Rothamsted Insect Survey. In: Wrigley, N. (ed.) *Statistical Applications in the Spatial Sciences*. Pion, London pp. 268-285.

Woiwod, I.P. (1986) Computer mapping in the Rothamsted Insect Survey. *Proceedings of the Seminar on Agrometeorology and Pest Forecasting, Fulmer Grange, UK, 24-28 June 1985 CTA/TDRI*, Wageningen, pp. 89-98.

Woiwod, I.P. (1991) The ecological importance of long-term synoptic monitoring. In: Firbank, L.G., Carter, N., Darbyshire, J.F. and Potts, G.R. (eds) *The Ecology Of Temperate Cereal Fields*, Blackwell Scientific Publications, Oxford, pp. 275-304.

Woiwod, I.P. and Hanski, I. (1992) Patterns of density dependence in moths and aphids. *Journal of Animal Ecology* 61, 619-629.

Woiwod, I.P. and Thomas, J.A. (1993) The ecology of butterflies and moths at the landscape scale. In: Haines-Young, R. (ed.) *Landscape Ecology in Britain*. IALE (UK), Department of Geography, University of Nottingham, Working paper No 21, pp. 76-92.

Woiwod, I.P., Tatchell, G.M. and Barrett, A.M. (1984) A system for the rapid collection, analysis and dissemination of aphid-monitoring data from suction traps. *Crop Protection* 3, 273-288.

Woiwod, I.P., Tatchell, G.M., Dupuch, M.J., Parker, S.J., Riley, A.M. and Taylor, M.S. (1990) *Rothamsted Insect Survey Twenty-first Annual Summary*. Lawes Agricultural Trust, Harpenden.

19 Long-term Studies and Monitoring of Bird Populations

J.J.D. Greenwood, S.R. Baillie and H.Q.P. Crick
British Trust for Ornithology, Thetford, Norfolk IP24 2PU, UK.

Introduction

Long-term studies are necessary in a number of areas of ecology (Likens, 1989; Risser, 1991), including avian ecology (Dunnet, 1991). Long-term population studies in particular are needed for the study of lifetime reproductive success and life history strategies of long-lived organisms (Clutton-Brock, 1986; Newton, 1989), and to understand the dynamics and regulation of natural populations (Lack, 1966; Perrins *et al.*, 1991). They are also needed for monitoring the effects of environmental changes, including those resulting from human action (Furness and Greenwood, 1993).

Population monitoring is long term by its very nature. If properly conducted, it includes the study of population dynamics and regulation, so it uses the methods of population dynamics, including the repeated measurement of population sizes, of reproductive rates, and of mortality rates at various life-history stages, year after year, as well as the identification of environmental processes and the within-population feedbacks that affect all of these. As a result, a well-designed monitoring programme, valuable in its own right, may have spin-offs for the broader study of population dynamics or for life history studies. The work of Crick *et al.* (1993a) on the seasonal variation in clutch size of British birds, using data from the British Trust for Ornithology (BTO) Nest Records Scheme, provides an example of such a contribution to life history theory.

The aim of this chapter, after briefly discussing what we mean by 'monitoring' and introducing the BTO, is to show, through a series of examples of BTO work, how long-term monitoring can contribute to conservation science and to ecology more generally. In particular, we stress the importance of measuring key demographic parameters of productivity and survival as well as population numbers.

What Is Monitoring?

Monitoring is sometimes still thought to be mere surveillance – that is, repeated survey using standard methods – but most modern conservation ecologists maintain that this is not enough (Hinds, 1984; Verner, 1986; Baillie, 1990, 1991; Pienkowski, 1990, 1991; Goldsmith, 1991; Greenwood and Baillie, 1991; Hellawell, 1991; Koskimies and Väisänen, 1991; Spellerberg, 1991; Usher, 1991; Koskimies, 1992). True monitoring involves surveillance plus the following.

1. A clear understanding of the objectives of the programme.
2. Assessment of any changes against some standard or target (and relative to the normal range of variation).
3. The gathering of data in such a way that the reasons for departure from the standard may be illuminated.
4. A mechanism by which the results of the monitoring are translated into action – in our case, conservation action.

For reasons of space, we shall largely neglect the last component in this chapter. This is not to minimize its importance, which is central.

The Common Birds Census (CBC) was set up by BTO over 30 years ago as a scheme for monitoring numbers of the commoner species of birds in Britain. It has a number of features that illustrate the four elements listed above. For example, because it grew out of the widespread concern over the effects of pesticides on wildlife in the late 1950s and early 1960s, it was targeted at farmland. Woodland was incorporated later because that was seen as another important wildlife habitat subject to substantial management and one in which it was therefore important to monitor the possible effects of human intervention. The scheme's objectives were therefore clear from the start. To criticize it, as some have done, on the grounds that it does not cover a comprehensive and random sample of the British countryside is to misunderstand these objectives. (Of course, objectives may be redefined as time goes by and we are now launching a scheme to census birds across the whole spectrum of habitats in Britain.) Also clear from the start was the need to assess changes against normal population levels, and the normal scale of variation around these levels, the last being particularly evident because the scheme was launched at a time when Britain experienced two very severe winters in succession, with dramatic effects on bird life (Dobinson and Richards, 1964). The need to gather the data such that they would provide insights into the reasons for population changes was not forgotten, so observers were asked to draw up detailed habitat maps of their census plots and to update this habitat information as necessary. This has contributed considerably to our understanding of the effects of farmland and woodland management on bird populations (O'Connor and Shrubb, 1986; Fuller *et al.*, 1989; Fuller and Warren, 1991; Fuller, 1992; Fuller and Crick, 1992; Lack, 1992), itself illuminating interpretations of population changes. It will often, of course, be true that the understanding provided from a monitoring programme is only partial, but it is nonetheless valuable: not only is it true in the real world of wildlife management that some understanding is better than none at all but the conclusions from the monitoring can be used to direct most cost-effectively the intensive, short-term research projects that complement the extensive, long-term work.

Monitoring numbers alone is not enough. The monitoring of reproductive and survival rates not only allows changes in these variables sometimes to serve as early warnings that the population is vulnerable (since they may change before numbers themselves change) but it also, and more importantly, allows better understanding of the factors responsible for changes in population size. For example, if in a declining species of migrant, reproductive rates remained high while survival decreased, this would point to problems on the wintering grounds rather than in the breeding area. Furthermore, the recording of reproductive and survival rates enables the dynamics of populations to be modelled. Such modelling not only deepens our understanding of the dynamics of a study population, but may help to highlight unexplained changes, identify the stages in the life cycle at which problems are occurring, point to factors that might be responsible, and reveal gaps in our knowledge. On this basis, conservation agencies can promote actions to reverse undesirable changes or, if knowledge is too incomplete for this, initiate more intensive research.

The limitations imposed if one monitors only numbers can be seen currently in North America, where a continent-wide scheme for monitoring numbers (but not reproduction and survival) has failed satisfactorily to explain perceived declines in populations of neotropical migrants, or even to establish the generality of the declines, although the problem has been a focus of much attention for many years (Hagan and Johnston, 1992; Böhing-Gaese *et al.*, 1993).

In order to be able to interpret the ornithological data in a way that allows the causes of changes to be illuminated, it is important also to gather environmental data in a form that is compatible with them.

The British Trust for Ornithology and Its Integrated Population Monitoring Programme

This chapter, although a general consideration of long-term studies for monitoring bird populations, deals almost exclusively with the work of the BTO for two good reasons. The first is that we are personally familiar with that work.

The second is that the Trust's work is in the lead, internationally, in the field of widespread bird population monitoring. The reasons for this happy state of affairs are various but can perhaps be summarized as 'collaboration'. The BTO was founded to promote collaborative survey work involving the network of birdwatchers throughout the country (Nicholson, 1983). As the work has developed, the Trust has had to employ staff to organize surveys, collate data, and analyse results. This has not been allowed to develop into a lop-sided partnership, with the volunteers merely acting as an unpaid workforce, because the volunteers themselves have an input at all stages through the Trust's Council and its committees. A second element of collaboration is that between the BTO and conservation bodies, particularly the UK government's conservation agencies (formerly the Nature Conservancy Council and now the four territorial agencies operating through the Joint Nature Conservation Committee) but also the voluntary conservation organizations. The BTO puts in the time and expertise of its volunteers and staff, amounting to hundreds (very possibly thousands) of man-years per annum; the statutory agencies put in their conservation expertise; the costs are shared.

Table 19.1 lists the various long-term schemes that are run and indicates which components of population dynamics – numbers, reproduction, or survival – each covers. Note the key importance of the Nest Records Scheme and Ringing Scheme in providing the demographic information to integrate with the census data. The overall programme involves not just the working together of people interested in different aspects of the work – census workers, nest recorders and bird ringers – but the integration of this work in a common framework. This is made possible by the various projects being run by a single organization; in almost all other countries ringing and bird censusing, for example, are run by different organizations and integration is difficult.

Note that many of our schemes also provide data for our habitat-related work, which may also require long-term approaches.

The value of drawing together data from the various schemes has led BTO to establish an Integrated Population Monitoring (IPM) programme (Fig. 19.1). The objective of this is to monitor British bird populations in relation to normal population levels and demographic rates and to establish 'alert limits' for departures from normality, such that the conservation authorities can be made aware of potential problems and take appropriate action. It is so far in its infancy, although some alerts have already been issued (e.g. Crick *et al.*, 1993c). Because the establishment of the IPM programme has led to some redefinition of the aims of component schemes, efforts have so far been concentrated on refining the methods of individual schemes and their analysis, so that they can make the most effective contributions to the overall programme. The examples quoted in this chapter are a mixture of case studies that had an immediate objective in their own right and others that were carried out to develop analytical methodology.

Table 19.1. Long-term surveys run annually by the British Trust for Ornithology with starting dates and geographical coverage (UK=United Kingdom; B&I=Britain and Ireland; E&W=England and Wales). 'Reproduction' means first-winter numbers (WPD), numbers of eggs and chicks per nest (NRS), or post-fledging numbers (CES). Note that routine annual monitoring through Ringing (including CES) is still under development and that BoEE was merged with the National Wildfowl Counts scheme (Wildfowl and Wetlands Trust) in 1993, to form an integrated Wetland Birds Survey.

Survey name	Target population	Start date and coverage	Data collected
Birds of Estuaries Enquiry	Wintering waders (all species)	1969; UK	Numbers
Wader Productivity Database	Wintering waders (all species)	1990; B&I	Reproduction
Common Birds Census	Breeding species of farmland and woodland (35% of species (=92% of individuals) covered by CBC and WBS together)	1961; B&I	Numbers
Waterways Bird Survey	Breeding species of linear waterways	1974; UK	Numbers
Nest Records Scheme	Breeding species (82 species routinely monitored)	1939; B&I	Reproduction
Constant Effort Sites Ringing	Breeding birds of wetland and scrub	1981; B&I	Numbers, Reproduction and Survival
Ringing	All birds	1907; B&I	Survival
Heronries Census	Breeding Grey Herons	1928; E&W	Numbers

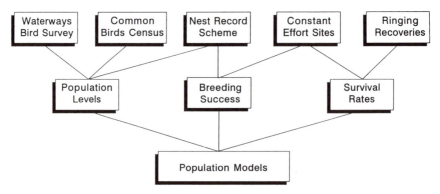

Fig. 19.1. Contributions of the various schemes to population models in the British Trust for Ornithology's Integrated Population Monitoring programme.

Those who are unfamiliar with the world of amateur ornithology some-times question the quality of the data provided by volunteers. Such doubts are unjustified: for the work that we ask them to do, the volunteers are at least as competent as paid staff. Much more significant are the difficulties of imposing rigorous sampling programmes on work that is carried out by unpaid volun-teers, though we are increasingly successful in this regard. There are also prob-lems with the statistical techniques available for interpreting the data, such as estimating survival rates from ringing recoveries (Lebreton and North, 1993), but in recent years there have been considerable advances. This chapter, will cover neither these technical aspects nor the details of individual schemes, which have been well covered elsewhere (Baillie, 1990, 1991; Marchant *et al.*, 1990; Stroud and Glue, 1991; Andrews and Carter, 1993; Greenwood *et al.*, 1993).

The Effects of Winter Weather on Bird Populations

The longest-running national bird census scheme in the world is the annual census of Grey Herons, *Ardea cinerea,* in England and Wales (Fig. 19.2). It is well-known not only for its length but also because it demonstrates clearly the adverse effects of severe winters on the size of the breeding population and subsequent recovery to an apparent equilibrium size. Because the effects of weather were established through the long-term data available, conservation-ists were not unduly alarmed by the crash in the population in the early 1960s: this could be explained by natural causes and would almost certainly be fol-lowed by recovery. The apparent long-term increase in the 'equilibrium' level may be an artefact of the statistical methods used. We are currently revising the data set before applying to it a new method of analysis (Thomas, 1993). Analy-sis of the ringing data for this population shows that cold winters markedly

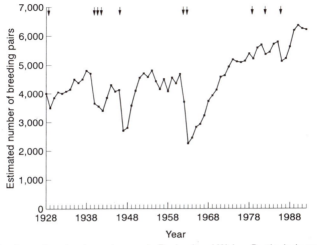

Fig. 19.2. Numbers of nests of grey herons in England and Wales. Particularly severe winters are marked by arrows. (For years prior to 1970, we have taken 'severe' winters from Stafford (1971); for subsequent years, 'severe' winters are those in which the mean temperature in any one winter month (given by Marchant *et al.*, 1990) fell below 1°C.) (BTO ·data, analysed by S.P. Carter).

depress survival of first-year birds (Fig. 19.3), though the effect on adult survival is not significant (North, 1979).

Widespread effects of weather on numbers were also shown in an analysis of CBC data for 14 species of resident passerines: numbers of most species are reduced by prolonged snowfall (frost and low temperatures seemed much less important; Greenwood and Baillie, 1991). An apparent positive effect of early spring rainfall on numbers censused, although possibly genuine, could have been an artefact, if birds concentrated their singing (important in detection) and other activity on the dry days in generally wet years (which are the days when fieldworkers make their censuses). Density-dependent effects were also clearly demonstrated in a number of these species in this analysis (as well as in several of 25 further species). Thus any monitoring programme should take weather and population level into account when assessing whether observed population changes exceed the alert limits that have been laid down.

Large-scale Changes in Numbers and Distribution of Wintering Waders

The Heron population figures were readily illuminated by incorporating environmental (meteorological) information. Even simply breaking population data down geographically can provide insights. Thus, Moser (1988) reviewed the relationship between the numbers of wintering grey plovers,

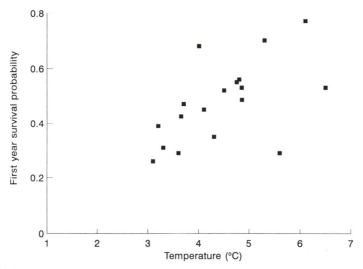

Fig. 19.3. Probability of survival of grey herons over the first year of life in relation to average temperatures in central England during January–March. Each point represents one year. (Redrawn from North, 1979.)

Pluvialis squatarola, on individual British estuaries and the size of the national population. If all estuaries were equal, one would expect the numbers on each to be a constant proportion of the national total as birds can move freely between estuaries. In fact (Fig. 19.4), the numbers on some estuaries (e.g. NW Solent) remained constant as the national population changed, changed more slowly than the national population (e.g. Chichester harbour), or reached an asymptote (e.g. the Mersey). These would appear to be pre-ferred estuaries, fully occupied (or almost so) even when the national popu-lation is low. On apparently less preferred sites (e.g. S. Solway), numbers varied proportionately more than did the national population. (There is no evidence for the alternative explanation, that conditions on estuaries such as the S. Solway improved over the years.)

The importance of wildlife sites is widely judged on the basis of numbers of animals there. It is, however, entirely possible for large, populous sites to be population sinks, maintained by continuing overspill from more productive areas. Using changes in distribution as Moser (1988) did, rather than simply distribution itself, allows us more insight into the true value of different sites. If information on reproduction and survival is also available, this can allow even more refined assessment of the quality of individual sites (Van Horne, 1983; Vickery *et al.*, 1992). Moser's (1988) work had another practical implication: that it is only the less preferred sites that are not already full, so that birds displaced from an estuary by industrial or other developments may have only suboptimal sites to go to as an alternative.

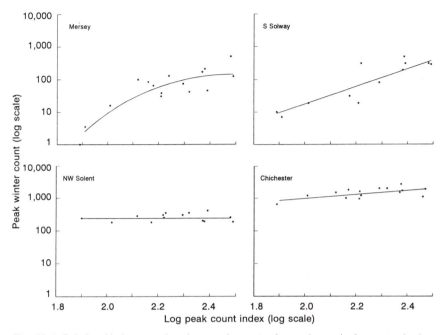

Fig. 19.4. Relationship between the winter peak counts of grey plovers on four estuaries in 1970–86, and the national winter peak count index. Note the logarithmic scale for both variables. (Redrawn from Moser, 1988.)

When environmental variation in time and space is laid alongside population variation in time and space, the results can be even more illuminating. The Birds of Estuaries Enquiry (BoEE) data again provide an example – Goss-Custard and Moser's (1988) analysis of the Dunlin, *Calidris alpina*. During 1977–86 there was a national decline of the British wintering population. This was most marked on estuaries where *Spartina anglica* had spread most (Fig. 19.5). Since *Spartina* occupies parts of the shore on which Dunlin would normally feed, and prevents them from feeding, it is reasonable to assume that the relationship is causal. An earlier decline which, unlike that of 1977–86, was matched elsewhere in western Europe showed no correlation with *Spartina* and was presumably the result either of factors affecting either the breeding grounds or all the western European wintering areas.

The Breeding Performance of Upland Birds

Because upland birds are sparsely distributed and sometimes inhabit remote areas, our knowledge of the changes in their numbers is less good than that of lowland species. As a result, information from the Nest Records Scheme has been particularly illuminating for a number of them. The merlin, *Falco*

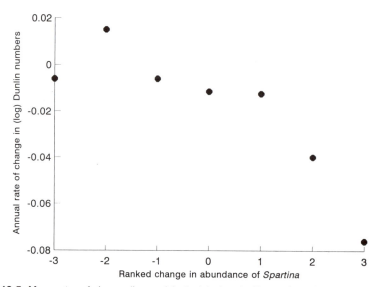

Fig. 19.5. Mean rates of change (log scale) of wintering dunlin numbers during 1977/78 to 1985/86 on estuaries grouped into seven categories according to change in abundance of *Spartina*. (Redrawn from Goss-Custard and Moser, 1988.)

columbarius, is of particular interest because its populations seem to have been affected by pesticides but, unlike those of other raptors such as spar-rowhawk (Marchant *et al.*, 1990) and peregrine (Ratcliffe, 1984; H.Q.P. Crick and D.A. Ratcliffe, in prep.), not to have recovered in recent years (Bibby and Nattrass, 1986), though there is some divergence of opinion in respect of the latter conclusion (A.F. Brown personal communication; S. Parr, personal communication). Drawing together data from various studies, Bibby and Nattrass (1986) suggested that numbers were declining particularly in areas where breeding success was poor. Analysis of nest record cards (Crick, 1993) showed that, in Britain as a whole, reproductive output appeared to have improved during the 1970s and 1980s but that it was still probably less than the 2.5 young per pair per year that Bibby (1986) had suggested was necessary to maintain numbers. Furthermore, there were differences between regions in the way in which reproductive rates had changed (Fig. 19.6). It is not clear why such regional differences occur but we have recommended to the conservation agencies that they should be taken into account in any intensive research aimed at understanding what determines the numbers of this species.

Another moorland species, the twite, *Acanthis flavirostris*, is the only passerine, apart from the endemic Scottish crossbill, *Loxia scotica*, that either breeds or winters in Britain in internationally important numbers (Batten *et al.*, 1990). A survey of the isolated population of the southern Pennines in 1990 showed it to be associated with grassy habitats at the moorland edge (Brown *et al.*, 1994). An analysis of nest records cards showed, however, that birds on

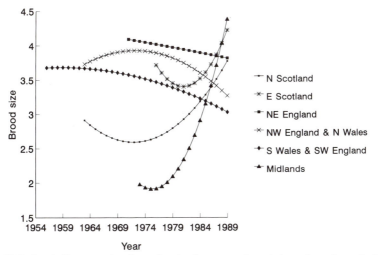

Fig. 19.6. Quadratic regression lines showing how mean brood sizes of merlin nests from which as least one chick fledged have changed in six regions (Meteorological Office regions). (Redrawn from Crick, 1993.)

heather appeared to produce more second broods and to be more successful than those on grassland, which suggests that, although the birds are more numerous on grassland, this is currently a suboptimal habitat (Brown *et al.*, 1994). Furthermore, mean brood sizes have recently declined in the south Pennines, but not in Scotland (Fig. 19.7), which flags up a cause for concern for the conservation authorities in England. Such alerts are a general objective of any monitoring programme.

Long-term Changes in Populations of Farmland Birds

Even quite simple analyses of carefully collected, long-term monitoring data can be instructive. Thus, if we take the data for species that are adequately censused by the CBC, we find that three species whose main habitat is farmland have increased during 1968–91 but 15 have decreased, whereas in woodland the equivalent figures are 13 and 12 respectively (R.J. Fuller *et al.*, in prep.). Plainly, though agrochemicals are now used with more care and discrimination than in the 1950s, the quality of farmland as habitat for birds has continued to decline. Given the range of species involved, it is inconceivable that farmland has not also declined as a habitat for wildlife generally, so this example illustrates how birds can be valuable indicators for wildlife in general, largely because they can be monitored on a wide scale very cost-effectively through the volunteer network.

Further evidence of the problems of wildlife on farmland come from comparisons of distribution of breeding birds made in two surveys two decades

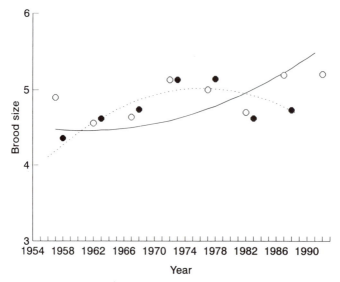

Fig. 19.7. Mean brood sizes of twite, from nests in which at least one egg hatched, during five-year periods in the southern Pennines (filled circles, dotted line) and in Scotland (open circles, continuous line). The lines were fitted by quadratic least-squares regression. (Redrawn from Brown *et al.*, 1994.)

apart (Table 19.2). Though comparisons must be made with care because the level of fieldwork effort was not assessed in the first atlas survey, there can be no doubt that farmland birds have fared badly compared with species in other habitats. Methods for the second atlas were more systematic, which will allow

Table 19.2. The numbers of species of British and Irish breeding birds that were found in more or in fewer 10 × 10 km squares in 1988–91 than in 1968–72, by main habitat. (From Gibbons *et al.*, 1993.)

Main habitat	More in 1988–91	Fewer in 1988–91
Farmland	4	24
Woodland	22	28
Lowland wetland	21	14
Upland	19	19
Coastal	11	13
Urban	5	2

$X^2 = 176$, 6 d.f., $P \ll 0.001$.

more reliable comparisons to be made in future, thus making the atlases an effective long-term monitoring tool (Gibbons *et al.*, 1993).

A preliminary analysis of the reproductive performance of some seed-eating birds (Crick *et al.*, 1991) showed that the daily rates of nest loss have declined in three species that have more or less maintained their populations (chaffinch, *Fringilla coelebs*; greenfinch, *Carduelis chloris*; and yellowhammer, *Emberiza citrinella*). In contrast, daily nest losses have increased in two species that have declined in numbers (linnet, *Acanthis cannabina* and reed bunting, *Emberiza schoeniclus*). Sample sizes of nest record cards for a third declining species, corn bunting, *Miliaria calandra*, were unfortunately too small for trends to be measured with any precision. Such multispecies studies provide pointers for further research. In this case, it would clearly be valuable to examine the data sets in more detail, especially those pertaining to breeding success, and perhaps to consider a wider range of seed-eaters. If detrimental changes are found in a suite of similar species, this is strong evidence of some general problem affecting them all.

Cross-species comparisons are particularly valuable in throwing light on species that are so scarce that it is difficult to obtain enough information about them directly. For example, the distribution of the cirl bunting, *Emberiza cirlus*, in England has contracted markedly since the 1940s, so that there now remains only one small population centred on south Devon (Evans, 1992; Gibbons *et al.*, 1993). Reasons suggested for the decline include increased competition with yellowhammer, loss of winter food resources, climatic change, and habitat change (Sitters, 1985; Evans, 1992; Evans and Smith, 1994). An analysis of nest record cards by H.Q.P. Crick, A.D. Evans, C. Dudley and K.W. Smith (in prep.) confirms that the cirl bunting is particularly susceptible to wet weather during the breeding season but that it nonetheless breeds as successfully in south-west England as do other buntings. Indeed, the limited data indicate that it was reproductively more successful than the yellowhammer in the rest of the UK, which weakens (but does not destroy) the hypothesis that its decline there was a result of competition from the yellowhammer. Both commoner buntings (yellow and reed) produce more chicks per nest in southwest England than elsewhere in the UK, suggesting that this area is generally good for buntings, rather than particularly good for cirl. Indeed, the continuing decline of the species parallels those of a number of other farmland seed-eaters, all of which may be affected by agricultural intensification, particularly weed control (Fuller *et al.*, 1991; Marchant and Gregory, 1994). Research into these more general problems may be more productive, even for illuminating the reasons for the cirl bunting's decline, than research focused on that species alone.

At an early stage in the development of IPM we conducted an integrated analysis of the data for the song thrush, *Turdus philomelos* (Baillie, 1990). The CBC index showed a decline in the population subsequent to the mid 1970s. A model fitted to the data for 1962–76, incorporating the effects of winter weather and population density on changes in numbers from year to year, explained 90% of the variation (Fig. 19.8). However, although part of the subsequent decline in numbers could be attributed to winter weather, the departure of the model from the data in later years showed that some other factor must be involved. There is no evidence of changes in measured components of reproduction, in adult survival, or in migration patterns (part of the population winters in France, where thrushes are hunted), so that recent problems may be ones that particularly affect first-year survival (Baillie, 1990).

Studies of another declining farmland bird, the lapwing, *Vanellus vanellus*, summarized and reviewed by Tucker and Galbraith (1994), show the value of bringing together data from a variety of schemes. Common Birds Census and Waterways Bird Survey (Marchant *et al.*, 1990), national censuses at intervals of 20–30 years (Shrubb and Lack, 1991), and the special survey of Breeding Waders of Wet Meadows (O'Brien and Smith, 1992) all show this species to have declined markedly, especially in areas of particularly intensified agriculture. Both short-term intensive studies (Galbraith, 1988; Baines, 1988, 1989) and analysis of nest record cards (Shrubb, 1990) indicate that lapwings produce most chicks when they can nest on bare earth or newly emerged crops and then take those chicks to nearby grassland (especially if unimproved, undrained and lightly grazed). The nest record cards show that, whereas clutch size has remained constant, brood size at hatching has declined, consequent upon greater losses of eggs associated with the switch to autumn cereals

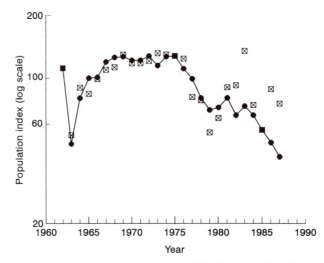

Fig. 19.8. Results of fitting a population model to the BTO Common Birds Census data for farmland song thrushes. The solid symbols and line show the observed indices of population size; the squares show values predicted from a model incorporating the number of freezing days in January and February and population size in the previous year. (Redrawn from Baillie, 1990.)

and higher stocking rates. Analysis of ringing recoveries shows that the survival rates of both first-year and adult birds fluctuate greatly (in response to winter weather) but that they have not declined in the long-term (Fig. 19.9; Peach *et al.*, 1994). This confirms that the decline in the population is a consequence of the decline in reproductive output. In most studies in western Europe the number of fledglings produced per pair was lower than that necessary to balance the mortalities measured by Peach *et al.* (1994).

Population Limitation in African-Palaearctic Migrants

Although some birds occupy different geographical ranges at different times of year, it is possible to gain some insight into their population dynamics through studies made at only one season. Thus key-factor analyses of data collected on the European breeding grounds have confirmed that changes in the populations of seven species of passerines that winter in Africa are driven by variation in survival rates rather than by reproductive output and that survival rates are density-dependent (Baillie and Peach, 1992). (These seven species were the only such migrants for which adequate long-term data exist; three had been studied professionally, for 13–18 years, and four through collaborative projects by volunteers, over 20–25 years.) If seven can be regarded as a sufficient sample, the concordance between these species should be of interest to students of population dynamics. It is certainly of interest to conservationists, for it suggests that the future of these birds may depend on habitat conditions in their non-breeding ranges in Africa. This certainly seems true for the sedge warbler, *Acrocephalus schoenobaenus*, in which rainfall on the wintering grounds determines both annual survival rates and the subsequent size of the breeding population (Fig. 19.10).

The Possible Effects of Magpies on Songbird Populations

The IPM programme is geared to detecting untoward changes in bird populations and identifying the possible causes. The data can be used retrospectively in the opposite way: to look for the possible effects of a perceived problem. For example, Gooch *et al.* (1991) analysed nest record and CBC data from 1966 to 1986 for the magpie, *Pica pica*, and for 15 species of songbirds, since it had been widely suggested that magpies, through nest predation, must have a detrimental effect on the populations of smaller birds. Magpie numbers increased by 4–5% per annum during this period; yet nesting success declined significantly in none of the songbirds. The increase in magpies varied from region to region and in different habitats but variation in songbird numbers and nesting success was not correlated with these variations. This suggests that the increase in magpies has not led to general declines in numbers or

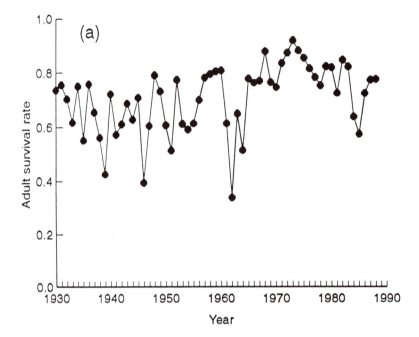

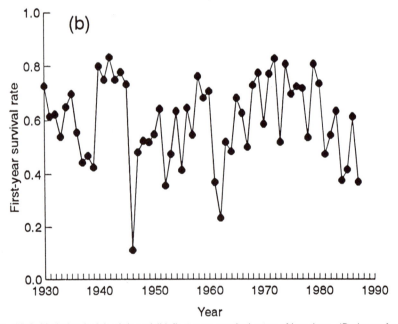

Fig. 19.9. Variation in (a) adult and (b) first-year survival rates of lapwings. (Redrawn from Peach *et al.*, 1994.)

nesting success of small birds, though such evidence cannot exclude the occurrence of possible effects on a more local scale.

Monitoring Environmental Change: Have We Got Our Priorities Right?

In recent years there has been considerable interest in environmental changes on the global scale brought about by the activities of humans. Politicians have made commitments involving huge sums of money to implement measures aimed at combatting some of the perceived problems. What has been the reaction of the scientific community in Britain? Some important and relevant research on climate and on marine ecosystems has been carried out. On land, there is the Terrestrial Initiative in Global Environmental Research (TIGER) programme and the Environmental Change Network (Tinker, Chapter 22). What we have not seen is any significantly increased support for any of the three long-term monitoring programmes run in Britain that are envied throughout the world, the Continuous Plankton Recorder Survey (Gamble, Chapter 20), the Rothamsted Insect Survey (Woiwood and Harrington, Chapter 18) and the various components of the BTO IPM programme. Indeed, all have come under strong financial pressure in recent years.

We have shown that the BTO's developing IPM programme, particularly because it incorporates demographic information and not just census data, has proven potential for conservation science and the study of population dynamics. Yet it is funded only by the Trust itself and by the conservation agencies. As a result, the level of funding is inadequate either to allow it to be developed quickly enough or to provide resources for analysing existing data. As an example, consider the monitoring of the timing of reproduction, particularly in relation to global warming. This example was used in the publicity launch of the Environmental Change Network, based on one bird species in one location. A preliminary analysis of BTO nest records for 82 species gathered across Britain during 1962–90 shows that the mean laying date has become significantly earlier in 33 species and later in only five; clutch sizes have apparently also generally increased (Crick *et al.*, 1993b). This result clearly demands closer investigation but the resources to do so are not currently available. This is an appropriate place in which to urge that the scientific community should build on the existing long-term schemes when confronting long-term problems. The current funding system is biased structurally in favour of short-term work. Furthermore, peer review tends to favour projects that promise to make decisive and rapid contributions to those issues that are intellectually fashionable at the time. As has been shown in this chapter, existing long-term programmes can provide intellectually rigorous and cost-effective monitoring and make considerable practical and long-lasting contributions to both pure and applied ecology.

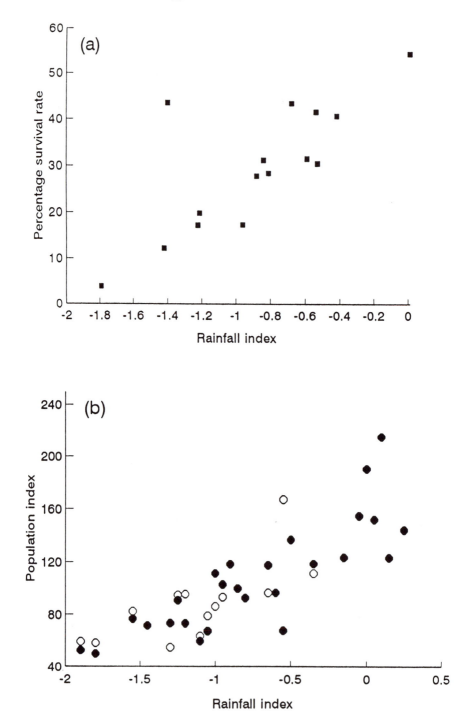

Acknowledgements

We thank the tens of thousands of volunteers for the millions of man-hours of expert and committed fieldwork on which our work is based. The Trust's work has been funded by a variety of agencies, chiefly by the Joint Nature Conservation Committee (on behalf of The Countryside Council for Wales, English Nature and Scottish Natural Heritage) and the Department of the Environment (Northern Ireland). We thank Tracey Brookes, Julie Sheldrake and Susan Waghorn for preparing the typescript and illustrations and several colleagues for access to unpublished results.

References

Andrews, J.A. and Carter, S.P. (eds) (1993) *Britain's Birds in 1990-91: The Conservation and Monitoring Review.* British Trust for Ornithology/Joint Nature Conservation Committee, Thetford.

Baillie, S.R. (1990) Integrated population monitoring of breeding birds in Britain and Ireland. *Ibis* 132, 151-166.

Baillie, S.R. (1991) Monitoring terrestrial breeding bird populations. In: Goldsmith, F.B. (ed.) *Monitoring for Conservation and Ecology.* Chapman and Hall, London. pp. 112-132.

Baillie, S.R. and Peach, W.J. (1992) Population limitation in Palaearctic-African migrant passerines. *Ibis* 134 (Suppl. 1), 120-132.

Baines, D. (1988) The effects of improvement of upland, marginal grasslands on the distribution and density of breeding wading birds (Charadriiformes) in Northern England. *Biological Conservation* 45, 221-236.

Baines, D. (1989) The effects of improvement of upland, marginal grasslands on the breeding success of Lapwings *Vanellus vanellus* and other waders. *Ibis* 131, 497-506.

Batten, L.A., Bibby, C.J., Clement, P., Elliott, G.D. and Porter, R.F. (1990) *Red Data Birds in Britain.* T & A D Poyser, London.

Bibby, C.J. (1986) Merlins in Wales: site occupancy and breeding in relation to vegetation. *Journal of Applied Ecology* 23, 1-12.

Bibby, C.J. and Nattrass, M. (1986) Breeding status of the Merlin in Britain. *British Birds* 79, 170-185.

Fig. 19.10. Effect of rainfall on the wintering grounds on (a) survival rate and (b) subsequent breeding populations of sedge warblers. Population sizes are indexed by the Common Birds Census (open circles) and the Waterways Bird Survey (closed circles). Each point represents a year. Note that different numbers of years' data are available for estimating survival rates and for the two census schemes. (Redrawn from Peach *et al.*, 1991.)

Böhning-Gaese, K., Taper, M.L. and Brown, J.H. (1993) Are declines in North American insectivorous songbirds due to causes on the breeding range? *Conservation Biology* 7, 76-86.

Brown, A.F., Crick, H.Q.P. and Stillman, R.A. (1994) The distribution, numbers and breeding ecology of twite *Acanthis flavirostris* in the south Pennines of England. *Bird Study* (in press).

Clutton-Brock, T.H. (1986) *Reproductive Success Studies of Individual Variation in Contrasting Breeding Systems.* University of Chicago Press, Chicago.

Crick, H.Q.P. (1993) Trends in breeding success of merlins (*Falco columbarius*) in Britain from 1937-198. In: Nicholls, M.K. and Clarke, R. (eds) *Biology and Conservation of Small Falcons.* Hawk and Owl Trust, London, pp. 30-38.

Crick, H.Q.P., Donald, P.F. and Greenwood, J.J.D. (1991) *Population Processes in Some British Seed-eating Birds.* BTO Research Report 80. British Trust for Ornithology, Thetford.

Crick, H.Q.P., Dudley, C. and Glue, D. (1993a) Breeding birds in 1991. *BTO News* 185, 15-18.

Crick, H.Q.P., Dudley, C. and Glue, D. (1993b) Nesting success in 1991. *BTO News* 187, 6-8.

Crick, H.Q.P., Gibbons, D.W. and Magrath, R.D. (1993c) Seasonal changes in clutch size in British birds. *Journal of Animal Ecology* 62, 263-273.

Dobinson, H.M. and Richards, A.J. (1964) The effects of the severe winter of 1962/63 on birds in Britain. *British Birds* 57, 373-434.

Dunnet, G.M. (ed.) (1991) Long-term studies of birds. *Ibis* 133 (Suppl. 1).

Evans, A.D. (1992) The numbers and distribution of cirl buntings *Emberiza cirlus* breeding in Britain in 1989. *Bird Study* 39, 17-22.

Evans, A.D. and Smith, K.W. (1994) Habitat selection of cirl buntings *Emberiza cirlus* wintering in Britain. *Bird Study* 41, 88-94.

Fuller, R.J. (1992) Effects of coppice management on woodland breeding birds. In: Buckley, G.P. (ed.) *Ecology and Management of Coppice Woodlands.* Chapman and Hall, London, pp. 169-192.

Fuller, R.J. and Crick, H.Q.P. (1992) Broad-scale patterns in geographical and habitat distribution of migrant and resident passerines in Britain and Ireland. *Ibis* 134, 14-20.

Fuller, R.J. and Warren, M.S. (1991) Conservation management in ancient and modern woodlands: responses of fauna to edges and rotations. In: Spellerberg, I.F., Goldsmith, F.B. and Morris M.G. (eds) *The Scientific Management of Temperate Communities for Conservation.* 31st Symposium of the British Ecological Society. Blackwell Scientific Publications, Oxford, pp. 445-471.

Fuller, R.J., Stafford, P. and Ray, C.M. (1989) The distribution of breeding songbirds within mixed coppiced woodland in Kent, England, in relation to vegetation age and structure. *Annales Zoologici Fennici* 26, 265-275.

Fuller, R.J., Hill. D.A. and Tucker, G.M. (1991) Feeding the birds down on the farm. *Ambio* 20, 232-237.

Furness, R.W. and Greenwood, J.J.D. (eds) (1993) *Birds as Monitors of Environmental Change.* Chapman and Hall, London.

Galbraith, H. (1988) Effects of agriculture on the breeding ecology of lapwings *Vanellus vanellus. Journal of Applied Ecology* 25, 487-503.

Gibbons, D.W., Reid, J.B. and Chapman, R.A. (eds) (1993) *The New Atlas of Breeding Birds in Britain and Ireland: 1988-1991.* T & A D Poyser, London.

Goldsmith, F.B. (ed.) (1991) *Monitoring for Conservation and Ecology*. Chapman and Hall, London.

Gooch, S., Baillie, S.R. and Birkhead, T.R. (1991) Magpie *Pica pica* and songbird populations. Retrospective investigation of trends in population density and breeding success. *Journal of Applied Ecology* 28, 1068–1086.

Goss-Custard, J.D. and Moser, M.E. (1988) Rates of change in the numbers of dunlin, *Calidris alpina*, wintering in British estuaries in relation to the spread of *Spartina anglica*. *Journal of Applied Ecology* 25, 95–109.

Greenwood, J.J.D. and Baillie, S.R. (1991) Effects of density-dependence and weather on population changes of English passerines: using a non-experimental paradigm. *Ibis* 133 (Suppl. 1), 121–133.

Greenwood, J.J.D., Baillie, S.R., Crick, H.Q.P., Marchant, J.H. and Peach, W.J. (1993) Integrated population monitoring: detecting the effects of diverse changes. In: Furness, R.W. and Greenwood, J.J.D. (eds), *Birds as Monitors of Environmental Change*. Chapman and Hall, London, pp. 267–342.

Hagan, J.M. and Johnston, D.W. (eds) (1992) *Ecology and Conservation of Neotropical Migrant Landbirds*. Smithsonian Institution Press, Washington.

Hellawell, J.A. (1991) Development of a rationale for monitoring. In: Goldsmith, F.B. (ed.) *Monitoring for Conservation and Ecology*. Chapman and Hall, London. pp. 1–14.

Hinds, W.T. (1984) Towards monitoring of long-term trends in terrestrial ecosystems. *Environmental Conservation* 11, 11–18.

Koskimies, P. (1992) Monitoring bird populations in Finland. *Vogelwelt* 113, 162–172.

Koskimies, P. and Väisänen, R.A. (1991) *Monitoring Bird Populations. A Manual of Methods Applied in Finland*. Zoological Museum, Finnish Museum of Natural History, Helsinki.

Lack, D. (1966) *Population Studies of Birds*. Clarendon Press, Oxford.

Lack, P. (1992) *Birds on Lowland Farms*. HMSO, London.

Lebreton, J.-D. and North, P.M. (eds) (1993) *Marked Individuals in the Study of Bird Population*. Birkhäuser Verlag, Basel.

Likens, G.E. (ed.) (1989) *Long-Term Studies in Ecology*. Springer-Verlag, New York.

Marchant, J.H. and Gregory, R.D. (1994) Recent population changes among seed-eating passerines in the United Kingdom. In: Hagemeijer, W. and Verstrael, T. (eds) *Bird Numbers 1992. Distribution, Monitoring and Ecological Aspects*. Proceedings of 12th International Conference of IBCC and EOAC. SOVON, Beek-Ubbergen (in press).

Marchant, J.H., Hudson, R., Carter, S.P. and Whittington, P. (eds) (1990) *Population Trends in British Breeding Birds*. British Trust for Ornithology, Tring.

Moser, M.E. (1988) Limits to the numbers of grey plovers *Pluvialis squatarola* wintering on British estuaries: an analysis of long-term population trends. *Journal of Applied Ecology* 25, 473–485.

Newton, I. (ed.) (1989) *Lifetime Reproduction in Birds*. Academic Press, London.

Nicholson, E.M. (1983) Origins and early days. In: Hickling, R. (ed.) *Enjoying Ornithology*. T & A D Poyser, Calton, pp. 15–28.

North, P.M. (1979) Relating grey heron survival rates to winter weather conditions. *Bird Study* 26, 23–28.

O'Brien, M. and Smith, K.W. (1992) Changes in the status of waders breeding on wet lowland grasslands in England and Wales between 1982 and 1989. *Bird Study* 39, 165–176.

O'Connor, R.J. and Shrubb, M. (eds) (1986) *Farming and Birds.* Cambridge University Press, Cambridge.

Peach, W.J., Baillie, S.R. and Underhill, L. (1991) Survival of British sedge warblers *Acrocephalus schoenobaenus* in relation to west African rainfall. *Ibis*, 133, 300-305.

Peach, W.S., Thompson, P.S. and Coulson, J.C. (1994) Annual and long-term variation in the survival rates of British Lapwings. *Journal of Animal Ecology*, 63, 60-70.

Perrins, C.M., Lebreton, J.D. and Hirons, G.J.M. (eds) (1991) *Bird Population Studies: Relevance to Conservation and Management.* Oxford University Press, Oxford.

Pienkowski, M.W. (1990) Foreword. In: Marchant, J.H., Hudson, R., Carter S.P. and Whittington, P. (eds) *Population Trends in British Breeding Birds.* British Trust for Ornithology, Tring, pp. v-viii.

Pienkowski, M.W. (1991) Using long-term ornithological studies in setting targets for conservation in Britain. *Ibis* 133 (Suppl. 1), 62-75.

Ratcliffe, D.A. (1984) The Peregrine breeding population of the United Kingdom in 1981. *Bird Study* 31, 1-18.

Risser, P.G. (ed.) (1991) *Long-term Ecological Research. An International Perspective. Scope* 47. Wiley, Chichester.

Shrubb, M. (1990) Effects of agricultural change on nesting lapwings *Vanellus vanellus* in England and Wales. *Bird Study* 37, 115-127.

Shrubb, M. and Lack, P.C. (1991) The numbers and distribution of Lapwings *V. vanellus* nesting in England and Wales in 1987. *Bird Study* 38, 20-38.

Sitters, H.P. (1985) Cirl buntings in Britain in 1982. *Bird Study* 32, 1-10.

Spellerberg, I.F. (ed.) (1991) *Monitoring Ecological Change.* Cambridge University Press, Cambridge.

Stafford, J. (1971) The heron population of England and Wales, 1928-1970. *Bird Study* 18, 218-221.

Stroud, D.A. and Glue, D. (eds) (1991) *Britain's Birds in 1989/90: The Conservation and Monitoring Review.* British Trust for Ornithology/Nature Conservancy Council, Thetford.

Thomas, G.E. (1993) Estimating annual total heron population counts. *Applied Statistics* 42, 473-486.

Tucker, G.M. and Galbraith, C.A. (eds) (1994) *The Ecology and Conservation of Lapwings Vanellus vanellus.* UK Nature Conservation No. 9. Joint Nature Conservation Committee, Peterborough.

Usher, M.B. (1991) Scientific requirements of a monitoring programme. In: Goldsmith, F.B. (ed.) *Monitoring for Conservation and Ecology.* Chapman and Hall, London, pp. 15-32.

Van Horne, B. (1983) Density as a misleading indicator of habitat quality. *Journal of Wildlife Management* 47, 893-901.

Verner, J.S. (1986) Future trends in management of non-game wildlife: a researcher's viewpoint. In: Hale, J.B., Best, L.B. and Clawson, R.L. (eds) *Management of Non-game Wildlife in the Midwest: a Developing Art.* Proceedings of a Symposium held at the 47th Midwest Fish and Wildlife Conference, Grand Rapids, Michigan, December 1985. pp. 149-71.

Vickery, P.D., Hunter, M.L., Jr and Wells, J.V. (1992) Is density an indicator of breeding success? *Auk* 109, 706-710.

Long-term Planktonic Time Series as Monitors of Marine Environmental Change

J.C. GAMBLE
Sir Alister Hardy Foundation for Ocean Science, The Laboratory, Citadel Hill, Plymouth PL1 2PB, UK.

Introduction

Marine ecosystems are subject to the same broad agencies of change as their terrestrial and freshwater counterparts. These include those induced by natural events, such as the consequences of the increased levels of radiatively active 'greenhouse' gases, as well as those caused by direct anthropogenic influence such as the localized effects of pollutants and the impact of fisheries. Despite the growing literature on the possible effects of global change on the marine environment (e.g. Glantz, 1992; Holligan and Reiners, 1992) including the consequences of CO_2 increases (e.g. Longhurst, 1991), the recognition of natural change is not so obvious. In contrast the literature abounds with the measurement of contaminant levels in the marine environment (e.g. Salomons *et al.*, 1988). The major question, however, must always be: to what extent does the measured change in the marine environment affect the functioning of the ecosystem? This question has been approached at every level of organization, from subcellular to ecosystem (Stebbing *et al.*, 1992). However, irrespective of the direct effects measured in a specific field or experimental investigation, the cumulative, long-term consequences are often the most important and invariably the most intractable. Thus systematic long-term monitoring of the environment is the only means of measuring such effects.

Long-term marine monitoring programmes have been carried out in many regions and are often associated with the proximity of a convenient marine laboratory. For instance sampling programmes took place at sites in the western English Channel in the proximity of the Marine Biological Association Laboratory at Plymouth (Southward, 1980), off the Northumbrian coast (Roff *et al.*, 1988) and close to the island of Helgoland (Radach *et al.*, 1990). Ironically the good intentions of such long-term programmes often rapidly founder and the life span of long-term projects is frequently shorter than many focused, fixed-period investigations (Duarte *et al.*, 1992). Long-term ecological research was considered a dispensable option until society became aware of the possibility of global climate change and the predicted dire consequences (see Glantz, 1992; Southwood, Chapter 1). It then became obvious that there was a fundamental need for long-term studies, if only as a baseline, against which to judge the effects of future changes.

The purpose of this review will be first, to outline briefly the nature and causes of marine environmental change, second, to look at some data collected by a specific long-term programme, the Continuous Plankton Recorder Survey, and finally to illustrate how this is used to indicate the consequences of change.

Causes of Change in the Marine Environment

Unlike much of the terrestrial environment, the physical structure and the ecological format of the marine system has not been altered by long-standing human interference and management. There is no true equivalent to the gross effects of urbanization and agricultural practices. Aquaculture systems, while increasingly extensive in inshore areas, have yet to have the equivalent impact in the marine environment as agriculture has had on land. Nevertheless the effects of human activity are probably among the greatest causes of change in the marine environment through the effects of fishing, oil and mineral extraction, and the discharge of pollutants.

The impact of fishing is probably the most difficult to assess since fishing mostly affects the ecosystem through the long-term effects of the removal of a significant biomass of, for the most part, the top predators in the system (Cushing, 1975). Further, most of the research effort has been directed towards the management of the fishery in a situation where the fishing capability can far exceed the capacity of the fish populations to sustain themselves. The uncertain relationship between stock size of many exploited fish populations and the level of recruitment to the fishery has further complicated management procedures. In fact, the causes of fluctuations in recruitment are not well understood and are thought to be rooted in events taking place in the early life stages of the fish during their planktonic larval phase.

Mineral and oil exploitation of the marine environment, although of increasing economic importance, might as yet be too localized to induce significant ecological change other than in the vicinity of the site of exploitation itself. These localized effects can be significant where, for instance, cuttings and other materials used in seabed drilling operations are deposited in a dense layer around the base of an oil-rig (Davies and Kingston, 1992). However, it is possible that the sustained effects of intensive hydrocarbon exploitation might ultimately have an overall effect over a wider area.

Pollution, through discharge, terrestrial runoff, and atmospheric input is a significant cause of ecological change in the marine environment. Toxins, high loads of suspended sediments and enhanced levels of inorganic nutrients have all been recognized as the causative agents of change in certain communities. However, the important question remains, how extensive are the effects of environmental contaminants both spatially and temporally.

While anthropogenic effects are indeed some of the most significant causes of change in the marine environment, undoubtedly the most pervasive and extensive agents are climatic (e.g. Cushing, 1982; Laevastu, 1993). The generation of weather systems is undoubtedly greatly influenced, if not determined, by the close association between atmospheric and oceanic processes; the sea surface temperature is one of the few readily measurable features which can be shown to be directly correlated with weather patterns (Cushing and Dickson, 1976). It is also well known that weather patterns and meteorological features show persistent spatial and temporal trends. For instance, the weather in the northeast Atlantic and over western Europe is much affected by the relative strengths of the low pressure system in the Iceland–Greenland region and the high pressure in the Azores–Bermuda region (Dickson *et al.*, 1988; Mann, 1993). Dominance of the former creates a southwesterly wind field in western Europe whereas the latter results in more northerly flows.

Changes in meteorological conditions undoubtedly affect marine ecosystems inasmuch that they are affected directly by alterations in light levels, temperature regime and wind-induced vertical mixing. But the continuum of the pelagic environment means that persistent anomalous wind fields can, among other effects, alter the trajectories of surface currents, affect the genesis of mesoscale eddy structures and modify upwelling patterns. Such changes to the hydrography of localities is frequently reflected in the nature of the plankton populations; before the advent of sensitive recording instrumentation, the presence of unusual planktonic 'indicator species' was usually taken to signify the presence of abnormal water masses (Russell, 1935; Fraser, 1955).

Planktonic Time Series

The pelagic environment is the most extensive of all the marine ecosystems encompassing the greatest volume and with an enormous variety and diversity of organism types. In trophodynamic terms the most important processes take place in the upper, photic layers where planktonic primary production is the principal fixed energy source for most other oceanic and subphotic layer shelf sea systems. Planktonic ecosystems therefore have certain ideal characteristics for monitoring change in the marine environment. In addition to their paramount significance in trophodynamic processes several other factors qualify plankton as appropriate monitors of environmental change and as a major avenue for understanding the processes affecting the survival and recruitment of commercially important fish species. These include the restriction of primary producers to the upper photic zone, the importance of upper layer stability for the effective growth and development of many groups of organisms, and the role of plankton as the food source of the larvae of most commercial fish species.

A major problem, however, lies with adequate sampling of the plankton. How can the effects of change be demonstrated in an ecosystem which has no definite boundaries, where the *milieu* is in constant flux, where the lives of the organisms are either spent entirely or partially within the system (as is the case with the larvae of many fish and benthic species) and where the distribution extends from the surface of the ocean to the depths of the deepest abyssal trench? Quite clearly a monitoring programme encompassing all these variables would be impossible and, for the most part, the strategy adopted has had to be one of great simplification based on pragmatism and cost effectiveness.

The types of long-term time series undertaken subdivide into two categories: single point surveys, usually in the vicinity of research stations, and spatial surveys aimed at encompassing a significant area. In the first case, the advantage lies in the operational cost as the vessel utilized for the survey is frequently small and the survey is fitted around other routine duties. However, such surveys can become increasingly elaborate, such as the Bermuda Atlantic (Lohrenz *et al.*, 1992) and the Hawaii Ocean Time Series (Karl and Winn, 1991) stations supported by the Joint Global Ocean Flux Study, where ocean-going research vessels are involved and appropriately sophisticated measurements made. These single point surveys can also be used to profile the water column or, at least, to provide a depth integrated value.

The great disadvantage of such systems is that they give no insight into advective processes affecting the distribution of planktonic organisms, nor do they provide information on the spatial patterns of planktonic distributions. Such information can only come from multistation monitoring systems or from the 'ship of opportunity' coverage of the Continuous Plankton Recorder Survey. Of the former systems two, the MARMAP project (Sherman, 1980) off the north-eastern USA and the CalCOFI monitoring programme off the Californian coast (Chelton *et al.*, 1982), are systematic station grids sampled on a routine basis. Other long-term surveys, such as in the sub-Arctic Pacific (Brodeur and Ware, 1992) have been described, but these are usually compilations of data from several sources and are not regular surveys. Since such surveys are made from research vessels with appropriate facilities for sampling the water column, their costs can be prohibitive and their spatial and, to a lesser extent, temporal coverage is limited.

In 1931 a unique plankton survey was established by the British marine biologist, Alister Hardy (1935) who subsequently (Hardy, 1939) wrote in explanation: 'the idea underlying the initiation of this ecological survey was that of attempting to apply methods similar to those employed in meteorology to the study of the changing plankton distribution, its causes and effects'. The genesis of this idea came from the desire to understand the causes of fluctuations in fish stocks, particularly herring, and it focused upon the specific trophic interactions with the planktonic carnivorous larval stages of fish elaborated 30–40 years later by Cushing (1975) in his match–mismatch hypothesis.

A most significant feature of this survey was that the sampling was carried out by commercial vessels while carrying out their normal business so reducing the costs of collecting the samples. It also, by the deployment of many sampling systems simultaneously on several routes, enabled Hardy to build up a spatial picture of the seasonal and longer-term changes in plankton populations. This was achieved by a special sampler (Fig. 20.1), the Continuous Plankton Recorder (CPR), which not only could survive the rigours of operational use on commercial 'ships of opportunity' but also permitted the varying pattern of the plankton distributions to be determined along the course of each tow. This was achieved by retaining the plankton on a mesh (filtering silk in Fig. 20.1) which moved across the stream of incoming sea water at a rate determined by the speed of movement through the water, and was then stored on a spool for later analysis. This simple, but robust, design gave the CPR Survey the potential to extend to wherever suitable, regular (and willing) commercial shipping routes occurred. At its peak the Survey extended across the North Atlantic as well as covering the shelf seas around the British Isles (Fig. 20.2).

Operational and sampling restrictions of the CPR are the main limitations and, although modern electronically instrumented undulating recorders that allow variable depth sampling now exist (Aiken, 1980; Williams and Aiken, 1990), these are still research instruments and have yet to be used in the repetitive operational mode of the CPR. This means that the CPR survey, throughout its existence, has sampled at a set depth (usually between 6 and 10 metres) within the upper mixed layer of the ocean. Furthermore, sampling is limited by the size of the filtering silk mesh (280 μm) and, until most recently, the lack of suitable self-contained instrumentation for monitoring additional environmental parameters. Despite such restrictions, the CPR Survey has provided much information about the ocean-wide distribution patterns, seasonality and long-term trends of Atlantic mesoplankton. The integrity of its long-term database is unquestionably due to its consistency of operation and inherent conservatism throughout its entire operation.

The question which immediately comes to mind when considering a Survey which was initiated over 60 years ago is whether it has achieved its purpose. Certainly we are still highly unsure as to what factors determine the recruitment success of fish populations but the consistent provision of data from the CPR Survey has provided an increasingly valuable insight into the

J.C. Gamble

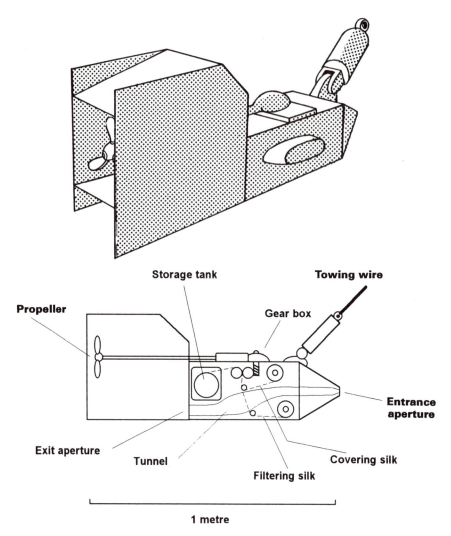

Fig. 20.1. Drawing and schematic of the Continuous Plankton Recorder currently in use.

distribution patterns and biogeography of mesoplankton in the sub-Arctic
North Atlantic. The early papers were largely descriptive in terms of specific
species, their distributions and regional differences in plankton community
structure. This is best illustrated by the respective distribution patterns of the
congeneric copepod species, *Calanus finmarchicus* and *C. helgolandicus*, in
the North Atlantic where the clear spatial differentiation between the two species played a major part in establishing their taxonomic identities (Fig. 20.3).
Subsequent studies on the vertical distribution of these species where they
co-occur established their distinct vertical separation in the water column

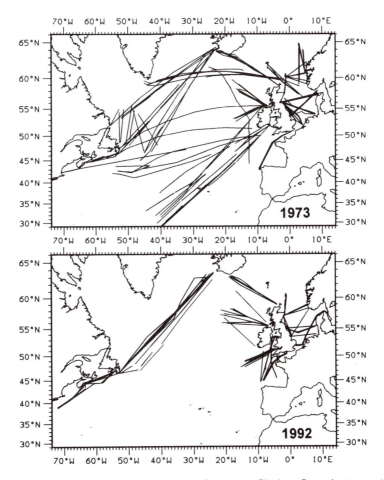

Fig. 20.2. Charts of the North Atlantic showing Continuous Plankton Recorder tow routes in 1973 and 1992.

(Williams, 1985). As the sampling schedules followed regular monthly patterns, data were accumulated on seasonal cycles of different plankton which, in turn, could be interpreted both in terms of interspecific seasonal trends or regional differences throughout the extent of the Survey (Fig. 20.4).

Some 20 years after the recommencement of the Survey, following the Second World War, the first attempts were made to evaluate long-term trends in the data (Colebrook, 1978). The preferred means of defining the trend in the data was to apply principal components analysis (PCA) to the standardized abundance of the sets of most representative zooplankton and phytoplankton in predetermined sampling areas. PCA is an ordination technique in which the original data set of intercorrelated variables are transformed into a new coordinate system where the axes are orthogonal linear combinations of the original

J.C. Gamble

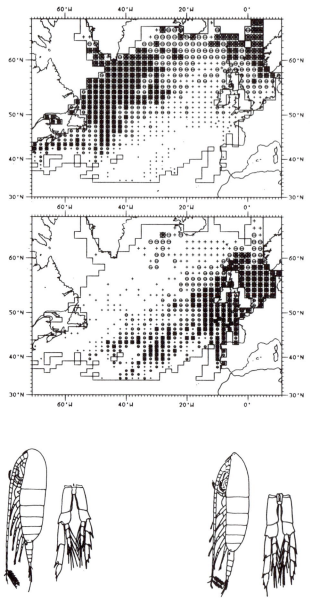

Fig. 20.3. Distribution patterns of two species of *Calanus* in the North Atlantic as determined from a compilation of records from the Continuous Plankton Recorder Survey collected over 25 years. Top map: *Calanus finmarchicus*; bottom map: *Calanus helgolandicus*.

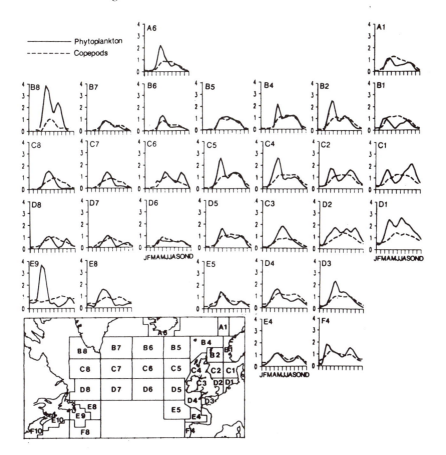

Fig. 20.4. Seasonal variations in the abundance of phytoplankton and copepods in the North Atlantic and North Sea. Annotations on each figure relate to the specific 'standard areas' indicated in the chart. Phytoplankton data are arbitrary units of greenness and copepod data are logarithmic means of numbers. (Adapted from Colebrook, 1979.)

variables. This effectively reduces a multiplicity of variables to many fewer, the principal components, which represent the covarying relationships among the original set of variables. In most cases most of the variance is explained by the first two components and, for the CPR data, experience has shown (Colebrook, 1978) that the first component was considered to be the best representation of the fluctuation and trends in abundance of the sampled plankton assemblage. This technique was used extensively by Colebrook throughout his career with the Survey and the results he described have become the classic representations of the nature of mesoplankton population trends in the North Atlantic (e.g. Colebrook, 1978, 1982, 1985; Colebrook *et al.*, 1984). The most well-known trend is that of the zooplankton in the North Sea and north-east Atlantic (Fig. 20.5) where the steady decline halted in the mid–late 1970s. In the North Sea the trend reversed to exceed in 1990 the original levels measured for the late 1940s (CPR Survey Team, 1992).

While this trend might well describe the population trends of the meso-plankton of the North Sea, the underlying cause of the trend is little understood. When compared to ocean-scale indices of climate (i.e. Atlantic ocean surface temperature) and hydrographic change (in this case Atlantic upwelling indices), it is possible to discern parallel trends which could imply correlation, if not cause (Dickson *et al.*, 1988). In fact, the extent of the observations of the CPR and the coherence of the data imply that the causative mechanism must operate at similar scales and hence is most likely to be concerned with contemporaneous oceanic and/or meteorological events. However, similarities between data sets and good correlations do not necessarily imply cause and effect.

Case Histories from the CPR Survey

The question therefore arises as to what, exactly, planktonic time series such as the Continuous Plankton Recorder Survey measure. Can such surveys, in fact be said to monitor the environment, or are they merely monitoring changes in the nature of the plankton populations? Obviously the latter is the true answer and the only environmental information is to be gained by surmise and inference from the observations on the plankton. This can be illustrated by three recent examples. The first relates to an increase in the abundance of dinoflagellates in the northern North Sea between 1988 and 1990 (Dickson *et al.*, 1992), the second to the distribution and abundance of the gelatinous, colony-forming flagellate, *Phaeocystis* sp., in the North Sea and western British shelf (Owens *et al.*, 1989) and the final concerns parallel trends between several trophic levels in an area of the North Sea adjacent to the NE coast of England (Aebischer *et al.*, 1990).

Dinoflagellates in the North Sea

The dinoflagellate bloom, *Ceratium* spp., occurred in the northern North Sea at a time when concern was being expressed about the consequences of eutrophication, particularly along the continental European seaboard of the North Sea. Initially this was suspected to be the cause although there was no direct evidence that such increases in nutrient levels had indeed taken place in the vicinity of the blooms. When examined in greater detail, it became apparent that the increase in the dinoflagellates coincided with a change in the vertical structure of the water column. The stratification due to the presence of low

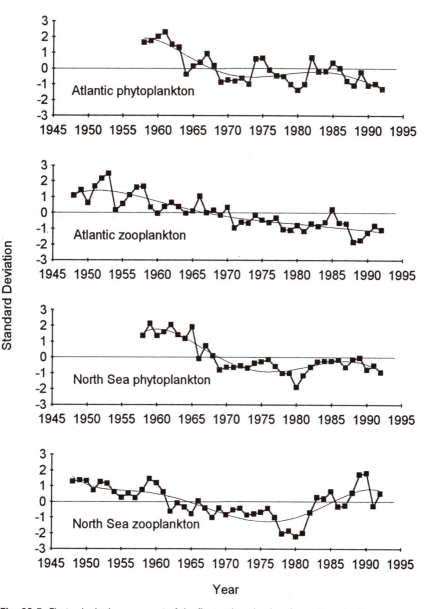

Fig. 20.5. First principal component of the fluctuations in abundance (standardized to zero mean and unit variance) of phytoplankton and zooplankton in the northeast Atlantic around the British Isles and the North Sea. The fitted curves (light continuous line) are fifth-order polynomial smoothing functions.

salinity water, was more intense than usual which resulted in enhanced stability and a prolonged development of the conditions which favoured the growth of the dinoflagellates (Fig. 20.6).

The cause of these circumstances had nothing to do with eutrophication but probably lay with alterations in persistent weather patterns affecting the hydrography of the region (Fig. 20.6). Indeed the hydrographic evidence suggested that an extensive westward excursion of low salinity water from the Baltic spread over the surface of the northern North Sea thus enhancing the stratification of the region. This is believed to have been caused by an anomalous weather pattern over western Europe which resulted in a more prevalent easterly air flow driving surface water out of the Baltic region into the North Sea.

The point to be made is that the initially presumed cause of this observation was not proven and the real cause was most probably an event remote from the site of the observation. In addition, it shows the responsiveness of plankton to anomalous weather conditions and, as such, is a pointer to their value in monitoring climate change. The links between weather and changes in the structure of plankton populations have been much emphasized in the interpretation of the CPR data. For instance, Dickson *et al* (1988) argued that increased northerly winds (caused by alterations in the relative strengths and positions of the Icelandic low and Azores high atmospheric pressure systems), affect the time of the stabilization of the water column, the depth of the critical layer for photosynthetic production and hence the timing of the onset of the spring phytoplankton bloom (the hypothesis of Sverdrup, 1953). On the other hand, Colebrook (1986) felt that there was better statistical evidence for the effects of persistent westerly wind patterns on the overwintering capabilities of many zooplankton species. These were reflected in their subsequent spring and summer abundance.

Distribution and abundance of Phaeocystis

This second example illustrates the benefit of an extensive survey which not only covers shallow waters but in the same time frame covers oceanic, deep-water environments. Thus we are able to answer questions such as to what extent do observations made in inshore areas relate to what is happening further offshore? In this case I shall take the example of the colonial flagellate, *Phaeocystis pouchetii* which has become of considerable nuisance value in many coastal regions of Europe, particularly in the inshore fringes of the Southern Bight of the North Sea (Cadée and Hegeman, 1986). Field data support the hypothesis that this alga has increased in abundance during the past two to three decades in these inshore areas and again the accusatory finger points at parallel changes in the environment including the increased loading of nutrients and contaminant levels (Cadée, 1990).

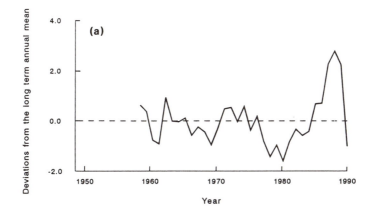

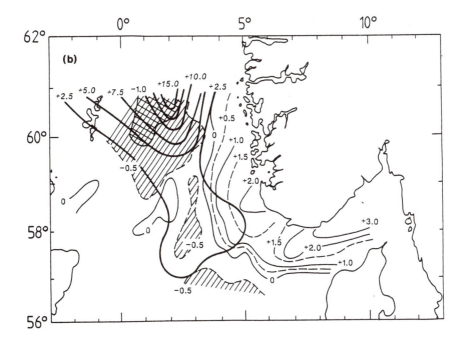

Fig. 20.6. The outburst of dinoflagellate genus *Ceratium* in the North Sea in the late 1980s. (a) Standardized annual mean abundance of *Ceratium* spp; (b) distribution of the differences in salinity in the surface (0–30 m) layer of the northern North Sea in summer 1987, compared with the long-term (1967–76) mean (finer isolines with differences in salinity, in parts per thousand, indicated; shaded and hatched areas signify less saline water). Contours (bold) of increased *Ceratium* abundance are superimposed ('+10' signifies a tenfold increase in abundance in 1982–88 compared with the long-term mean abundance, 1968–81). (Adapted from Dickson *et al.*, 1992.)

Although too small and fragile to be trapped by the meshes of the CPR, *Phaeocystis* can be recorded by the prevalence of gelatinous material on the filter silk which can be scaled to several levels of abundance. Owens *et al.* (1989) made a comparison of the changes in abundance of the species in several areas of the North Sea and on the western continental shelf of the British Isles, and although there were regional differences in abundance, there was no evidence in any of the individual areas of a significant, and sustained increase during the last 15–20 years (Fig. 20.7). This is in marked contrast with the coastal observations (Cadée and Hegeman, 1986) and the conclusion from the CPR Survey was that the recent increases in number of these flagellates did not occur in offshore areas.

The point this emphasizes is that single station observations, or observations in restricted localities, should not be used to extrapolate into the wider environment. Spatial surveys such as the CPR Survey are the only means whereby such broad generalizations can be made and supported.

Trends between trophic levels in the North Sea

With this final example I wish to discuss the use of the data from the CPR Survey to interpret parallel trends between trophic levels (Fig. 20.8). This relates to the similarity between the trends of phyto- and zooplankton populations as measured by the CPR and various indices relating to the long-term reproductive success of a piscivorous gull, the kittiwake (*Rissa tridactyla*). The similarities are clearly evident, in particular the timing of the reversal of the decreasing trend around 1980 in the phytoplankton, zooplankton and the three kittiwake indices (Aebischer *et al.*, 1990). When compared also with the trend in the recruitment of herring (*Clupea harengus*) in the North Sea, it is tempting to consider that there is some direct trophic linkage. Although it is logical that a trophic link can be claimed from phytoplankton–zooplankton–herring–kittiwake, it is not likely that this would be the direct cause in the similarity between the trends particularly when the evidence suggested that changes occurred simultaneously in all time series and without the lags that might be expected in responses between trophic levels. It is more likely that there is some underlying common cause suggested in this case to be meteorological as evidenced by the similarity in the trend in the westerly weather index. This again relates to Colebrook's (1985) earlier work where he felt that over-wintering success of many important copepod species was in some way related to the intensity of westerly weather conditions.

This suggests that parallel trends in populations of marine organisms can be related to pervasive factors operating over large areas and over lengthy time periods. Although this point has been made on several occasions (e.g. Dickson *et al.*, 1988; CPR Team, 1992), the work of Taylor and Stephens (1980) (updated by Taylor *et al.* (1992) would appear to show, beyond all doubt, that the passage of weather systems had a profound effect on the mixed layer

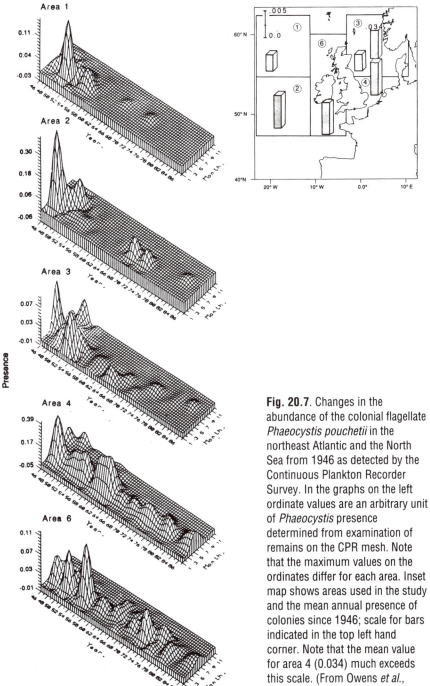

Fig. 20.7. Changes in the abundance of the colonial flagellate *Phaeocystis pouchetii* in the northeast Atlantic and the North Sea from 1946 as detected by the Continuous Plankton Recorder Survey. In the graphs on the left ordinate values are an arbitrary unit of *Phaeocystis* presence determined from examination of remains on the CPR mesh. Note that the maximum values on the ordinates differ for each area. Inset map shows areas used in the study and the mean annual presence of colonies since 1946; scale for bars indicated in the top left hand corner. Note that the mean value for area 4 (0.034) much exceeds this scale. (From Owens *et al.*, 1989.)

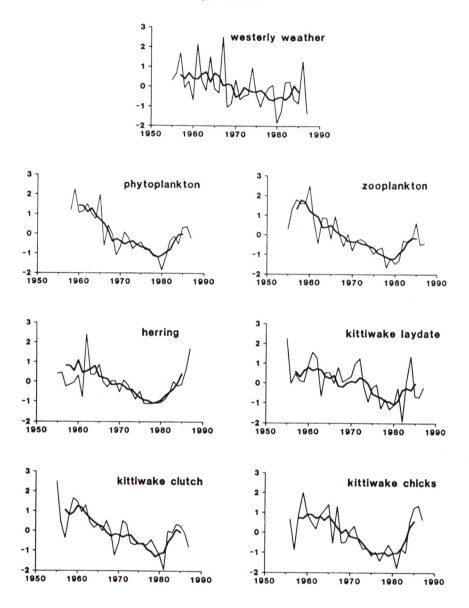

Fig. 20.8. Standardized (zero mean, unit variance) time series (thick lines) and 5-year running means (thin lines) for abundance of phytoplankton, zooplankton and herring, for kittiwake laying date, clutch size and chick production, and for frequency of westerly weather, from 1955 to 1987. Ordinates in standard deviation units. (From Aebischer *et al.*, 1990.)

plankton populations. This was demonstrated indirectly through an investigation of the shift in position of the 'North Wall' of the Gulf Stream as the current moves east-northeast away from the coastline of the USA. There was a very close correlation between the variation in the Gulf Stream position and the variation in the abundance of copepods collected by the CPR in the Eastern Atlantic and North Sea (Fig. 20.9). As these individual points were simultaneously correlated (i.e. there were no obvious phase differences), it was argued that the connection between the two sets of measurements was not oceanic, i.e. it was not due to changes within the water column being transmitted across the Atlantic in the current system (this would take several months) but was through the atmosphere. This hypothesis has recently received support from long-term studies on the plankton in Lake Windermere, where there were also significant correlations with the Gulf Stream position (Mr A.H. Taylor, personal communication; see also Maberly *et al.*, Chapter 21).

Once again the hypothesis rests upon a climatic solution. Only weather systems can have similar long-term effects on both marine and fresh water communities and recent models generated by Taylor (personal communication) show the effects of altering the position of the Gulf Stream on the eastward trajectories of pressure systems. Again the suspicion is that the increased prevalence, or otherwise, of storms affects the time of onset of the stratification which, in turn, controls the timing of the spring bloom.

Conclusion

The results presented above show that data from marine planktonic monitoring surveys can therefore be used to illustrate changes in marine ecosystems although the surveys themselves should be sufficiently extensive if it is wished to separate localized from pervasive, basin-wide effects. This latter point has been illustrated from a few examples of the results of the CPR Survey where, in two of the chosen examples, climatological events were seen to be the main agency of change. It has been argued that upper layer plankton are well suited as monitors of climatic change. In the first instance they exist close to the interface between the ocean and the atmosphere and secondly the rapid turnover times, particularly of the planktonic primary producers in the system, are within the time constants of many meteorological processes. Hence it can be expected that the response of a planktonic population should be one of the earliest indications of the effects of climatic change, although the means of recognizing such a response from the inherent natural variation in such populations is by no means certain.

The problem of recognition of cause and effect, together with the cost–benefit attributes of 'ship-of opportunity' versus 'multistation' versus 'fixed point' survey techniques will inevitably always be the main questions directed at marine planktonic monitoring surveys. Much has been said in favour of the

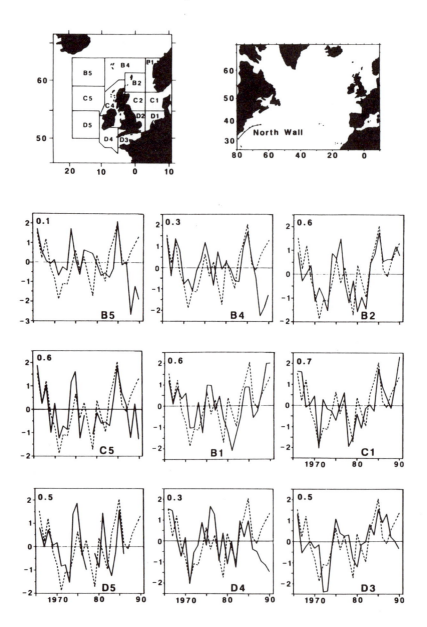

Fig. 20.9. Annual means of the latitude of the north wall of the Gulf Stream (1966 to 1990, broken line) compared with the logarithm of the number of copepods in areas of the Continuous Plankton Recorder Survey (solid line). Each graph has been standardized to have zero mean and unit variance and is accompanied by the correlation coefficient between the two time-series. Charts showing the position of the areas and of the north wall of the Gulf Stream are included. (From Taylor *et al.*, 1992.)

spatiotemporal attributes of the CPR Survey but single point sampling stations have certain advantages. In single point surveys it is, for instance, usually easier to do more routine measurements of environmental data as well as carry out more complex and a wider variety of procedures on the organisms themselves.

One of the best known single station surveys was Station E1 in the English Channel off Plymouth where plankton, fish populations and nutrients were measured for several decades (Russell *et al.*, 1971) and where a long frequency cycle, the Russell Cycle (Cushing and Dickson, 1976), was described. There is no single explanation for the changes observed at this station although it is considered that climatic changes affecting the temperature regime in an area close to a biogeographical boundary between cold and warm-water species can alter the competitive advantage of one species when compared with another. In this case it is the alteration between the closely related clupeoid fish species, herring (*Clupea harengus*) and pilchard (*C. pilchardus*), the chaetognaths, *Sagitta elegans* and *S. setosa* and several other species (Southward, 1980) but it is clear that there is evidence of fluctuations in the herring and pilchard fisheries at intervals over the past 400 years (Southward *et al.*, 1988). Such results from intensive surveys at a single site indicate that trends in decadal time series, such as the CPR Survey, must be regarded with caution as they too could be part of a much longer frequency period of change.

Duarte *et al.* (1992), while emphasizing the need for more long-term marine monitoring systems, have shown that on average they, ironically, have the shortest life of all projects. Good intentions founder rapidly. I hope, however, that I have demonstrated the worth of long-term time series in the planktonic realm. They can identify change and, more importantly, if sufficiently extensive, can discern the nature of the change and differentiate the localized effect from pervasive events. It is the possible increase in the frequency and magnitude of the pervasive events due to global climate change which reinforces the paramount importance of persisting with long-term monitoring of marine planktonic systems.

References

Aebischer, N.J., Coulson, J.C. and Colebrook, J.M. (1990) Parallel long-term trends across four marine trophic levels and weather. *Nature* 347, 753–755.

Aiken, J. (1980) A marine environmental recorder. *Marine Biology* 57, 238–240.

Brodeur, R.D. and Ware, D.M. (1992) Long-term variability in zooplankton biomass in the subarctic Pacific Ocean. *Fisheries Oceanography* 1, 32–38.

Cadée, G.C. (1990) Increased bloom. *Nature* 346, 418.

Cadée, G.C. and Hegeman, J. (1986) Seasonal and annual variation in *Phaeocystis pouchetii* (Haptophyceae) in the westernmost inlet of the Wadden Sea during the 1973 to 1985 period. *Netherlands Journal of Sea Research* 20, 29–36.

Chelton, D.B., Bernal, P.A. and McGowan, J.A. (1982) Large-scale interannual physical and biological interaction in the California Current. *Journal of Marine Research* 40, 1095–1125.

Colebrook, J.M. (1978) Continuous Plankton Records: zooplankton and environment, north-east Atlantic and North sea, 1948–1975. *Oceanologica Acta* 1, 9–23.

Colebrook, J.M. (1979) Continuous Plankton Records: seasonal cycles of phytoplankton and copepods in the North Atlantic Ocean and the North Sea. *Marine Biology* 51, 23–32.

Colebrook, J.M. (1982) Continuous Plankton Records: phytoplankton, zooplankton and environment, north-east Atlantic and North Sea, 1958–1980. *Oceanologica Acta* 5, 473–480.

Colebrook, J.M. (1985) Continuous Plankton Records: overwintering and annual fluctuations in the abundance of zooplankton. *Marine Biology* 84, 261–265.

Colebrook, J.M. (1986) Environmental influences on long-term variability in marine plankton. *Hydrobiologia* 142, 309–325.

Colebrook, J.M., Robinson, G.A., Hunt, H.G., Roskell, J., John, A.W.G., Bottrell, H.H., Lindley, J.A., Collins, N.R. and Halliday, N.C. (1984) Continuous Plankton Records: a possible reversal in the downward trend in the abundance of plankton of the North Sea and the north-east Atlantic. *Journal du Conseil* 41, 304–306.

CPR Survey Team (1992) The 1992 Continuous Plankton Records: the North Sea in the 1980s. *ICES Marine Science Symposia* 195, 243–248.

Cushing, D.H. (1975) *Marine Ecology and Fisheries*. Cambridge University Press, Cambridge.

Cushing, D.H. (1982) *Climate and Fisheries*. Academic Press, London.

Cushing, D.H. and Dickson, R.R. (1976) The biological response in the sea to climatic changes. *Advances in Marine Biology* 14, 1–122.

Davies, J.M. and Kingston, P.F. (1992) Sources of environmental disturbance associated with offshore oil and gas developments. In: Cairns, W. (ed.) *North Sea Oil and the Environment: Developing Oil and Gas Resources, Environmental Impacts and Responses*. Elsevier, Amsterdam, pp. 417–440.

Dickson, R.R., Kelly, P.M., Colebrook, J.M., Wooster, W.S. and Cushing, D.H. (1988) North winds and production in the eastern north Atlantic. *Journal of Plankton Research* 10, 151–169.

Dickson, R.R., Colebrook, J.M. and Svendsen, E. (1992) Recent changes in the summer plankton of the North Sea. *ICES Marine Science Symposia* 195, 232–242.

Duarte, C.M., Cebrian, J. and Marba, N. (1992) Uncertainty of detecting sea change. *Nature* 356, 190.

Fraser, J.H. (1955) The plankton of the waters approaching the British Isles. *Marine Research Scotland* 1955, No.1, 1–12.

Glantz, M.H. (ed.) (1992) *Climate Variability, Climate Change, and Fisheries*. Cambridge University Press, Cambridge.

Hardy, A.C. (1935) The Continuous Plankton Recorder. A new method of survey. *Rapports et Procès-verbaux des Réunions. Conseil International pour L'Exploration de la Mer* 95, 36–47.

Hardy, A.C. (1939) Ecological investigations with the Continuous Plankton Recorder: object, plan and methods. *Hull Bulletins of Marine Ecology* 1, 1–47.

Holligan, P.M. and Reiners, W.A. (1992) Predicting the responses of the coastal zone to global change. *Advances in Ecological Research* 22, 212–255.

Karl, D.M. and Winn, C.D. (1991) A sea of change: monitoring the oceans' carbon cycle. *Environmental Science and Technology* 25, 1976–1981.

Laevastu, T. (1993) *Marine Climate, Weather and Fisheries*. Fishing News Books, Oxford.

Lohrenz, S.F., Knauer, G.A., Asper, V.L., Tuel, M., Michaels, A.F. and Knap, A.H. (1992) Seasonal variability in primary production and particle flux in the northwestern Sargasso Sea: US JGOFS Bermuda Atlantic Time-Series Study. *Deep-Sea Research II* 39, 1373–1391.

Longhurst, A.R. (1991) Role of the marine biosphere in the global carbon cycle. *Limnology and Oceanography* 36, 1507–1526.

Mann, K.H. (1993) Physical oceanography, food chains, and fish stocks: a review. *ICES Journal of Marine Science* 50, 105–119.

Owens, N.J.P., Cook, D., Colebrook, M., Hunt, H. and Reid, P.C. (1989) Long term trends in the occurrence of *Phaeocystis* sp. in the north-east Atlantic. *Journal of the Marine Biological Association of the United Kingdom* 69, 813–821.

Radach, G., Berg, J. and Hagmeier, E. (1990) Long-term changes of the annual cycles of meteorological, hydrographic, nutrient and phytoplankton time series at Helgoland and at LV ELBE 1 in the German Bight. *Continental Shelf Research* 10, 305–328.

Roff, J.C., Middlebrook, K. and Evans, F. (1988) Long-term variability in North Sea zooplankton off the Northumberland coast: productivity of small copepods and analysis of trophic interactions. *Journal of the Marine Biological Association of the United Kingdom* 68, 143–164.

Russell, F.S. (1935) On the value of certain plankton animals as indicators of water movements in the English Channel and the North Sea. *Journal of the Marine Biological Association of the United Kingdom* 20, 309–332.

Russell, F.S., Southward, A.J., Boalch, G.T. and Butler, E.I. (1971) Changes in biological conditions in the English Channel off Plymouth during the last half century. *Nature* 234, 468–470.

Salomons, W., Bayne, B.L., Duursma, E.K. and Forstner, U. (eds) (1988) *Pollution of the North Sea: An Assessment*. Springer-Verlag, Berlin.

Sherman, K. (1980) MARMAP, a fisheries ecosystem study in the NW Atlantic: fluctuations in ichthyoplankton zooplankton components and their potential impact on the system. In: Diemer, F.P., Vernberg, F.J. and Mirkes, D.Z. (eds) *Advanced Concepts in Ocean Measurements for Marine Biology*. University of South Carolina Press, South Carolina.

Southward, A.J. (1980) The western English Channel – an inconsistent ecosystem? *Nature* 285, 361–366.

Southward, A.J., Boalch, G.T. and Maddock, L. (1988) Fluctuations in the herring and pilchard fisheries of Devon and Cornwall linked to change in climate since the 16th Century. *Journal of the Marine Biological Association of the United Kingdom* 68, 423–445.

Stebbing, A.R.D., Dethlefsen, V. and Carr, M. (eds) (1992) *Biological Effects of Contaminants in the North Sea*. Inter-Research, Amelinghausen.

Sverdrup, H.U. (1953) On conditions for the vernal blooming of phytoplankton. *Journal du Conseil* 18, 287–295.

Taylor, A.H. and Stephens, J.A. (1980) Latitudinal displacements of the Gulf Stream (1966 to 1977) and their relation to changes in temperature and zooplankton abundance in the N.E. Atlantic. *Oceanologica Acta* 3, 145–149.

Taylor, A.H., Colebrook, J.M., Stephens, J.A. and Baker, N.G. (1992) Latitudinal displacements of the Gulf Stream and the abundance of plankton in the north-east Atlantic. *Journal of the Marine Biological Association of the United Kingdom* 72, 919–921.

Williams, R. (1985) Vertical distribution of *Calanus finmarchicus* and *C. helgolandicus* in relation to development of the seasonal thermocline in the Celtic Sea. *Marine Biology* 86, 145–149.

Williams, R. and Aiken, J. (1990) Optical measurements from underwater towed vehicles deployed from ships-of-opportunity in the North Sea. *SPIE Environment and Pollution Measurement Sensors and Systems* 1269, 186–194.

21

The Sensitivity of Freshwater Planktonic Communities to Environmental Change: Monitoring, Mechanisms and Models

S.C. Maberly, C.S. Reynolds, D.G. George, E.Y. Haworth
and J.W.G. Lund .
*Institute of Freshwater Ecology and Freshwater Biological
Association, Windermere Laboratory, Far Sawrey,
Ambleside, Cumbria LA22 0LP, UK.*

Introduction

The planktonic communities of freshwaters comprise bacteria, fungi and eu-karyotic algae drawn from several phyla, and animal representatives of the Crustacea, Rotifera, Protozoa and other phyla, including certain species of Insecta and Mollusca during their larval stages. As in other ecosystems, the community of open water is structured about its primary producers, its consumers and its decomposers. Most biologists will recognize that the striking feature of the plants of the plankton (phytoplankton) and the first consumers in the dependent foodchain (zooplankton) is their universally small size (10^{-6}–10^{-3} m for the phytoplankton and 10^{-5}–10^{-2} m for the zooplankton; Reynolds, 1993), apparently a requirement for sustained planktonic existence.

What may be less readily appreciated is that there is a wide range of more specialized adaptations within each broad functional group. Among the primary producers, for example, there are species which are aquatic analogues of invasive colonists, ruderals and persistent canopy-formers, each requiring distinctive morphological, physiological and behavioural adaptations, as interpreted from careful observations and experiments (Reynolds, 1988). Because the generation times of phytoplankton are measured in days, the response of the species composition of the assemblage can become manifest within a week or so (Reynolds, 1993). Monitored at the appropriate frequency, these signals provide insights into the governing functions and constraints; in aggregate, weaker interannual differences in the response signals may identify longer-term shifts in the ambient environmental conditions.

Of what we believe is a total of about 5000 accepted species of algae and cyanobacteria (blue–green algae) in the freshwater phytoplankton, almost 500 have been recorded from routine samples in three lakes in the English Lake District – Windermere, Esthwaite Water and Blelham Tarn – collected over the past 50 years. Of these, only some 40–50 are sufficiently 'common' or regularly 'abundant' to have been reliably observed or subjected to searching experimentation. This chapter gives examples of long-term changes detected by the monitoring programme, uses the records to illustrate the mechanisms believed to be involved and describes models which test and apply our understanding of these processes.

History of the Database

It should be emphasized that the record, although one of the longest and most complete for freshwater phytoplankton populations available (Strayer *et al.*, 1986; Elliott, 1990), was not designed as a long-term database. The changes which it has shown were not anticipated at the outset: their detection has been fortuitous, though the clues come entirely from the information it was chosen to collect.

The Freshwater Biological Association has records for some biological variables which extend back into the 1930s (Elliott, 1990). However, the main monitoring programme for algae and water-chemistry began with samples from the North and South Basins of Windermere, initiated in 1945 by Dr J.W.G. Lund. The intention was to provide crucial field evidence in support of his now-renowned observational and experimental study (Lund, 1949, 1950; Lund *et al.*, 1963) on the population dynamics of the vernally blooming diatom *Asterionella formosa* Hass. However, Lund and co-workers also counted other algae of interest and these counts form the base of the long-term records of algal composition.

The early addition to the programme of regular sampling of Esthwaite Water was intended to maintain the observations of C.H. Mortimer on physicochemical stratification during his absence on military service. Lund decided to add Blelham Tarn to the programme some months later when, while skating on this small lake, his curiosity was aroused by the distribution of planktonic algae under ice. Grasmere was added to the routine programme in 1969 when it became known that a sewage works was to be constructed there. Grasmere is the only site in the programme where the collection of long-term records was started with the purpose of monitoring rather than testing scientific ideas.

The sampling programme has involved the weekly (latterly fortnightly) collection of water, integrated over the surface layer, from fixed points within the five lake basins mentioned above. The samples have been used to determine species composition of the phytoplankton, their total biomass as chlorophyll *a* (since 1964) and nutrient concentrations. Descriptions of the chemical methods used are given in Sutcliffe *et al.* (1982) and Heaney *et al.* (1988). *In situ* depth-profiles of temperature, dissolved oxygen and light penetration were also measured. Samples have been stored for zooplankton composition and abundance. This programme has been maintained, albeit with minor modifications to methods and sampling frequency. Some determinands have been added, such as pH, which reminds us that you cannot necessarily design long-term data-collection to be able to answer the questions you or others will eventually ask.

Environmental Change and Plankton Response

Modelling the response of phytoplankton to light

One example of the use of a long-term database is to test our understanding of the effect of environmental parameters on plankton performance. For example, light is a key environmental resource which affects the productivity of freshwaters. Most aquatic habitats receive low amounts of light because of reflection at the air–water interface and, more particularly, through attenuation by scattering and absorption within the water column. Analysis by Neale *et al.* (1991) of the annual early-spring growth of the diatom *A. formosa* in Windermere through the period 1964–88 has shown that the average weekly rate of increase is closely regulated by the average weekly availability of surface light (Fig. 21.1a). The rate of increase can be successfully modelled as a function of the availability of light in the water column and laboratory measurements of photosynthesis and growth rate parameters using equation 21.1 (Fig. 21.1b):

$$\mu = (D\mu_{max}[1 - \exp(-\alpha_g I_{sml}/\mu_{max})] - 24\alpha_0) \tag{21.1}$$

where: μ is the specific growth rate ($\log_e$ day^{-1}); μ_{max} is the maximum specific growth rate per hour of daylength (D), α_g is the slope of the growth–irradiance curve at zero irradiance, ($\log_e$ day^{-1} (cal cm^{-2} s^{-1})$^{-1}$), I_{sml} is the mean daytime irradiance (400–700 nm) in the surface-layer (cal cm^{-2} s^{-1}) and μ_0 is the hourly maintenance rate ($\log_e$ h^{-1}). The close agreement between modelled and long-term average rate in the lake suggests that one aspect of our understanding of the mechanisms which control phytoplankton abundance is broadly correct. However, Neale *et al.* (1991) note that an alternative model based on more

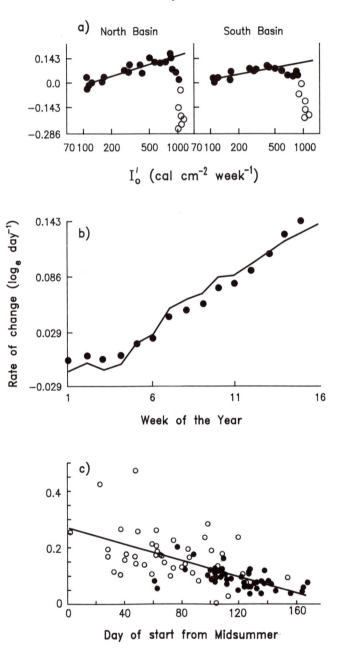

basic photosynthesis parameters was unsuccessful, thus highlighting areas where more research is needed.

In an analysis of interannual variation in rate of increase of *A. formosa* in the South Basin of Windermere between 1946 and 1990, Maberly *et al.* (1994) showed that the rate of increase in the spring and in the early autumn are not different statistically when expressed in terms of the number of days from midsummer (Fig 21.1c). This implies that light availability rather than temperature is important in controlling rates of increase since temperatures are typically in the range of 4-8°C in the spring and 14-19°C in the autumn.

Loss Processes

Loss processes can be equally important in controlling phytoplankton biomass and species composition because different types of phytoplankton are differentially sensitive to particular losses such as grazing, sedimentation, or parasitism by fungi (Reynolds *et al.*, 1982). One other direct source of loss which affects all phytoplankton is flushing of a lake basin. This source of loss is particularly important in lakes with a short retention time, such as Grasmere, a small lake in the northwest catchment of Windermere. The relatively large catchment, high annual precipitation and small lake volume determine that the average retention time of the lake is only 24-37 days. However, owing to variation in intensity and temporal distribution of rainfall, the instantaneous retention time can vary between 8 and >65 days (Reynolds and Lund, 1988). High rates of discharge that produce retention times of 21 and 8 days are equivalent to exponential loss rates of 0.05 and 0.12 $\log_e$ day^{-1} respectively. These are equal to, or greater than, net rates of growth of phytoplankton in nature, and so preclude, or seriously restrict, the development of biomass. An example can be seen in the phytoplankton chlorophyll *a* concentration for this site in years with wet or dry springs and autumns (Fig. 21.2). High rates of flushing in early spring (e.g. 1978) suppress or delay the vernal increase in biomass (Fig. 21.2a). Conversely, floods in the autumn (1977) cause a rapid reduction in

---

Fig. 21.1. The effect of light on rate of change of cell concentration for *Asterionella formosa* in Windermere. (a) Rate of change against mean time-integrated surface irradiance (I'_o) between 1964 and 1988. Open symbols indicate silicate concentrations < 0.5 g SiO$_2$ m^{-3}. The regression line for week 1–17 (North Basin) and week 1–13 (South Basin) is shown. (After Neale *et al.* 1991.) (b) Comparison of rate of change in the North Basin as a weekly mean between 1964 and 1988 determined by regression analysis (solid line) with modelled values using equation 21.1 (●). (After Neale *et al.* 1991.) (c) Rate of change for individual years between 1946 and 1990 in the spring (●) and autumn (○) expressed as the number of days between the start of growth and midsummer. (After Maberly *et al.* 1994.)

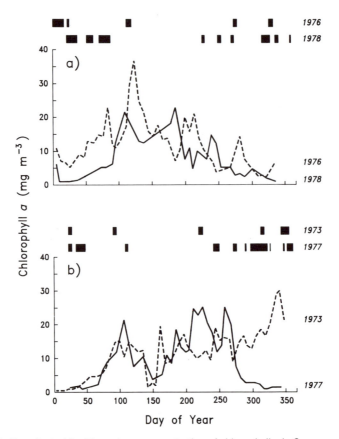

Fig. 21.2. The effect of flushing rate on concentration of chlorophyll *a* in Grasmere for: (a) dry (1976) and wet (1978) springs; and (b) dry (1973) and wet (1977) autumns. Bars at the top of each panel indicate periods when the flushing rate exceeded 0.05 log$_e$ day^{-1}. (Derived from data in Reynolds and Lund, 1988.)

biomass (Fig. 21.2b), as do floods throughout the year. In terms of species composition, the lake lacks substantial populations of cyanobacteria even though the concentrations of phosphate and nitrate appear to be sufficient to support their growth. Cyanobacteria are typically slow-growing and achieve dominance in late summer by virtue of making efficient use of scarce resources and being relatively unpalatable to zooplankton. In a lake such as Grasmere, the rapid flushing will tend to wash-out such species and their overwintering propagules (Reynolds and Lund, 1988).

Nutrient Concentrations

The nutrient-regime of a lake is a major factor which determines its 'trophic-status', productivity, biomass and species composition. The major change detected in the environment of the Cumbrian lakes since 1945 is the increase in nutrient loading resulting from an increased human population in the catchment, particularly summer visitors, increased discharge of waste water and use of detergents and increased use of agricultural fertilizers (Lund, 1972). This 'eutrophication' is evident in the South Basin of Windermere where winter concentrations of the key nutrients, PO_4-P (Fig. 21.3a) and NO_3-N (Fig. 21.3b), have increased. In contrast, soluble reactive silica expressed as SiO_2 shows year-to-year variation, but no indication of a long-term change (Fig. 21.3c). The method for measuring nitrate has altered over the period of the database, with the method used earlier giving lower values than that used subsequently (Heaney *et al.*, 1988). Corrected and uncorrected values are presented which show that knowledge of the method and its limitations is essential for the correct interpretation of past data (see also Leigh *et al.*, Chapter 14, for an example from a field experiment). There is some evidence for low winter concentrations of PO_4-P in years 1988, 1990 and 1991, although it is not clear why. Since April 1992 phosphate stripping has been introduced at the sewage works which discharges into the South Basin of Windermere to try to reduce the phosphate concentrations, although processes of internal loading from reserves in the sediment are likely to make this a slow process (Reynolds, 1992). Monitoring is now taking place to determine the consequences of this reduction in the external load of phosphorus.

Long-term Changes in Chlorophyll Concentration

Information on total phytoplankton biomass in the surface waters of Windermere, Esthwaite Water and Blelham Tarn, as the concentration of chlorophyll *a*, has been collected since 1964. Essentially the same method of extraction in hot methanol has been used since the work was initiated by Dr J.F. Talling. In an analysis of the data between 1964 and 1989, Talling (1993) showed a long-term increase in biomass. These data for Windermere South Basin are presented here in a simplified form for the period 1965–92, as the annual maximum (Fig. 21.4a) and annual average, weighted to take into account the non-regular sampling (Fig. 21.4b). Both show a statistically significant increase over the study period. Based on our understanding of factors which control phytoplankton biomass the expectation is that the chlorophyll concentration may be linked to the availability of limiting nutrients such as PO_4-P and NO_3-N. However, although within the South Basin of Windermere there is an overall link between increasing biomass and nutrients, there is no statistically significant relationship between chlorophyll and annual concentrations of PO_4-P,

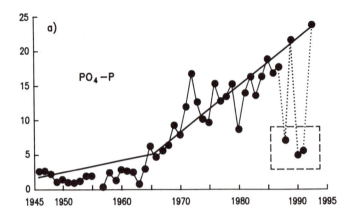

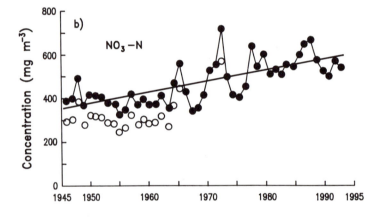

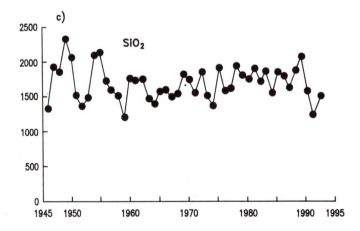

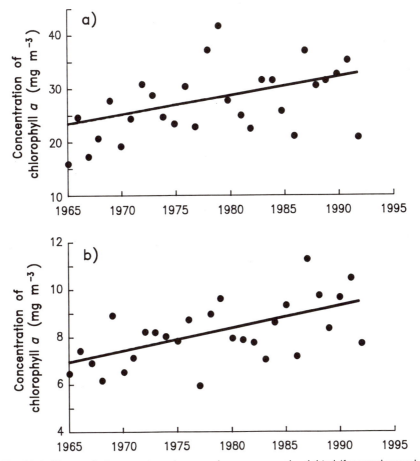

Fig. 21.4. Changes in the annual maximum and average annual weighted (for equal annual frequency) concentration of phytoplankton chlorophyll *a* (mg m⁻³) between 1965 and 1992 in the South Basin of Windermere, (a) Maximum chlorophyll *a* with regression (standard error in parenthesis): chlorophyll *a* = −699 (259) + 0.367 (0.130) × year; adjusted r^2 = 0.20, $P < 0.01$. (b) average chlorophyll *a* with regression: chlorophyll *a* = −165 (50) + 0.088 (0.025) × year; adjusted r^2 = 0.29, $P < 0.01$. (Data derived, in part, from Talling, 1993.)

Fig. 21.3. Time-trend between 1946 and 1992 in the South Basin of Windermere of average concentrations between January and March of : (a) PO_4-P; (b) NO_3-N and (c) SiO_2. For NO_3-N uncorrected concentrations (○) and corrected concentrations (●) depending on the method used following Heaney *et al.* (1988) are given. Linear regression of annual concentrations (mg m⁻³) (*y*) against year (*x*) yielded (standard errors in parentheses): for PO_4-P, *y* = −372 (123) + 0.192 (0.063) × *x* + 0.483 (0.100) × (year after 1965), adjusted r^2 = 0.89, $P < 0.001$ (boxed values excluded); for NO_3-N, *y* = −8975 (1444) + 4.80 (0.73) × *x*, adjusted r^2 = 0.47, $P < 0.001$, and for SiO_2 there was no significant trend ($P = 0.9$).

NO_3-N and SiO_2. Thus the probability (P) values for the relationship between the January to March concentrations of nutrients and the maximum and average chlorophyll concentration, respectively, are 0.27 and 0.45 for PO_4-P, 0.07 and 0.13 for NO_3-N and 0.33 and 0.90 for SiO_2. The lack of a relationship between biomass and SiO_2 is not surprising since there is no evidence for a long-term trend for this nutrient (Fig. 21.3c). Furthermore, the concentration of SiO_2 is essential only for the diatoms which dominate the spring maximum, whereas the annual maximum is usually in the summer in this lake (Talling, 1993). Although none of these relationships is statistically significant, changes in NO_3-N appear to be more closely linked to average and maximum chlorophyll concentration whereas PO_4-P is more usually considered limiting in temperate freshwaters.

Changes in Species Composition

The study of changes in phytoplankton as total biomass hides the fact that species composition may change. Some species have increased in recent times, particularly some cyanobacteria, and it is inferred from known nutritional correlations that the cause of this is increased nutrient levels. For example, the cyanobacterium *Aphanizomenon flos-aquae* has increased in abundance in both the North and particularly the more nutrient-rich South Basin of Windermere (Fig. 21.5). In Blelham Tarn, which was affected by increased nutrient loading much earlier than Windermere, several algae whose distribution is known to be in nutrient-rich water became established in the lake (Lund, 1978). Comparing the average abundance in the period 1945–60 with the period 1961–76, Lund showed increases in the cyanobacteria *A. flos-aquae* and *Planktothrix mougeotii* (Bory ex Gom.) Anagnostidis and Komárek (= *Oscillatoria agardhii* var. *isothrix* (Skuja) V. Poljaski) and recorded *Microcystis aeruginosa* for the first time in 1974. The flagellates *Cryptomonas* spp. and *Trachelomonas* spp. also increased, whereas other species, notably *Dinobryon divergens*, typical of oligotrophic water, declined. Some species have been present throughout the sampling period, and in these cases long-term patterns of change may be discernible. For example, Maberly *et al.* (1994) have described the annual seasonal cycle of the diatom *A. formosa* in the South Basin of Windermere and showed that long-term changes had occurred for eight of the 19 descriptors of seasonal change which were studied. The causes of these changes are currently under investigation. Further quantitative examples of long-term change in species composition in the Windermere catchment are given in Lund (1972), Talling and Heaney (1988), and Heaney *et al.* (1988).

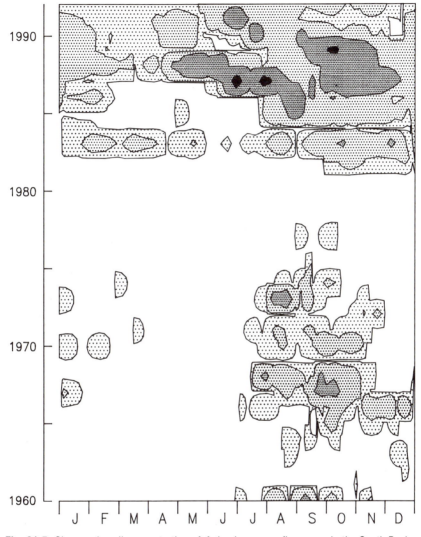

Fig. 21.5. Changes in cell concentration of *Aphanizomenon flos-aquae* in the South Basin of Windermere between 1960 and 1992. Between 1960 and 1977 the species was counted as filaments cm^{-3} (cells cm^{-3} = filament cm^{-3} × 40) and between 1978 and 1990 as mm filament cm^{-3} (cells cm^{-3} = mm cm^{-3} × 400). Cell concentration (cell cm^{-3}): ☐ < 10; ▨ 10–99; ▨ 100–999; ▨ 1000–9999; ■ > 10,000.

The Effects of Weather

In recent years it has become clear that changes in the timing and intensity of wind-induced mixing have a profound effect on the seasonal succession of lake plankton. Reynolds (1980) gives a general view of the processes involved and recent publications by the same author (Reynolds, 1990, 1993) relate these processes to more general ecological theory. The classic view of seasonal succession is essentially directional and implies a predetermined order in which different functional groups dominate a community. Such processes of succession are typically characterized by increasing species diversity and increasing production until a characteristic 'climax' community has evolved. In terrestrial ecosystems, such natural patterns of succession are seldom subjected to sudden natural reversals. In communities of freshwater phytoplankton the expected pattern of succession is frequently interrupted, and sometimes reversed, by external perturbations.

In deep, thermally stratified lakes, episodes of intense wind mixing are usually responsible for such sudden reversals. Each lake can be regarded as a recording climatic instrument capable of responding to quite small regional variations in the input of solar and wind energy (George, 1989). Year-to-year changes in species dominance are thus intimately related to changes in the intensity and timing of wind-induced mixing. The implications of these weather effects for the chemistry and biology of a lake are discussed below.

The chemical consequences of year-to-year changes in the weather

When a lake becomes thermally stratified, very little oxygen moves across the thermocline from the epilimnion (upper layer) to the hypolimnion (lower layer). The concentration of oxygen in the hypolimnion is then controlled by the reducing properties of the sediment and the rate at which organic particles decompose in the water column. Most productive lakes develop a pattern of deep-water anoxia that remains relatively constant from year to year. In some lakes, however, the concentration of oxygen in the hypolimnion changes from year to year as changing weather conditions generate different patterns of phytoplankton succession.

A good example of weather-related anoxia is shown by the South Basin of Windermere where the deep water only becomes anoxic when weather conditions favour the growth of the non-vacuolate cyanobacterium *Tychonema bourrellyi* (Lund) Anagnostidis and Komárek (=*Oscillatoria bourrellyi* Lund) (Heaney, 1986; Mills *et al.*, 1990). This alga is denser than water and requires a relatively high level of turbulence to keep it in suspension, so large crops of *T. bourrellyi* typically appear in the lake when the summer is relatively windy. Once established, the species can remain dominant for several weeks and then give rise to very low oxygen tensions in the hypolimnion when it eventually sinks and decays (Heaney, 1986; Mills *et al.*, 1990). Figure 21.6 shows the

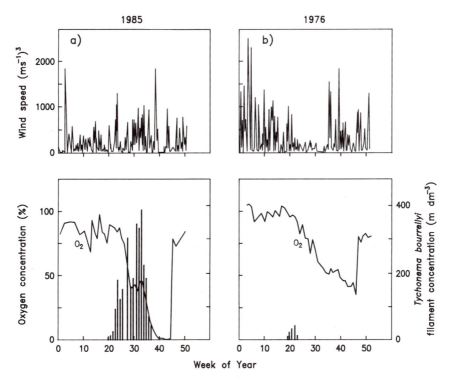

Fig. 21.6 Wind speeds (upper graphs) and growth of the non-vacuolate cyanobacterium *Tychonema bourrellyi* (bars, lower graphs) and the development of deep-water anoxia (lines, lower graphs) in the South Basin of Windermere in relation to wind speed: (a) 1985 – a relatively windy summer and (b) 1976 – a relatively calm summer. Wind speeds measured at Sellafield and expressed as daily average values raised to the power of 3.

extent to which periodic growths of this rather unusual species of *Tychonema* can influence the chemical characteristics of the late-summer hypolimnion in Windermere. In 1976 (Fig. 21.6b) the summer was relatively calm, the growth of *T. bourrellyi* was suppressed and so the deepest water in the lake never became anoxic. In 1985 (Fig. 21.6a), the early summer was calm but high winds in July and August favoured the growth of *T. bourrellyi* which greatly enhanced the consumption of oxygen in the late-summer hypolimnion. These year-to-year changes in the oxygen content of the hypolimnion have a pronounced effect on the vertical distribution of fish (Mills *et al.*, 1990) and fundamentally alter the chemical reactions that lead to the regeneration of nutrients from the sediment in deep water. This is particularly the case for phosphate which is released from the sediment when the surface layers become anoxic, potentially causing a feedback loop to increase productivity. In calmer years, *T. bourrellyi* is usually replaced by buoyant or motile species such as *Aphanizomenon flos-aquae* that remain in suspension. Much of this summer biomass

is then lost through the outflow, or decays in the epilimnion, so the organic
load to the hypolimnion is greatly reduced.

The biological consequences of year-to-year changes in the weather

At one time, blooms of cyanobacteria were regarded as an almost inevitable
consequence of lake enrichment. In recent years, however, it has become
clear that, in most lakes, very dense surface blooms only appear if the weather
conditions are favourable (Steinberg and Hartman, 1988; George *et al.*, 1990).
In temperate lakes, cyanobacteria typically appear as the 'climax' community
in waters that support a succession of other species in spring and early summer (Reynolds, 1984). Most genera of cyanobacteria grow relatively slowly but
their growth rate can be accelerated by high temperatures and prolonged
periods of relative calm. The effect of calm periods on the development and
dispersion of a bloom of cyanobacteria can be illustrated by contrasting the
development of *Aphanizomenon flos-aquae* in Esthwaite Water in 1985 and
1989. The summer of 1985 was cool and windy, particularly in July and August
(Fig. 21.7a). The *Aphanizomenon* crop appeared late in the year and its
growth was soon checked by intense mixing in late August. In 1989 the early

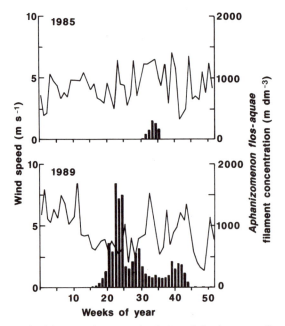

Fig. 21.7. The growth of the vacuolate cyanobacterium *Aphanizomenon flos-aquae* (bars)
in Esthwaite Water in relation to wind speed (lines): (a) 1985– a relatively windy summer
and (b) 1989– a relatively calm summer. Wind speeds measured at Sellafield and
expressed as mean weekly values.

summer was not only warm but unusually calm (Fig. 21.7b). A dense bloom of *Aphanizomenon* was recorded in the lake towards the end of May and the species remained relatively abundant throughout the summer.

Recent long-term studies of plankton dynamics in Esthwaite Water (George *et al.*, 1990) suggest that many of these year-to-year changes are quasicyclical in nature. Figure 21.8 shows the year-to-year changes in the late summer crop of *Aphanizomenon* in Esthwaite Water between 1956 and 1972. Although the lake became increasingly nutrient-rich during this 18-year period, the critical 'bloom forming' factor was the year-to-year variation in the late summer mean wind speeds. Very high *Aphanizomenon* numbers were

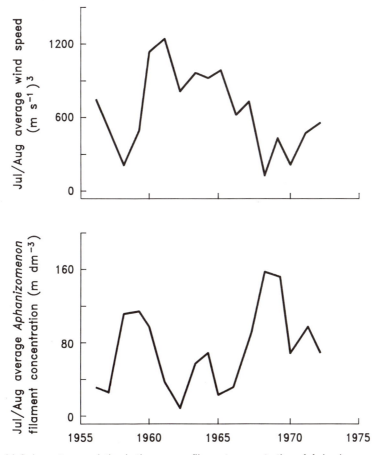

Fig. 21.8. Long-term variation in the average filament concentration of *Aphanizomenon flos-aquae* between May and September in Esthwaite Water in relation to the average wind speeds in July and August. Modified from George *et al.* (1990). Wind speeds measured at Sellafield and expressed as average values raised to the power of 3.

recorded in the late 1950s and the late 1960s when wind speeds were, on average, lower and there were prolonged periods of calm. This '10-year' cycle of calm summers matches the reported temperature cycle in Windermere (George and Harris, 1985) and appears to be related to the incidence of storms in the Western Atlantic.

Extending the Record: the Use of Algal Remains in Sediments

In many lakes, sediments are laid down in an orderly manner in deep water, and the remains of diatom frustules and chrysophyte scales and cysts therein remain preserved and identifiable. Analysis of the species assemblage present at different depths can be calibrated to time using radioisotope markers to provide a history of phytoplankton change. For four of the Cumbrian lakes, the sedimentary record can, uniquely, be cross-checked against the long data-series of phytoplankton composition. Allowing for limitations of the sedimentary record in terms of time-resolution caused by low rates of sedimentation and hence sediment thickness, there is a good agreement between the two sources of data. For example, changes in the abundance of the diatoms *Asterionella formosa* and *Fragilaria crotonensis* between 1946 and 1990 in the South Basin of Windermere are clearly recorded in the sediments (Fig. 21.9).

The sedimentary record thus enables us to interpret past environments; which for these lakes extends back *c.* 14,000 years to the last ice-age. One result of this is that the pH of a lake can be inferred over long-periods, with obvious implications for current concerns over acid-deposition, by relating the known distribution of present day taxa to pH. For example, changes in the postglacial diatom profile in the sediment of Ullswater, a mesotrophic lake of similar size and shape to Windermere, indicates that the pH was about 1 unit higher immediately after the retreat of the last ice-sheet and then declined steadily in the subsequent 5000 years to approximately present-day values of about pH 7.0 (Haworth and Allen, 1984; Haworth, 1990; Haworth, in prep.).

Conclusions

The desirability of the programme on which this chapter is based has been reviewed on numerous occasions but the reluctance to stop it has always triumphed, at one time contrary to official UK Science Policy (Jones, 1990). It is interesting to see how far and how quickly attitudes have shifted. The value of long-term monitoring, producing data such as those considered here, is no longer questioned: they alone can attest, within the limits of acuity of the data collected, to what has really happened. Furthermore, by combining this information with data from experiments in the laboratory and directly in lakes, the

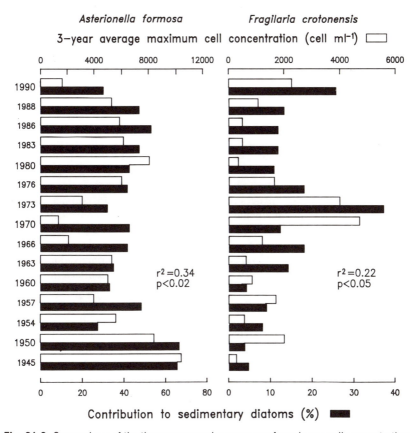

Fig. 21.9. Comparison of the three-year running average of maximum cell concentration for *Asterionella formosa* and *Fragilaria crotonensis* in the South Basin of Windermere with their contribution to the diatom assemblage in the sediments, derived from analysis of 5 mm thick slices from a sediment core collected in 1990 and analysed by Dr S. Sabater.

mechanisms which drive any changes may be determined and models derived which facilitate the prediction of future events.

Acknowledgements

We are grateful to Drs J. F. Talling FRS, S.I. Heaney and S. Sabater for giving us advice and access to their data. Dr J. F. Talling made constructive comments on an earlier manuscript. Richard Poles helped with the statistical analysis of the data. Christine Butterwick extracted and calculated the data for *Aphanizomenon* in Windermere. Trevor Furnass drew the figures.

References

Elliott, J.M. (1990) The need for long-term investigations in ecology and the contribution of the Freshwater Biological Association. *Freshwater Biology* 23, 1-5

George, D.G. (1989) The thermal characteristics of lakes as a measure of climate change. *Conference on Climate and Water*, Helsinki, Finland. The Publications of the Academy of Finland, pp. 402-412.

George, D.G. and Harris, G.P. (1985) The effect of climate on long-term changes in the crustacean zooplankton biomass of Lake Windermere, U.K. *Nature* 316, 536-539.

George, D.G., Hewitt, D.P., Lund, J.W.G. and Smyly, W.J.P. (1990) The relative effects of enrichment and climate change on the long-term dynamics of *Daphnia* in Esthwaite Water, Cumbria. *Freshwater Biology* 23, 55-70.

Haworth, E.Y. (1990) Algal changes during the Holocene: records from a maritime, montane region of Britain. In: Burchardt, L. (ed.) *Evolution of Freshwater Lakes*. Proceedings of the IXth Symposium Phycological Section of the Polish Botanical Association , UAM Poznán, pp. 33-36.

Haworth, E.Y. and Allen, P.V. (1984) Temporary changes in the composition of a post-glacial diatom profile from Ullswater. *Proceedings of the 7th International Diatom Symposium (Philadelphia 1982)*, pp. 431-442.

Heaney, S.I. (1986) Unusual *Oscillatoria* populations in Windermere - is the lake changing? *British Phycological Journal* 21, 330.

Heaney, S.I, Lund, J.W.G., Canter, H.M. and Gray, K. (1988) Population dynamics of *Ceratium* spp. in three English Lakes, 1945-1985. *Hydrobiologia* 161, 133-148.

Jones, J.G. (1990) Long-term research and the future: resumé. *Freshwater Biology* 23, 161-164.

Lund, J.W.G. (1949) Studies on *Asterionella formosa* Hass. I The origin and nature of the cells producing seasonal maxima. *Journal of Ecology* 37, 389-419.

Lund, J.W.G. (1950) Studies on *Asterionella formosa* Hass. II Nutrient depletion and the spring maximum. *Journal of Ecology* 38, 1-35.

Lund, J.W.G. (1972) Eutrophication. *Proceedings of the Royal Society of London B* 180, 371-382.

Lund, J.W.G. (1978) Changes in the phytoplankton of an English Lake, 1945-1977. *Hydrobiological Journal, Kiev* 14, 6-21.

Lund, J.W.G., Mackereth, F.J.H. and Mortimer, C.H. (1963) Changes in depth and time of certain chemical and physical conditions and of the standing crop of *Asterionella formosa* Hass. in the north basin of Windermere in 1947. *Philosophical Transactions of the Royal Society of London B* 246, 255-290.

Maberly, S.C., Hurley, M.A., Butterwick, C., Corry, J.E., Heaney, S.I., Irish, A.E., Jaworski, G.H.M., Lund, J.W.G., Reynolds, C.S. and Roscoe, J.V. (1994) The rise and fall of *Asterionella formosa* in the South Basin of Windermere: analysis of a 45-year series of data. *Freshwater Biology* 31, 19-34.

Mills, C.A., Heaney, S.I., Butterwick, C., Corry, J.E. and Elliott, J.M. (1990) Lake enrichment and the status of Windermere Charr, *Salvelinus alpinus* (L.). *Journal of Fish Biology* 37 (Supplement A), 167-174.

Neale, P.J., Talling, J.F., Heaney, S.I., Reynolds, C.S. and Lund, J.W.G. (1991) Long time-series from the English Lake District: irradiance-dependent phytoplankton dynamics during the spring maximum. *Limnology and Oceanography* 36, 751-760.

Reynolds, C.S. (1980) Phytoplankton assemblages and their periodicity in stratifying lake systems. *Holarctic Ecology* 3, 141-159.

Reynolds, C.S. (1984) *The Ecology of Freshwater Phytoplankton.* Cambridge University Press, Cambridge.

Reynolds, C.S. (1988) Functional morphology and the adaptive strategies of freshwater phytoplankton. In: Sandgren, C.D. (ed.) *Growth and Reproductive Strategies of Freshwater Phytoplankton.* Cambridge University Press, Cambridge, pp. 388-433.

Reynolds, C.S. (1990) Temporal scales of variability in pelagic environments and the response of phytoplankton. *Freshwater Biology* 23, 25-53.

Reynolds, C.S. (1992) Eutrophication and the management of planktonic algae: what Vollenweider couldn't tell us. In: Sutcliffe, D.W. and Jones, J.G. (eds.) *Eutrophication: Research and Application to Water Supply*, Freshwater Biological Association, Ambleside, pp. 4-29.

Reynolds, C.S. (1993) Scales of disturbance and their role in plankton ecology. *Hydrobiologia* 249, 157-171.

Reynolds, C.S. and Lund, J.W.G. (1988) The phytoplankton of an enriched, soft-water lake subject to intermittent hydraulic flushing (Grasmere, English Lake District). *Freshwater Biology* 19, 379-404.

Reynolds, C.S., Thompson, J.M., Ferguson, A.J.D. and Wiseman, S.W. (1982) Loss processes in the population dynamics of phytoplankton maintained in closed systems. *Journal of Plankton Research* 4, 561-600.

Steinberg, E.W. and Hartmann, H.M. (1988) Planktonic bloom-forming Cyanobacteria and the eutrophication of lakes and rivers. *Freshwater Biology* 20, 279-287.

Strayer, D., Glitzenstein, J.S., Jones, C.G., Kolasa, J., Likens, G.E., McDonnell, M.J., Parker, G.G. and Pickett, S.T.A. (1986) Long-term ecological studies: an illustrated account of their design, operation and importance to ecology. *Occasional Publication of the Institute of Ecosystem Studies* 2.

Sutcliffe, D.W., Carrick, T.R., Heron, J., Rigg, E., Talling, J.F., Woof, C. and Lund, J.W.G. (1982) Long-term and seasonal changes in the chemical composition of precipitation and surface waters of lakes and tarns in the English Lake District. *Freshwater Biology* 12, 451-506.

Talling, J.F. (1993) Comparative seasonal changes, and inter-annual variability and stability, in a 26-year record of total phytoplankton biomass in four English lake basins. *Hydrobiologia* 268, 65-98.

Talling, J.F. and Heaney, S.I. (1988) Long-term changes in some English (Cumbrian) lakes subject to increased nutrient inputs. In: Round, F.E. (ed.) *Algae and the Aquatic Environment.* Biopress Ltd, Bristol, pp. 1-29.

Monitoring Environmental Change Through Networks

P.B. TINKER
*Department of Plant Sciences, University of Oxford, South
Parks Road, Oxford OX1 3RB, UK.*

Introduction

The activities of monitoring and research stand in a curious relationship to
each other. Laboratory scientists are often unwilling to accept monitoring as a
valid scientific activity, and, in particular some ten years ago, monitoring had a
very poor image, with phrases such as 'mindless monitoring' being used
widely. Indeed, it is possible for monitoring to be done without proper objectives, but all forms of science can be done badly. Science depends on the acquisition of data, and monitoring is an important form of this. Usually, it is taken to
mean measurement of one or more variables in a situation which has not been
experimentally perturbed, yet there are an infinite number of degrees of perturbation, and in that sense monitoring is only one extreme of a range of meth-
odologies. Thus agricultural scientists lay down experiments in the field, and
then monitor information from them, whereas ecologists may monitor information from a series of undisturbed sites or habitats. In some situations experimental perturbation may be totally impossible, e.g. in observational
astronomy. Bradshaw (1987) has emphasized the virtues of the comparative
approach, and in one sense monitoring is simply a comparison over time, and
possibly over space. Properly used, monitoring is a basic tool of science,
though ideally monitoring and experimentation should be combined, so that
each supports the other. Especially in ecological research it is often difficult to
determine where the boundary lies between research and monitoring, and in
this chapter it is taken for granted that they are mutually complementary
(Risser *et al.*, 1991).

   The essence of proper monitoring is to have a well-defined purpose. In
this sense a single-issue monitoring operation is highly focused and its purposes are clear. An excellent example is the acid rain monitoring of the UK

Department of the Environment (DOE), which has established a network of sites to measure acidity in rain. The sites were selected to produce the most useful data for the DOE's purposes, and the network has proved its value abundantly. Similarly, the operation which detected the 'ozone hole' over Halley Bay in the Antarctic was directed uniquely towards measuring the amount of ozone in the atmosphere there. However, this project was fortunate to detect such a striking effect as the ozone hole, because such narrowly focused work may well produce little of general interest. There may be good reasons for a less sharply focused approach, if a number of variables may be changing simultaneously, the relationships between these changes define the system, and the system itself is not well understood. In such general-purpose monitoring the choice of variables is always difficult to make, and it is inevitable that it will often be found, by subsequent work, that a better choice was possible. This has led to the belief by some that it is best to start with a modelling programme, so that the necessary variables can be identified. However, if subsequent measurements prove the initial models to be mistaken, the variables may still be inappropriate. In practice, monitoring and modelling should go hand in hand, with constant adjustments and interplay, just as for other forms of research.

There is often a need to justify the 'long-term' (say, over 20 years) nature of monitoring. This aspect causes genuine problems to science administrators, as it implies an open-ended commitment of part of the available funding stream into the future, with only a general statement of objectives and likely achievements. Nevertheless, the long-term aspect is crucial. Some processes have long time scales, such as the development of a forest. Some very important processes occur briefly, but rarely over time, such as water erosion or fire, and a long period is needed to capture representative data. Some processes are subtle or complex, and a long and accurate data set is needed to allow them to be explained and modelled. Finally, the system may not be well understood at the outset of monitoring, and it is essential to allow time to adjust and improve the list of variables and the protocols for their measurement.

The Purposes of Monitoring

The possible reasons for monitoring are almost endless, but there are some particular issues. Monitoring for pollutants is particularly frequent because of the very large economic consequences, e.g. the extraction of sulfur from power station flue gases. There is now very heavy emphasis on the monitoring of heavy metals in the environment, because their build-up may be almost irreversible (Burton, 1986; McGrath, 1993).

The other major reason for monitoring is now 'Global Change' (Correll and Anderson, 1991). This postulates that the effects of man's activities are now becoming pervasive on a global scale, with consequences which are at least partly unpredictable. The International Geosphere–Biosphere Pro-gramme (IGBP) defines three main causes: atmospheric composition change, climatic change and land use change (IGBP, 1992). Some of the consequences are truly global, for example the increase in carbon dioxide in the atmosphere, whereas others are essentially local in nature, but repeated so often that their total impact becomes global. The latter include processes such as erosion or salinization of soil, or loss of biodiversity. This global interest is giving rise to requirements for global monitoring systems, which are discussed later.

Broadly, the data obtained from general-purpose monitoring networks may be used in three ways. The first is scientific: to parameterize and validate ecological models of vegetation, land cover and biota at different scales, including ecosystems and biomes, and hydrological or atmospheric compo-sition models. Second, they provide information about changes in the environ-ment of direct value to agencies and governments for regulatory or policy purposes. Third, they provide information that can be used in studies or mod-els of socioeconomic factors. The question is whether a large network can meet all these needs simultaneously, or whether there will be conflict between the different demands; this question has not yet been answered.

The issue of biodiversity is prominent, and poses particular problems for monitoring (Solbrig *et al.*, 1994). The majority of the world's species are pres-ently not identified or described. For those that are, their populations have to be monitored to determine whether they are decreasing or increasing, and ultimately whether there is a danger of extinction. This normally demands careful and tedious observational work and the more scarce and dispersed a species is, the more difficult it is to monitor it.

The Spatial and Time Dimensions

A great deal of monitoring has been done on single sites, or on a few sites that were not logically related to each other in any way. For example, a report on biological monitoring (Burton, 1986) reviews a large number of studies on heavy metal pollutants, most of which were on a very few sites, and measured only a few variables. Such studies must have a very limited purpose, and may be difficult to relate to analogous studies elsewhere.

A complete monitoring study requires a knowledge of how variables change in both time and space, to allow models to be built that can define how the change is occurring, and how it may develop and extend in the future. There are essentially three ways of sampling in time. First, one may set up sites to take regular observations and to analyse these, and a main problem is then to determine the appropriate time intervals. This can be difficult if change is irregular, as in accidental pollution episodes in a river. Second, it may be poss-ible to use 'space-for-time' methods (Pickett, 1989) in which appropriate selections of sites on some form of chronosequence allows the secular change to be inferred – as in studying the vegetation succession in a forest by using a series of regrowth patches of different ages. If such sites are well distributed, both the space and time dimensions can be studied. Third, one can, in the right circumstances, analyse samples from a substrate that contains a defined time sequence. This can be in peat, ice, sediments, tree rings or preserved specimens (MARC, 1985). The Rothamsted soil sample collection (Johnston, Chapter 2) is an outstanding example of the last. When I was at Rothamsted, we considered whether we could drill down into the chalk underlying the Rothamsted Classical experiments to determine the rate at which nitrate was leached out of them over the preceding decades. In practice, we found that the chalk was too irregular for this, but it would have been a very interesting method of following the experiments' nitrogen dynamics if it had proved practicable. However, all these alternatives have pitfalls and uncertainties. The most reliable method is clearly to establish sites at the necessary density, and continue to measure the variables for as long as necessary.

The best arrangements for sampling in space are rather difficult to define. The most obvious way is to sample on a grid basis, as was done in the soil sampling to provide the basis for the soils map of England and Wales, and for the Soil Geochemical Atlas of England and Wales (McGrath and Loveland, 1992). Such data can be used for interpolation, including the use of geostatistical techniques. The 5 km grid sampling that was used for the Geochemical Atlas could be applied over the whole country for other characteristics which change only exceedingly slowly and are therefore sampled rarely, but it would be quite uneconomic for variables which may fluctuate rapidly.

The next possibility is to use stratified sampling, as with the Countryside Survey of Britain, produced by the Institute of Terrestrial Ecology (Bunce *et al.*, 1992). For this, the whole of Great Britain, was classified into 32 land classes on the basis of easily accessible environmental information. This classification was then used as the basis for sampling 1 km squares selected randomly within each land class, measurements being done by direct observation. An analogous approach is used in ecological work when sampling is arranged in defined agroclimatic zones, vegetation types or biomes, and the results are taken to apply to the whole area of the appropriate class. The Countryside Survey is repeated roughly every five years, so that a secular pattern is also obtained which is adequate for many variables at this scale.

The advent of remote sensing, by aircraft or satellite, has greatly expanded the possibility of obtaining spatially explicit data. In a few cases the direct satellite measurements may be fully sufficient, for example sea-surface temperature can be sensed directly, with relatively few assumptions. However, in many cases, and particularly on land, only a small number of variables can be measured, and most of these demand a lengthy train of inference for their derivation. Some variables, such as populations of animals or insects, are unlikely to be measurable at all by such methods, except as extrapolations of ground-based data on the basis of vegetation type, weather and soils. The ability to extrapolate from a point of precisely known ground truth will be a most important advantage for sparse networks such as the Environmental Change Network (see below). The Data and Information System of IGBP has just begun to release global maps of vegetation cover, based on the AVHRR instrument data, as 10-day averages (Rasool, personal communication). This has a fairly low resolution for local applications, but will be most valuable for work at global, regional and national scales.

If observations are taken at random, but the location of each observation is noted, the resulting information can still be made spatially explicit. A good example of this is in the maps produced by the Biological Records Centre at the Monks Wood station of the Institute of Terrestrial Ecology (Harding, 1990). These rely on the observations of a large group of amateur but highly skilled biologists, as is also the case for the British Trust for Ornithology (see Greenwood *et al.*, Chapter 19).

However, the most general method for spatially distributed monitoring is for a series of carefully arranged monitoring sites, i.e. a network. Any group of sites can be used for comparative purposes, but it is essential to define what is being compared. The sites must be located in well-defined positions, with respect to climate, soil, terrain, ecosystem and other descriptors. Ideally the sites will be located on all main classes in the landscape. In the limit, there will be sufficient sites to allow isolines to be drawn for a mapping of the variables. Geostatistical techniques can also be applied to support this process. However, few if any networks have yet made a complete analysis of both variation in time and in space.

Some Current Networks for Monitoring

A number of monitoring networks have of course been in existence for a long time. The Acid Rain network was referred to above, and there is also the Harmonized Water Archive, based in the Institute of Hydrology, which is derived from a network of sites on rivers. Examples of international networks operating in Europe at present are the United Nations Economic Commission for Europe scheme, coordinated from Finland, and the ENCORE (European Network of Catchments Organized for Research in Ecosystems) sites for catchment research coordinated by the European Community (Hornung, 1992) (Fig. 22.1). A large Long-Term Ecological Research (LTER) network has been operational in the USA for a number of years, but this was designed for research, and it is only recently that much attention has been given to coordinated monitoring. The network provides an excellent set of sites for this, though the density over a huge area like the USA is low. Most networks at present concentrate partly or wholly on natural or seminatural systems, but it is essential that managed ecosystems be given full parity, as it is there that the impacts of change will have the largest effects in economic and social terms. A

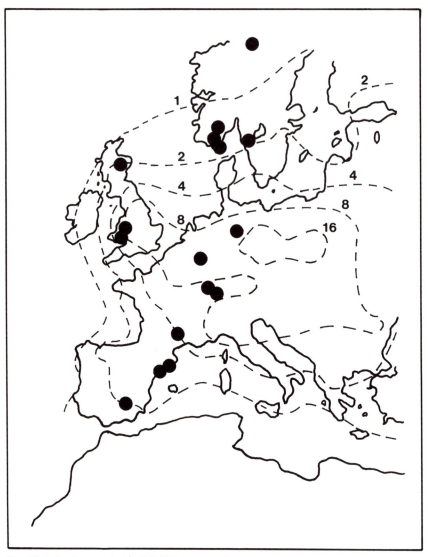

Fig. 22.1. Catchments in the ENCORE network. Catchments are marked ●. The dashed isolines of SO$_2$ concentration (μg Sm^{-3}) are obtained from the EMAP network, and show that the ENCORE sites are well-distributed with regard to atmospheric pollution.

rough classification of sites/systems, given in Table 22.1, shows the wide variety of monitoring systems that exist.

A very useful brief review of existing monitoring networks is in the record of a meeting organized at Fontainebleau in 1992 (Heal *et al.*, 1993). It is clear that at present there is no single network that gives full global cover. The Glo-

Table 22.1. Types of monitoring systems.

Site	Purpose	Example
Single	Single	Many
Single	Multiple	Rothamsted Experimental Station; Windermere; Hubbard Brook
Multiple	Single	Acid Rain Network
Multiple	Multiple	ENCORE; Long-Term Ecological Research Network; Environmental Change Network
Multiple	Global	CO_2 Flask Networks; Global Terrestrial Observing System; Biosphere Observatories

bal Environment Monitoring System (GEMS) of the United Nations Environment Programme (UNEP) does valuable work in coordination and integration of data, but does not operate its own sites. The largest group of potential monitoring sites is in the United Nations Education Science and Cultural Organisation's Man and Biosphere (UNESCO/MAB) system of Biosphere Reserves. These are on sites selected for their conservation interest by individual countries, and the monitoring carried out is variable in quantity, quality and methodology. The observatories also have a remit for the study of sustainable development, within an innovative 'zonal' layout. Despite the fact that these sites are not fully coordinated and integrated now, they are a very valuable resource for monitoring activities.

There are also at present very well-established networks in tropical areas of the world for agricultural research. These are largely commodity-based, and are usually organized from one of the international centres operated by the Consultative Group for International Agricultural Research (Greenland *et al.,* 1987). A great deal of experience has been built up with these networks. Above all, this experience points to the fact that all members of a network must obtain perceived benefit from it – otherwise their involvement will be both lukewarm and temporary. Most of these networks arc for experimentation rather than for observation, but the methodology for agricultural and environmental research is in many ways similar, and this experience should be useful in erecting networks in the tropics for environmental monitoring, which could often, with advantage, use agricultural centres as sites.

The Environmental Change Network

The Environmental Change Network (ECN) has recently been set up in the UK, and this is certainly the most ambitious national multipurpose monitoring network launched in this country, and possibly in the world. This network is explicitly designed for general-purpose monitoring in relation to the major anthropogenic changes that are occurring now or likely to occur in the next decades.

The main objectives are as follows:

1. To provide a network of sites from which to obtain comparable long-term data sets. This will require measurement, at regular intervals, of variables identified as being of major environmental importance.
2. To integrate and analyse the data in order to define possible environmental change and improve understanding of the causes of such change.
3. To use the data for modelling and prediction of future change.
4. To provide a range of sites with good instrumentation and reliable information, for research purposes.

The purposes underlying these objectives vary quite widely. One specific origin of the ECN was the recommendation by a House of Lords Committee (HMSO, 1984) that a set of monitoring stations should be provided to detect changes in pesticides in the environment. It was felt in the National Environment Research Council (NERC) that this specific need should be combined with a much more general monitoring programme. A group of agencies with clear interests in the environment was formed, with the author acting as chairman, and over a long series of discussions this group developed a model for the ECN that satisfied all the members. The data sets will be important to Government for policy and regulatory purposes, for informing the actions of agencies, commercial companies and local authorities, and as a basis for modelling and prediction in the scientific community.

The mechanism in this case is for a relatively modest number of carefully selected sites placed in the major ecosystems of the country and for similar data sets to be collected at all of them. As far as possible, all measurements will be of the same variables, at the same frequency, and carried out according to the same protocols. The intention is that the data sets shall be fully comparable and rigorous.

Membership and sites

The Network covers the whole of the UK, and has a long list of agencies as sponsors (Table 22.2). It includes the Biotechnology and Biological Sciences Research Council (BBSRC), which has designated Rothamsted as a site. Sites with a history of previous measurements have been preferred, to give the largest possible amount of background data to support the new data sets, and

Table 22.2. Sponsors and sites in the ECN.

Sponsor	Site	Type
Biotechnology and Biological Sciences Research Council	Rothamsted North Wyke	Arable Lowland Grassland
Department for Agriculture for Northern Ireland	Hillsborough	Lowland Grassland
Department of the Environment	General Support	
English Nature/NERC	Moor House	Moorland/Grassland
Forestry Commission	Alice Holt	Woodland
Ministry of Agriculture, Fisheries and Food	Drayton	Mixed farming
National Rivers Authority	Many sites	Rivers
Natural Environemnt Research Council	Merlewood (CCU)	
Oxford University/NERC	Wytham	Old woodland and grassland
Scottish Office Agriculture and Fisheries Dept	Glensough Sourhope	Upland grassland Upland grassland

Rothamsted is a valuable asset to the ECN, because the existing data sets are so exceptionally good. An initial search selected potential terrestrial sites, which have been steadily added to. By the end of 1994 there will be 10 terrestrial and 37 freshwater (river and lake) sites (NERC, 1994). A Planning Committee under Professor W. Heal initially selected 24 potential terrestrial sites that ideally should be included, and it is intended that this number will eventually be operational. The underlying principle is that each sponsor supports one or more sites, or puts in equivalent resources for general use. Data are available to all, though ownership remains with the agency controlling the site.

The varied purposes for which the data can be used are mirrored in the agencies who support the ECN. Because each agency is expected to contribute one or more sites and to support the work there, the sites reflect the agencies' remits and interests. Both natural and managed ecosystems are represented in the list of sites (Table 22.2), and sponsors include research councils (BBSRC and NERC), agencies (English Nature and Forestry Commission) and Government departments (MAFF, DOE, Scottish Office and Department of Agriculture for Northern Ireland). The accession of river sites *via* the National Rivers Authority is a great advantage as these will allow the interaction of land use with water quantity and quality to be addressed. This land/water dimension has been increased with the addition of the lake sites and subsequently several other agencies have joined in.

Organization

The organization of the ECN is as simple as possible, while maintaining a strong common identity (Fig. 22.2). This organization has been designed for continuity, because a long-term monitoring network that collapses after a few years is of no value. The Steering Committee (chaired by Professor W. Heal) consists of one representative of each member organization. All sign a common Memorandum of Understanding, but NERC has a special position as 'managing agent' for the rest. NERC provides funds for the 'Central Coordination Unit', including two site managers at the main sites of Moor House and Wytham (Oxford). The latter are also focal sites for the NERC Terrestrial Initiative in Global Environmental Research, and there is a valuable linkage with this programme.

All major decisions are taken by the Steering Committee, always by consensus. No public bodies can bind themselves to provide funding for more

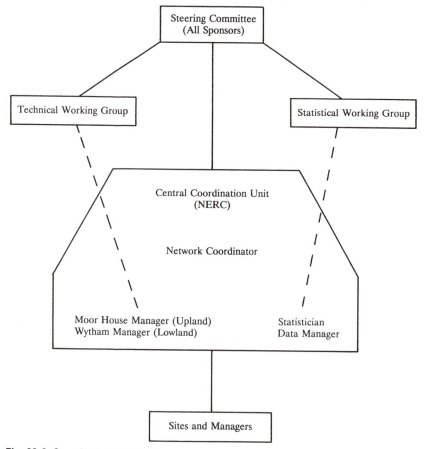

Fig. 22.2. Organization of the ECN.

than three years ahead in practice, and it was recognized that no member organization could be constrained to remain in the ECN against its will. All agree to follow the ECN protocols and methods in monitoring, and to provide data obtained on their site(s) for the use of all members. In return, they have access to all other data in the ECN. There is no way of guaranteeing continuity – short of a large private financial trust – and a collaborative group like the ECN must be designed to provide its members with such value in its integrated data stream that they will need and wish to remain members because each gains added value from the rest.

Two very important subcommittees of the Steering Committee, the Technical Working Group and the Statistics and Data Handling Working Group, deal with technical issues and advise the Steering Committee and the Network Coordinator at present J.M. Sykes.

The Central Coordination Unit and data handling

If it is to function well, a network needs a managing team. The Central Coordination Unit (CCU) contains the Network Coordinator, who has responsibility for the smooth running of the ECN. He is supported by a statistician, a data manager and two main site managers (Moor House and Wytham), who also have responsibility for advice to sites in the uplands and lowlands, respectively.

The statistical underpinning is seen as essential. All data obtained need to be assessed for quality, and then integrated between sites. Pre-existing data sets need to be integrated with new ones. Above all, the data sets from the sites need to be related to the other sources of spatially identified information, including that from other data-acquisition programmes currently in the Terrestrial and Freshwater Directorate of NERC (Table 22.3). The ultimate goal of this initiative, with additions from other organizations, should be a national terrestrial database.

The Minimum Dataset

Much effort has been expended on selecting the variables to be measured, and the protocols for this. Basically, they are intended to meet the priority interests of all the sponsors. There is now a Minimum Dataset (Table 22.4) that all sites will measure – if technically possible – but any agency that wishes to measure additional variables on its site(s) is, of course, free to do so. It is certain that as time goes on, good reasons will emerge to change the variables or the methodology. Very considerable care will be needed then, because continuity is essential. Constant change – even if for the better – will damage this continuity, and make it more difficult to use the data. The problem is analogous to that faced on the long-term experiments at Rothamsted, where it became essential to

Table 22.3. Programmes in the Natural Environment Research Council for obtaining spatially explicit terrestrial information, with dates when conducted or started.

Programme	Year started or conducted
Countryside Survey	1978, 1984, 1991
Remotely sensed land cover map	1993
Biological Records Centre datasets	from 1964
Ground and Surface Water Archives	from 1935
Environmental Change Network	from 1992

change the cultivars grown (see Leigh *et al.*, Chapter 14), but this thereby introduced a discontinuity in the data. Such problems are even more difficult in a network with a wide range of diverse interests.

The features of the network described here are designed to reduce operational problems to a minimum by standardizing the procedures as far as possible (clearly some variables cannot be measured at some of the sites). This does demand that the managers at individual sites do not introduce changes in procedure, even with good reason. The tendency for 'procedural drift' has been a serious problem in many joint programmes before. It is for this reason that such emphasis is placed on the effectiveness of overall management within the ECN.

The costs of monitoring are always an issue, and each new variable to be measured – or measured more frequently – will increase the cost. As the maintenance of a site carries basic overheads, the broader the objectives to be met there, the lower the individual costs will be. Most of the sites were already in use by their owners, for research, development or monitoring. The ECN status thus has often meant simply a change of variables, methodology and protocol, and the integration with other ECN systems at modest cost. Very approximately, it is believed the costs of one site to its sponsor will be around £50,000 per year.

The output from the ECN should be highly cost-effective, and should fully repay the investment of the participants. It is innovative in several features of its design, and may have lessons for the monitoring networks which will surely develop on a European or global scale.

The Global Terrestrial Observing System (GTOS)

In the general upsurge of interest in monitoring, global change concerns demand a global monitoring system. This need was expressed at the 2nd World Climate Conference, which called for a Global Climate Observing Sys-

Table 22.4. The ECN Minimum Dataset to be collected at each site using standardized protocols.

Information class		Data collected
Vegetation	1.	Permanent quadrant samples on 3- and 9-year cycles with field estimates
Vertebrates	1.	Birds; following national common bird census scheme
	2.	Bats; using fixed transects and ultrasonic detectors
	3.	Rabbits; using fixed transects
Invertebrates	1.	Moths; light traps following existing national scheme
	2.	Butterflies; transects following existing national scheme
	3.	Spittle bugs; annual population estimates and colour morphs in quadrat samples
	4.	Beetles; pitfall traps for Carabidae and Opilionids in summer
Soil biology	1.	Tipulids; extraction of larvae from soil cores twice per year
Site management	2.	Changes in management, land use
Meteorology	1.	Automatic weather station
Surface water	1.	Continuous recording of stream/river discharge, temperature, conductivity, pH and turbidity
	2.	Weekly samples for content of selected ions
Atmospheric chemistry	1.	Diffusion tubes for NO_2
	2.	Weekly rain collection for pH and ion content
Soil sampling	1.	Classification and mapping
	2.	At 20-year intervals for total contents of nine trace elements, Ni, extractable Fe, Al; water-soluble N and P; mineralogy, particle size analysis, bulk density
	3.	At 5-year intervals for pH; exchangeable cations, Al; total N, P, S, organic C
	4.	Soil solution chemistry; six suction lysimeters at base of A and B horizons; analysed alternate weeks for ions, dissolved organic carbon, total N and pH

tem (GCOS), including a terrestrial component (GTOS). So far there has been much attention to the use of remote sensing, because cost will prevent the use of direct measurement for most variables in the atmosphere and the oceans. However, for the reasons explained for the ECN, remote sensing is less useful for terrestrial situations. A planning meeting in 1992 agreed that there was a need for a set of fixed sites in carefully chosen locations – related to major biomes and covering both managed and natural land cover (Heal *et al.*, 1993). Currently five international agencies have agreed to accept joint responsibility for GTOS (UNEP, UNESCO, Food and Agriculture Organization (FAO), World Meteorological Organization (WMO) and the International Council of Scientific Unions (ICSU)) and a task force is now working on the detailed planning. The basis for the work has recently been published (UNEP, 1994).

Conclusion

There is now a solid consensus among both scientists and policy makers that sound and up-to-date environmental data sets are essential if we are to understand, predict and react to the various forms of change in the biological and physical world around us. Many of these changes result from, and interact with, social and economic change in the world, in industry and land use, so that it is essential to define land use and development on the monitored sites. Despite the feeling that remote sensing is the only technology that can deliver data at the necessary speed and cost, each group that has considered the issue has concluded that ground-level monitoring is essential to provide fully reliable data. These may, of course, be used as ground truth for remote sensing, and must be integrated with all other forms of data, but at present there is no substitute for well-established and managed networks of sites where rigorous quality control can be maintained.

The outline of the ECN given here has intentionally stressed the organizational and managerial aspects, rather than the purely scientific ones. Considerable scientific insight is required to identify the variables that will be most useful in such a system and to develop practical protocols for their measurement. However, the greatest danger for such a general-purpose network is in managerial breakdown, technical fragmentation or inability to handle and deliver large quantities of information. Continuity, persistence and sustainability are the main demands upon a *long-term* network, and these must be designed into it.

References

Bradshaw, A.D. (1987) Comparison – its scope and limits. In: Rorison, I.H., Grime, J.P., Hunt, R. and Lewis, D.H. (eds), *Functions of Comparative Plant Ecology*. Academic Press, London, pp. 3–22.

Bunce, R.G.H., Barr, C.J. and Fuller, R.M. (1992) Integration of methods for detecting land use change, with special reference to the 'Countryside Survey 1990'. In: Whitley, M.C. (ed.) *Land Use Change, Causes and Consequences*, ITE Symposium No. 27, HMSO, London, pp. 69–78.

Burton, M.A.S. (1986) Biological monitoring of environmental contaminants (plants). *MARC Report No.32*. Monitoring and Assessment Research Centre, University of London.

Correll, R.W. and Anderson, P.A. (1991) *Global Environmental Change*, NATO ASI Series, Springer-Verlag, Berlin.

Greenland, D.J., Crasswell, E.T. and Dag, M. (1987) International networks and their contribution to crop and soil management research. *Outlook on Agriculture* 16, 42-50.

Harding, P.T. (1990) National species distribution survyes. In: Goldsmith, B.J. (ed.) *Monitoring for Conservation and Ecology*. Chapman and Hall, London, pp. 133-154.

Heal, O.W., Menaut, J.-C. and Steffen, W.L. (1993) Towards a Global Terrestrial Observing System (GTOS): detecting and monitoring change in terrestrial ecosystems. Report from workshop at Fontainebleau, France, July 1992. *MAB Digest 14*, UNESCO, Paris.

HMSO (1984) Agricultural and environmental research. *Select Committee of the House of Lords for Science and Technology, 4th Report*. HMSO, London.

Hornung, M. (1992) The European network of catchments organized for research on ecosystems (ENCORE). In: Teller, A., Mathy, P. and Jeffers, J.N.R. (eds), *Responses of Forest Ecosystems to Environmental Change*, Elsevier, London, pp. 315-324.

IGBP (1992) Global Change and Terrestrial Ecosystems – the Operational Plan. *IGBP Report No. 21*, International Geosphere–Biosphere Programme, Stockholm.

MARC (1985) Historical Monitoring. *Report No.31*. Monitoring and Assessment Research Centre, University of London.

McGrath, S.P. (1993) Soil quality in relation to agricultural uses. In: Eijsackers, H.J.P. and Hamers, T. (eds.) *Integrated Soil and Sediment Research: A Basis for Proper Protection*. Kluwer Academic Publishers, Dordrecht, pp. 187-200.

McGrath, S.P. and Loveland, P.S. (1992) *The Soil Geochemical Atlas of England and Wales*. Blackie Academic and Professional, Chapman and Hall, Andover.

NERC (1994) *Strategy for Terrestrial and Freshwater Science and Technology, 1994-2000*. Natural Environment Research Council, Swindon, UK.

Pickett, S.T.A. (1989) Space-for-time substitution as an alternative to long-term studies. In: Likens, G.E. (ed.) *Long-term Studies in Ecology*. Springer-Verlag, New York, pp. 110-135.

Risser, P.G., Melillo, J.M. and Gosz, J.R. (1991) Current status and future of long-term ecological research. In: Risser, P.G. (ed.) *Long-term Ecological Research, SCOPE*. Wiley, Chichester, pp. 275-285.

Solbrig, O.T., van Emden, H.M. and van Oordt, P.G.W.J. (1994) *Biodiversity and Global Change*. Revised edition. CAB International, Wallingford, UK.

UNEP (1994) Report of the first meeting of the *ad hoc* Scientific and Technical Planning Group for the Global Terrestrial Observing System (GTOS) *Gems Report Series* 24. United Nations Environmental Programme, Nairobi, Kenya.

Index